Sustainable Bioenergy

Sustainable Bioenergy

Advances and Impacts

Edited by

Mahendra Rai
Department of Biotechnology, SGB Amravati University, Amravati, Maharashtra, India

Avinash P. Ingle
Department of Biotechnology, Engineering School of Lorena, University of Sao Paulo, Lorena, SP, Brazil

ELSEVIER

Elsevier
Radarweg 29, PO Box 211, 1000 AE Amsterdam, Netherlands
The Boulevard, Langford Lane, Kidlington, Oxford OX5 1GB, United Kingdom
50 Hampshire Street, 5th Floor, Cambridge, MA 02139, United States

British Library Cataloguing-in-Publication Data
A catalogue record for this book is available from the British Library

Library of Congress Cataloging-in-Publication Data
A catalog record for this book is available from the Library of Congress

ISBN: 978-0-12-817654-2

For Information on all Elsevier publications
visit our website at https://www.elsevier.com/books-and-journals

Publisher: Brian Romer
Acquisition Editor: Raquel Zanol
Editorial Project Manager: Aleksandra Packowska
Production Project Manager: R. Vijay Bharath
Cover Designer: Greg Harris

Typeset by MPS Limited, Chennai, India

Contents

List of Contributors

Abubakar Sani Sambo Faculty of Engineering and Environmental Sciences, Usmanu Danfodiyo University, Sokoto, Nigeria

Andrea Komesu Department of Marine Sciences, Federal University of São Paulo (UNIFESP), Santos, Brazil

Arnaldo Sarti Department of Biochemistry and Chemical Technology, Institute of Chemistry, São Paulo State University-Unesp, Araraquara, Brazil; São Paulo State University (Unesp), Bioenergy Research Institute (IPBEN), Araraquara, Brazil

Atthapon Srifa Department of Chemical Engineering, Faculty of Engineering, Mahidol University, Nakhon Pathom, Thailand

Avinash P. Ingle Department of Biotechnology, Engineering School of Lorena, University of São Paulo, Lorena, Brazil

Bruna Sampaio de Mello Department of Biochemistry and Chemical Technology, Institute of Chemistry, São Paulo State University-Unesp, Araraquara, Brazil

Debashis Sut Department of Energy, Tezpur University, Tezpur, India

Dehua Liu Key Laboratory for Industrial Biocatalysis, Department of Chemical Engineering, Ministry of Education of China, Institute of Applied Chemistry, Tsinghua University, Beijing, P.R. China

Hwei Voon Lee Nanotechnology and Catalysis Research Center (NANOCAT), University of Malaya, Kuala Lumpur, Malaysia

Indarchand Gupta Department of Biotechnology, Institute of Science, Aurangabad, Maharashtra, India

João Moreira Neto Department of Engineering, Federal University of Lavras (UFLA), Lavras, Brazil

Johnatt Allan Rocha de Oliveira Institute of Health Sciences, Nutrition College, Federal University of Pará (UFPA), Belém, Brazil

Joon Ching Juan Nanotechnology and Catalysis Research Center (NANOCAT), University of Malaya, Kuala Lumpur, Malaysia; Monash University, Sunway Campus, Subang Jaya, Malaysia

José Luz Silveira Department of Energy, Laboratory of Optimization of Energy Systems (LOSE), School of Engineering, Guaratinguetá and Institute of Bioenergy Research (IPBEN-UNESP), Sao Paulo State University (UNESP), São Paulo, Brazil

Kajornsak Faungnawakij National Nanotechnology Center (NANOTEC), National Science and Technology Development Agency (NSTDA), Pathum Thani, Thailand

Kelly J. Dussán Department of Biochemistry and Chemical Technology, Institute of Chemistry, São Paulo State University-Unesp, Araraquara, Brazil; São Paulo State University (Unesp), Bioenergy Research Institute (IPBEN), Araraquara, Brazil

Lee Eng Oi Nanotechnology and Catalysis Research Center (NANOCAT), University of Malaya, Kuala Lumpur, Malaysia

Lina Gogoi Department of Energy, Tezpur University, Tezpur, India

Luana Cardoso Grangeiro Department of Biochemistry and Chemical Technology, Institute of Chemistry, São Paulo State University-Unesp, Araraquara, Brazil

Lucas Tadeu Fuess Biological Processes Laboratory (LPB), São Carlos School of Engineering (EESC), University of São Paulo (USP), São Carlos, Brazil

Luiza Helena da Silva Martins Department of Natural Sciences and Technology, State University of Pará (UEPA), Belém, Brazil

Mahendra Rai Nanobiotechnology Lab, Department of Biotechnology, Sant Gadge Baba Amravati University, Amravati, Maharashtra, India

Mahmoud Muhammad Garba Sokoto Energy Research Centre, Usmanu Danfodiyo University, Sokoto, Nigeria

Min-Yee Choo Nanotechnology and Catalysis Research Center (NANOCAT), University of Malaya, Kuala Lumpur, Malaysia

Mohammad Barati Department of Applied Chemistry, Faculty of Chemistry, University of Kashan, Kashan, Iran

Narendra Naik Deshavath Centre for the Environment, Indian Institute of Technology Guwahati, Guwahati, India

Nilutpal Bhuyan Department of Energy, Tezpur University, Tezpur, India

Noorsaadah Abdul Rahman Department of Chemistry, University of Malaya, Kuala Lumpur, Malaysia

Pawnprapa Pitakjakpipop National Metal and Materials Technology Center (MTEC), Pathum Thani, Thailand

Pramod Ingle Nanobiotechnology Lab, Department of Biotechnology, Sant Gadge Baba Amravati University, Amravati, Maharashtra, India

Pravin G. Suryawanshi Department of Chemical Engineering, Indian Institute of Technology Guwahati, Guwahati, India

Raquel Von Hohendorff University of the Vale do Rio dos Sinos—UNISINOS, São Leopoldo, Brazil

Ravichandra C. Patil Centre for Energy, Indian Institute of Technology Guwahati, Guwahati, India

Regina Franciélle Silva Paulino Department of Energy, Laboratory of Optimization of Energy Systems (LOSE), School of Engineering, Guaratinguetá and Institute of Bioenergy Research (IPBEN-UNESP), Sao Paulo State University (UNESP), São Paulo, Brazil

Rupam Kataki Department of Energy, Tezpur University, Tezpur, India

Sâmilla Gabriella Coêlho de Almeida Department of Biochemistry and Chemical Technology, Institute of Chemistry, São Paulo State University-Unesp, Araraquara, Brazil

Saidu Muhammad Maishanu Sokoto Energy Research Centre, Usmanu Danfodiyo University, Sokoto, Nigeria

Shiv Prasad Centre for Environmental Sciences and Climate Resilient Agriculture (CESCRA), ICAR-Indian Agricultural Research Institute, New Delhi, India

Tahereh Nematian Department of Applied Chemistry, School of Chemistry, College of Science, University of Tehran, Tehran, Iran

Vaibhav V. Goud Centre for Energy, Indian Institute of Technology Guwahati, Guwahati, India; Department of Chemical Engineering, Indian Institute of Technology Guwahati, Guwahati, India

Venkata Dasu Veeranki Department of Bioscience and Bioengineering, Indian Institute of Technology Guwahati, Guwahati, India

Vinicius O.O. Gonçalves Institute of Chemistry, Department of Physical Chemistry, Federal University of Rio de Janeiro, Rio de Janeiro, Brazil

Weerawut Chaiwat Environmental Engineering and Disaster Management Program, School of Interdisciplinary Studies, Mahidol University, Kanchanaburi, Thailand

Wilson Engelmann University of the Vale do Rio dos Sinos—UNISINOS, São Leopoldo, Brazil

Wipark Anutrasakda Department of Chemistry, Faculty of Science, Chulalongkorn University, Bangkok, Thailand

Xiaoying Sun Institute of New Energy Technology, Academy of Military Science of Chinese, Beijing, P.R. China

Xingkai Cui Key Laboratory for Industrial Biocatalysis, Department of Chemical Engineering, Ministry of Education of China, Institute of Applied Chemistry, Tsinghua University, Beijing, P.R. China

Xuebing Zhao Key Laboratory for Industrial Biocatalysis, Department of Chemical Engineering, Ministry of Education of China, Institute of Applied Chemistry, Tsinghua University, Beijing, P.R. China

Preface

Energy is considered as the backbone of a nation and plays an important role in the socioeconomic development of every country. The ever-increasing demand for energy fuels and their use by a huge portion of the global population for industrialization and motorization has resulted in the continuous and rapid depletion of the limited stocks of fossil fuels. It is estimated that the limited fossil fuel resources will be exhausted in the next 40–50 years. In addition, the burning of various fossil fuels such as coal, gas, and oil are majorly responsible for the emission of greenhouse gases into the atmosphere. This is considered as a significant anthropogenic source of climate change and air pollution, and it leads to global warming, which is a major concern worldwide. Moreover, global warming and air pollution are responsible for the rise in sea level and loss of biodiversity. Collectively, these problems create a pressing need to search for novel and sustainable alternative energy sources that have the potential to fulfill the increasing demand for fossil fuels and simultaneously overcome the issues related to environmental pollution. In this context, the sustainable production of bioenergy including bioethanol, butanol, biohydrogen, biogas, biooils, and others is found to be an emerging alternative that can overcome different issues associated with the use of fossil fuels like the depletion of fossil resources and the emission of greenhouse gases.

Considering the present environmental and energy issues, this book has broadly focused on recent advances and developments in the production of various sustainable bioenergies and their impacts on mankind. The book is divided into two sections consisting of 14 chapters. *Section I* presents recent advances and developments in the production of various sustainable bioenergies, comprising chapters on different lignocellulosic biomasses, their structures and compositions, rural biorefinery in India for the production of biofuels and other valuable chemicals, issues and challenges in the production of biofuels in Africa, biohydrogen production as a sustainable biofuel, recent advances in biojet fuel and biooil production, new trends in biogas production and utilization, current challenges and advances in butanol production, various catalysts for the conversion of nonedible oil into biogasoline, an overview of third-generation biofuels, and nanobiocatalytic processes for the production of biodiesel from algae. *Section II* is devoted to the impacts of bioenergy production on mankind and includes chapters on general and socioeconomic impacts of biofuel production and the regulatory challenges in nanotechnology for the sustainable production of biofuel in Brazil. The text in each chapter is supported by numerous clear and informative tables and figures. Each chapter contains relevant references to published articles, which offer a potentially large amount of primary information and further links to a nexus of data and ideas.

All the chapters in this book have been written by one or more specialist or expert in the concerned topic, and are highly informative and detailed. In this way, we would like to offer a rich guide for researchers in this field and undergraduate or graduate students in various disciplines like agriculture, food science, biotechnology, biofuel and bioenergy as well as in allied subjects. In addition, this book is useful for people working in various industries, regulatory bodies, and global fuel and energy organizations.

The editors are thankful to all the contributors for their outstanding efforts to provide updated information on the subject matter of the respective chapters. Their efforts will certainly enhance and update the knowledge of readers interested in sustainable bioenergies. We express our sincerest thanks to the publisher and the authors of the chapters whose research works have been cited in this book. We are also thankful to Raquel Zanol, Aleksandra Packowska, Sheela Bernardine, and the team at Elsevier for their generous cooperation and efforts in producing this book.

We hope that the book will be useful for all readers in terms of finding the required information on the latest research and advances in the field of sustainable bioenergy production.

Mahendra Rai and Avinash P. Ingle

Chapter 1

Lignocellulosic feedstocks for the production of bioethanol: availability, structure, and composition

Narendra Naik Deshavath[1], Venkata Dasu Veeranki[2] and Vaibhav V. Goud[3]

[1]Centre for the Environment, Indian Institute of Technology Guwahati, Guwahati, India, [2]Department of Bioscience and Bioengineering, Indian Institute of Technology Guwahati, Guwahati, India, [3]Department of Chemical Engineering, Indian Institute of Technology Guwahati, Guwahati, India

1.1 Introduction

Owing to fossil fuel depletion, it is necessary to explore alternative renewable resources for the production of transportation fuels.[1] In this context, the production of biofuels from lignocellulosic biomass is found to be an emerging trend that can overcome the depletion of fossil fuel resources.[2] Lignocellulosic materials are one of the most promising potential feedstocks for the production of biofuels, such as bioethanol, butanol, biohydrogen, biogas (biomethane), and biooils. Among these, bioethanol production from lignocellulosic biomass has gained significant momentum due to its benefits as a substitute for fossil fuels.[2] Bioethanol can be a partial or direct substitute for petroleum products, and thus can reduce fossil fuel dependency. Currently, the United States and Brazil are using bioethanol as a transportation fuel in the proportions: E5 (5% ethanol: 95% gasoline), E15 (15% ethanol: 85% gasoline), E85 (85% ethanol: 15% gasoline), and E100 (100% ethanol: 0% gasoline).[2] Generally, bioethanol acts as an octane enhancer in unleaded gasoline, which also has the capability to replace methyl-tertiary-butyl-ether.[3] Due to the presence of oxygen atoms in ethanol, it is used as an oxygenated compound for the clean combustion of gasoline, thereby decreasing greenhouse gas emissions, which ultimately reduces the global warming

Sustainable Bioenergy. DOI: https://doi.org/10.1016/B978-0-12-817654-2.00001-0

effect. Due to these reasons, bioethanol is considered as a potential and environmentally sustainable fuel.[2,3]

Several potential lignocellulosic materials are abundantly available in nature (around 1.5 billion dry tons/year),[4] including agricultural crop residues (e.g., corn stover, sugarcane bagasse, wheat straw, rice straw, and sorghum stalks), forest residues (e.g., aspen wood, pine wood, poplar wood, and bamboo), and grass varieties (e.g., *Miscanthus giganteus*, switchgrass, and elephant grass).[3,5,6] Among these, a considerable amount of lignocellulosic biomass is being generated through agricultural practices.[6] These are generally used as animal feed and/or fuel by direct combustion. The consumption of these agricultural crop residues as a feedstock for animals is low. Therefore a large amount of agricultural crop residue is being disposed of as a waste product. Approximately 600–900 million tons of rice straw is being produced every year,[6] of which a minor portion is being used as a feedstock for animals and the rest is disposed of by burning. This practice contaminates the ambient air because the burning of lignocellulosic biomass in open fields emits greenhouse gasses (like CO_2, CH_4, and NO_2) in large volumes, which could be a principal contributor to global warming. In the United States, about 64–139 million metric tons of corn stover is generated per year, of which only 1%–5% of corn stover is collected and used for animal feed and industrial processing; the remaining corn stover is left in the fields as waste.[7]

According to the 2017 statistics of the Food and Agriculture Organization of the United Nations (FAOSTAT), around 181.8 million tons of agricultural residues (derived from rice, wheat, corn, and sugarcane crops) were burnt in Brazil, China, India, and the United States in 2016, which produced ~15.77 million tons of CO_2 (Table 1.1). In addition to this, around 19.6 million tons of wheat and corn residues were burnt in the European Union (EU-28), which produced ~1.7 million tons of CO_2 (Table 1.2). Open field burning of agricultural residues is already banned in many countries in western

TABLE 1.1 Amount of major agricultural crops residues was burned in the year of 2016 and their CO_2 emissions.

Agriculture crop residues (million dry ton/year)					Million ton/year
Country	Corn	Rice	Wheat	Sugarcane	CO_2 emission
The United States	35.11	0.07	7.10	0.24	3.73
India	10.20	23.63	12.09	3.22	4.25
China	38.98	16.75	9.74	1.09	5.75
Brazil	14.96	1.07	0.87	6.65	2.03

TABLE 1.2 Amount of corn and wheat residues wasted through burning and their CO_2 emission in EU-28 countries.

S. no.	Region	Corn ($\times 10^4$ t)	Wheat ($\times 10^4$ t)	CO_2 emission ($\times 10^4$ t)
1	Austria	19.53	12.60	2.78
2	Belgium	5.21	8.25	1.16
3	Bulgaria	40.69	47.70	7.64
4	Croatia	25.01	6.72	2.74
5	Cyprus	NA	0.36	0.03
6	Czech Republic	8.64	33.59	3.65
7	Denmark	0.57	23.32	2.06
8	Estonia	NA	6.58	0.57
9	Finland	NA	8.60	0.74
10	France	148.70	222.50	32.08
11	Germany	41.63	128.07	14.67
12	Greece	17.11	24.51	3.60
13	Hungary	119.82	42.22	14.00
14	Ireland	NA	2.72	0.23
15	Italy	66.07	76.50	12.32
16	Latvia	NA	19.16	1.66
17	Lithuania	1.23	34.84	3.12
18	Luxembourg	0.01	0.55	0.05
19	Malta	NA	0.13	0.01
20	The Netherlands	0.84	5.09	0.51
21	Poland	59.54	95.36	13.39
22	Portugal	8.86	1.53	0.90
23	Romania	257.85	85.41	29.67
24	Slovakia	18.48	16.66	3.04
25	Slovenia	3.64	1.26	0.42
26	Spain	38.46	83.12	10.51
27	Sweden	NA	17.94	1.55
28	The United Kingdom	NA	72.92	6.30

NA, Data not available.

Europe, and some other countries, like India, have considered it seriously. Therefore these huge amounts of agricultural crop residues can be converted into bioethanol, thereby providing multiple benefits, such as enhancing indigenous energy sources, strengthening sustainable energy, and boosting the rural economy and environmental systems.

1.2 Structure and composition of lignocellulosic biomass

It is important to understand the chemical and structural composition of the various components of lignocellulosic materials before proceeding to bioethanol production, as there are several factors that majorly hamper the formation of monomeric sugars from lignocellulosic biomass and their subsequent conversion into bioethanol. The typical structure of lignocellulosic biomass is mainly composed of cellulose, hemicellulose, and lignin. In addition, lignocellulosic biomass also contains extractives, pectin, protein, and metal or ash in fractions.[8] Cellulose and hemicellulose are polymeric carbohydrates that yield different types of hexose and pentose sugars upon hydrolysis. While, lignin is a three-dimensional methoxylated polyphenolic compound that principally covers cellulose and hemicellulose.[9] Therefore lignin restricts the depolymerization of cellulose and hemicellulose during lignocellulosic biomass hydrolysis. The percentages of these lignocellulosic biomass constituents vary from species to species, for instance, hardwood biomass has a higher amount of cellulose than softwood biomass, whereas different agricultural feedstocks have more hemicellulose than that of hardwood.[5,10] Moreover, agricultural residues and grass have lower lignin contents than softwood and hardwood biomass. These polymers (cellulose, hemicellulose and lignin) are linked together with various types of bonds (hydrogen bonds, covalent bond, ether bonds) and form a heteromatrix structure.[9] The complexity and rigidity of heteromatrix structures also vary with respect to the composition of lignocellulosic constituents, which mainly depends on the type, species, and even source of the biomass.[8] Table 1.3[11–38] shows the chemical composition of different lignocellulosic biomass.

1.2.1 Cellulose

Except for hemicellulose and lignin, the structural properties of cellulose are the same in all types of lignocellulosic biomass and it has been found to be a predominant scaffolding constituent of plant cell walls with a generic formula of $(C_6\,H_{12}O_5)n$.[9] However, the degree of polymerization and the length of the cellulose chain can vary based on the type of lignocellulosic biomass. Cellulose is a homopolymer made up of glucose subunits that are linked together through β-1,4-glycosidic bonds.[39,40] This linear polysaccharide is linked together through different inter- and intramolecular hydrogen bonds,

TABLE 1.3 Chemical composition of different lignocellulosic biomass.

Forest feedstocks	Cellulose	Hemicellulose	Lignin	References
Hardwood				
Bamboo	43	23.1	26.2	[11]
Aspen	49	18.2	25.6	[12]
Poplar	49.7	12.3	16.4	[13]
Eucalypt wood	43.4	15.3	25.5	[14]
Oak	41.3	28	24.3	[15]
Softwood				
Pine wood	42	14.7	26.3	[16]
Spruce	38.5	4.6	50.6	[17]
Douglas fir	40.9	18.41	29.30	[18]
Agriculture feedstocks				
Rice husk	30.1	13.5	22.4	[19]
Rice straw	36.1	24.7	16.8	[20]
Wheat straw	40.5–42.8	25–26	18–19	[21,22]
Corn stover (stalk, leaves, and husk)	35.6–41.2	26.1–37.4	10.7–21	[23,24]
Corncobs	32.6–37.4	31.7–34.2	16.9–19.2	[25,26]
Sorghum stalks	31–36.8	22-29.7	13.9–22	[27,28]
Banana waste (rachis and pseudostem)	20.1–26.1	9.6–11.2	10.1–10.8	[19]
Sugarcane bagasse	32–45	20–32	17–22	[29,30]
Pineapple leaf	46.17	19.28	2.8	[31]
Cotton stalks or stems	39.5	22.3	19.8	[32]
Coconut coir	34.8	18.3	35.8	[33]
Sunflower stalks	35.8	19.7	14.9	[34]
Sunn hemp	37	21.3	13.8	[28]
Grassy feedstocks				
Switchgrass (*Panicum virgatum* L.)	29.9	25.9	14.8	[35]
Elephant grass (*Pennisetum purpureum*)	22.6	20.9	19.4	[36]
Citronella spent biomass[a]	28.1	32.2	18.7	[37]
M. giganteus	45.3	27.1	9.8	[38]

[a]*After extraction of essential oil from citronella.*

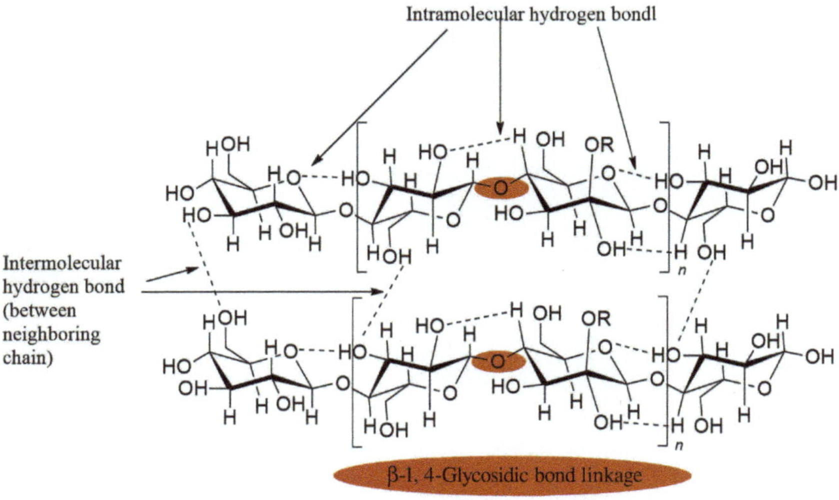

FIGURE 1.1 Chemical structure of cellulose chains.

which enable them to be packed into microfibrils[39] (Fig. 1.1). Such polymers are insoluble in water as linear polysaccharides containing hydroxyl groups are bonded with each other, which develops a hydrophobic scenario. Noticeably, the typical cellulose structure exists in two forms, that is, crystalline and amorphous. A major portion of cellulose structure present in crystalline nature, whereas a small percentage of unorganized chains are present in the amorphous form.[39,40]

1.2.2 Hemicelluloses

Hemicellulose is the second most abundant polymeric carbohydrate in nature and a major component of lignocellulosic biomass. Unlike homologous cellulose, hemicelluloses are heteropolymeric in nature, which is made up of different types of compounds, like pentoses (xylose, rhamnose, and arabinose), hexoses (glucose, mannose, and galactose), and organic acids (acetic, 4-*O*-methylglucuronic, D-glucuronic, and D-galacturonic acids). The backbone of hemicellulose consists of either a homopolymer (xylan, glucan, and galactan) or a heteropolymer (glucomannan), which are made up of their own respective monomeric sugar moieties that are mostly linked by β-1,4 glycosidic bonds and occasionally β-1,3-glycosidic bonds.[10] Moreover, several types of sugars and organic acid substituents are interlinked with xylan, glucan, galactan, or glucomannan backbones that prevent the formation of crystalline structures.[10,41] Instead they form long chains that can associate with cellulose through hydrogen bonding.[8,42]

Hemicellulose forms a complex network with cellulose by holding the microfibrils together, probably at specific connections rather than along the entire length of the microfibril.[43]

Xylans and glucomannans are the most relevant hemicelluloses present in different lignocellulosic materials. Among these, xylans are ubiquitously found and are the main hemicellulosic constituent of secondary cell walls.[10] About 20%–30% xylan is present in lignocellulosic sources, such as hardwoods and herbaceous plants. However, some tissues of various grasses contain about 50% xylans.[44] Xylans are generally available in large quantities as byproducts of agro-industries and the paper and pulp industries. Moreover, galactoglucomannans and glucomannans are the main hemicellulosic components of the secondary cell wall of softwoods. In contrast, mannan-type hemicelluloses are found in small quantities in hardwoods.[10] Based on biological origin, different hemicellulose structures can be found in nature.

1.2.3 Xylan backbone containing hemicelluloses

1.2.3.1 Arabinoglucuronoxylans

Arabinoglucuronoxylans (AGXs) or arabino-4-*O*-metylglucuronoxylans are major constituents of agricultural crop residues (nonwoody biomass); accounting for 25%–30%, whereas it is a minor constituent of softwoods; typically around 5%–10% of the dry mass. AGXs are made up of D-xylopyranose units linked with β-1,4-glycosidic bonds to xylan (homopolymer) backbones. Xylans have additional substitutes, such as an arabinofuranose ring attached through α-1,3-glycosidic bonds, 4-*O*-methyl-D-glucuronic acid, and/or glucuronic acid residues crosslinked through α-1,2-glycosidic bonds and the acetyl group found at the xylan C_2 position (Fig. 1.2).[41] The molar ratio of arabinose/glucuronic acid/xylose is 1:2:8.[45] The average degree of polymerization of AGX ranges between 50 and 185. Conversely to hardwood xylans, AGXs might be less acetylated but may contain low amounts of galacturonic acid and rhamnose.[45]

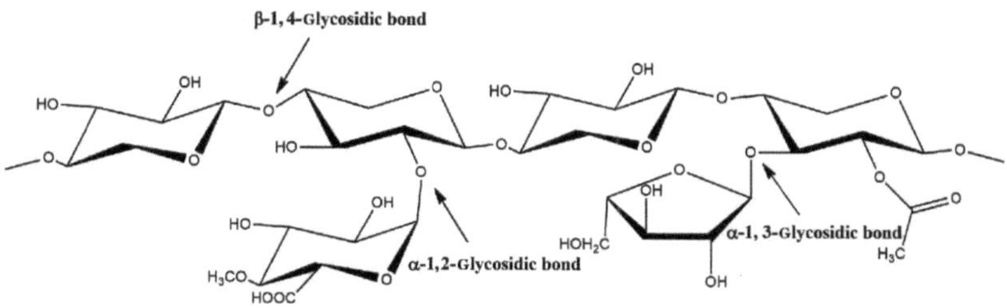

FIGURE 1.2 Chemical structure of arabinoglucuronoxylans.

1.2.3.2 Glucuronoxylans

Glucuronoxylans (GXs) or *O*-acetyl-4-*O*-methylglucuronoxylans are major hemicellulose constituents of hardwood biomass; accounting for 15%–30% of their dry mass.[46] The average degree of polymerization range of GXs is between 100 and 200.[45] The backbone of GX is made up of β-D-xylopyranose units that are linked with β-1,4 glycosidic bonds. Some xylose residues are substituted with acetyl groups at C_2 and/or C_3 which accounted to be 8%-17%.[46] In addition, the uronic acid group (4-*O*-methylglucuronic acid) is also substituted on the C_2 position of xylose units through α-1,2-glycosidic bonds (Fig. 1.3). One out of 10 xylose units has a uronic acid group. 4-*O*-Methylglucuronic side chains are more resistant to acids than the xylopyranose units and acetyl groups. Moreover, apart from main structural units, GXs may also contain lower quantities of galacturonic acid and L-rhamnose.

1.2.3.3 Arabinoxylans

Arabinoxylans (AXs) constitute the main hemicellulose structures of the cell walls of cereal grains. The backbone of AXs and their substitutions or side chains are synthesized by three essential membrane enzymes which belong to glycosyltransferases: (1) the backbone of AX, that is, xylan, is synthesized by AX synthase; (2) the substitution of arabinose onto the xylan backbone is constructed by arabinosyl transferases; and (3) ferulate or ferulic acid residue is attached to the arabinose unit by the action of AX feruloyl transferase.[47] Therefore the backbone of AX is composed of β-D-xylose units linked with β-1,4 glycosidic bonds. The structure of AXs is similar to that of hardwood xylans, but the xylan backbone is highly substituted with α-L-arabinose on the C_2 and/or C_3 position. Some α-L-arabinofuranose units are esterified (at the O-5 position of arabinose) with ferulic acid.[48,49] Esterified ferulic acids may also form intra- and intermolecular cross-linkages with the xylan backbone. Therefore the interaction of physical and/or covalent bonds with other cell wall compounds hampers xylan extractability. In addition, α-D-glucuronic acid units or its 4-*O*-methyl derivatives attach to the xylan backbone at the 2-*O* position.[48,49] Substitution of the *O*-acetyl groups may also occur in AXs[48,49] (see Fig. 1.4).

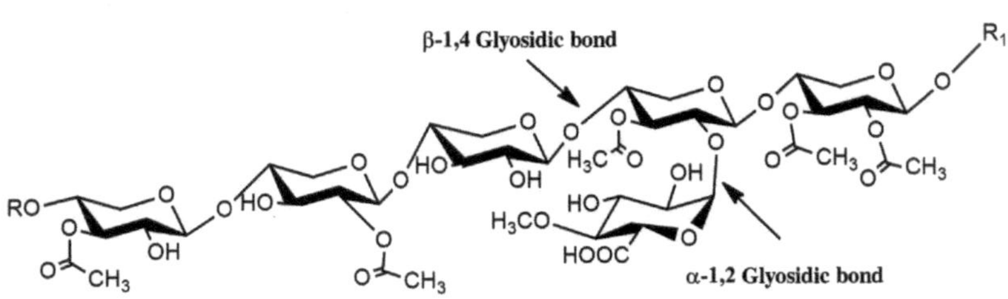

FIGURE 1.3 Chemical structure of glucuronoxylans.

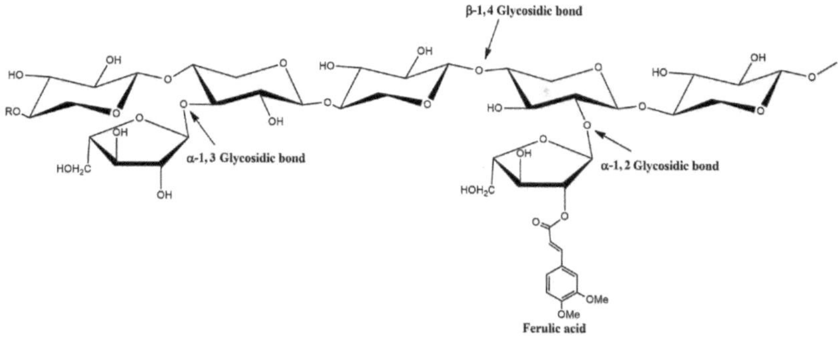

FIGURE 1.4 Chemical structure of arabinoxylans.

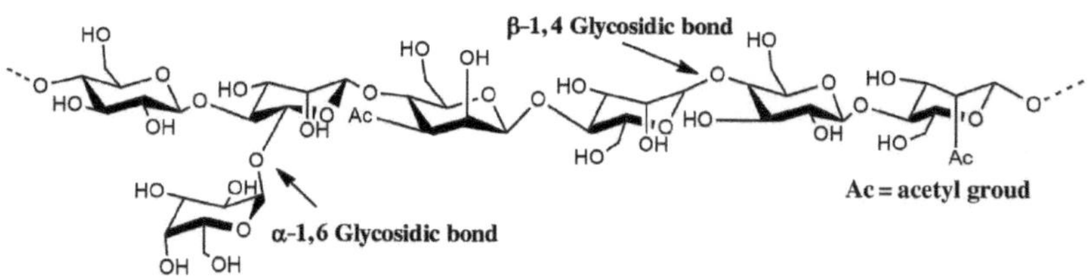

FIGURE 1.5 Chemical structure of glucogalactomannan.

1.2.4 Glucan and glucomannan backbone containing hemicelluloses

Softwood hemicelluloses contain a glucomannan backbone that is formed through the random distribution of β-D-mannopyranosyl and β-D-glucopyranosyl units linked with β-1,4 glycosidic bonds. This backbone partially is substituted by the α-*d*-galactopyranose and acetyl groups. In addition to this, the primary cell wall of softwood biomass also constitutes xyloglucans (XGs) that are made up of β-1,4-linked glucose (glucan) backbone with α-linked xylose substitutions.

1.2.4.1 Galactoglucomannans

Galactoglucomannans or *O*-acetyl-galactoglucomannans are major hemicellulosic constituents of softwoods, accounting for 20%–25% (w/w) of their dry mass.[45] A typical structure of glucogalactomannan (GGM) consists of a glucomannan backbone, which is made up of β-D-glucopyranose (Glup) units and β-D-mannopyranose (Manp) units attached with β-1,4 glycosidic bonds.[50] This backbone is partially substituted by α-D-galactopyranose units through α-1,6-glycosidic linkages, whereas the acetyl groups attach at the C_2–OH or C_3–OH positions of glucose and mannose (Fig. 1.5). GGM contains around 6% acetyl groups, which approximates to one acetyl group for every 3–4 hexose units (Glup or Manp).[46] Based on the galactose content, glucomannan can be categorized into two types: (1) galactose-rich glucomannans, which are normally referred to as "galactoglucomannans" and (2) galactose poor glucomannans, which simply called as glucomannans. The molar ratios of galactose/glucose/mannose were found to be 1:1:3 in galactoglucomannans, whereas they are 0.1−0.2:1:3−4 in glucomannans. However, the typical overall molar ratios of galactose/glucose/mannose in softwoods are 0.5:1:3.5.[45] The average degree of polymerization (DP) of GGM ranges between 40 and 100.[46]

Unlike softwoods, the glucomannan content of hardwoods is low, that is, approximately 1%−4% of the dry biomass. The typical glucomannan structure of hardwood biomass is made up of β-D-Glup units and β-D-Manp units (with β-1,4 glycosidic bonds) with 1:1−2 molar ratios.

1.2.4.2 Xyloglucans

XGs are found to be major hemicellulosic constituents of the primary cell walls of hardwood biomass. A typical structure of XGs is composed of a β-1,4-linked D-glucose backbone, and three out of four glucose units are substituted at the *O*-6 position with D-xylose. Moreover, around half of the α-D-xylose units in XG of dicotyledons are attached to β-D-galactose or L-arabinose units, and some of the β-D-galactose units are substituted with α-L-fucosyl units.[51] L-Arabinose and D-galactose residues can be attached to xylose residues, thus forming di- or triglycosyl side chains. In addition, *O*-linked acetyl groups are also present in XGs.[52,53] XGs are also found in minor fractions in the cell walls of grasses and the structure also varies; for instance, less branching with only a small number of β-D-galactose units substituted with xylose and no L-arabinose and α-L-fucosyl side chains present (Fig. 1.6). With the help of hydrogen bonds, XGs bind with cellulose microfibrils, which provides structural integrity to the cellulose network.[51−53]

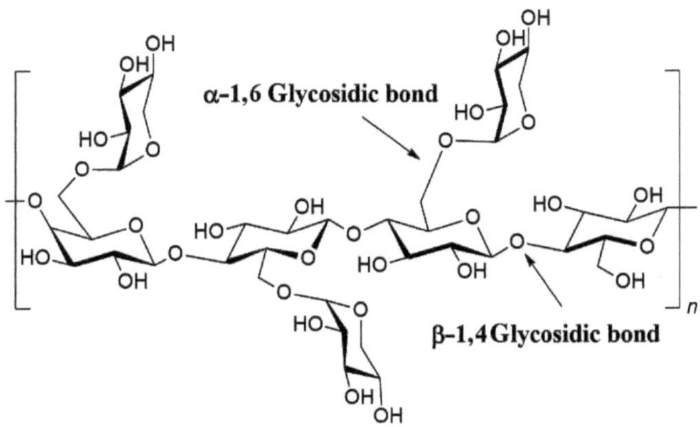

FIGURE 1.6 Chemical structure of xyloglucans.

1.2.4.3 Arabinogalactans

Arabinogalactans (AGs) are a form of slightly water-soluble polysaccharide, which are present in hardwood biomass, like Siberian larch wood (*Larix sibirica*).[54] The structure of AG in Siberian larch wood is similar to that of other types of larch wood.[55] In both cases, this polymer displays a highly branched framework. The backbone of AG consists of galactopyranose units linked with α-1,3-glycosidic bonds with branching points at the C-6 atom in the majority of these residues. This backbone is substituted with galactopyranosyl residues (di-*O*-substitution) through α-1,6-glycosidic bonds. Similarly, arabinopyranosyl or arabinofuranosyl residues are also substituted on the backbone of AG through α-1,6-glycosidic bonds (Fig. 1.7).

1.3 Lignin

Lignin is highly aromatic in nature and the third most abundant polymer in the world. Unlike cellulose and hemicellulose, lignin does not contain any carbohydrates (sugars) in its polymeric structure. Lignin is a three-dimensional methoxylated phenylpropane compound that is uniquely responsible for the structural rigidity of lignocellulosic biomass, which typically covers hemicellulose and cellulose.[9] Elaborately, lignin contains guaiacyl, *p*-hydroxyphenyl, and syringyl units that are polymerized by ether bonds or carbon–carbon linkages.[56] The chemical structure and composition of lignin are significantly influenced by origin.[57] Based on the type, species, origin and genetic variability, chemical composition, and structural properties of lignin, there are differences mainly with regards to the content

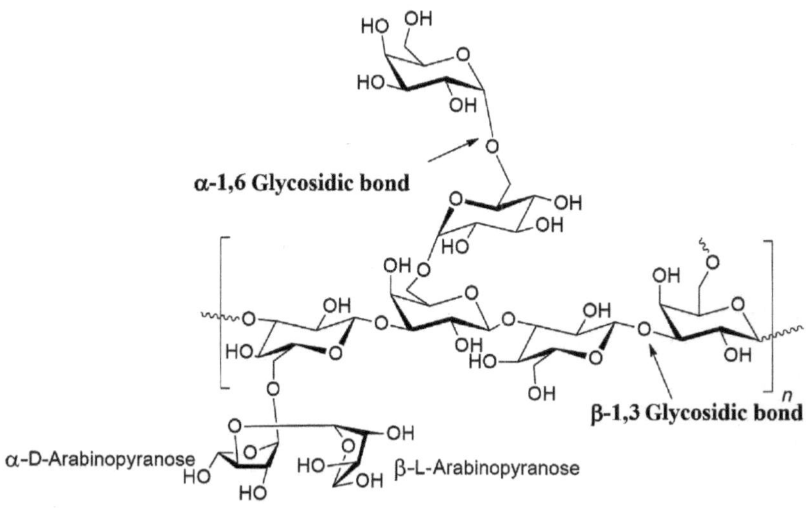

FIGURE 1.7 Chemical structure of arabinogalactan.

of *p*-coumaryl, coniferyl, and syringyl alcohols.[9] Coniferyl alcohol occurs in all types of lignins, and it is understood to be the dominant monomer in softwoods. Hardwood lignin contains both coniferyl alcohol and syringyl alcohol units, estimated to be roughly in equal proportions, while agricultural crops and grasses may also contain *p*-coumaryl alcohol units.[43,58]

1.4 Theoretical bioethanol potential of lignocellulosic biomass

Several technologies are being investigated to convert cellulose and hemicellulose into monosaccharides and subsequently into bioethanol through fermentation. In general, the theoretical bioethanol potential of lignocellulosic materials mainly depends on its holocellulose content, which commonly accounts for 50%–65% (w/w) of their dry mass (Table 1.3). Based on the stoichiometric equation shown in Eq. (1.i), the maximum theoretical yield (g_p/g_s) of ethanol from 1 g of substrate (C_6 or C_5 sugar) is 0.51 g. Therefore the theoretical bioethanol potential of lignocellulosic materials can be calculated using Eqs. (1.1) and (1.2).

$$C_6H_{12}O_6 \rightarrow 2C_2H_5OH + 2CO_2 \tag{1.i}$$

$$\text{Theoretical bioethanol potential}(\text{kg}/t) = C_s(\text{kg}/t) \times \frac{0.51\ \text{kg ethanol}}{\text{kg sugar}} \tag{1.1}$$

$$V(L/t) = \frac{\text{Eq. (1.1)}}{\rho} \tag{1.2}$$

where C_s is the concentration of sugars (C_6 or C_5) present in the lignocellulosic biomass, 0.51 is the theoretical ethanol yield (gp/gs) constant, and ρ is the ethanol density (0.789 kg/L at 20°C). For instance, substituting the holocellulose (374 kg of glucan, 211 kg of xylan, and 29 kg of arabinan) content (614 kg/dry ton) of corn stover into Eqs. (1.1) and (1.2) will give, around 313.14 kg or 396.8 L of bioethanol.

1.5 Factors affecting bioethanol production

In a practical way, there are several factors that principally hamper the hydrolysis of holocellulose into monomeric sugars and its subsequent conversion into ethanol; for instance, sugar decomposition during the pretreatment process and incomplete enzymatic hydrolysis due to the interference of recalcitrant lignin. Apart from fermentable sugars (glucose, xylose, arabinose, etc.), products such as furans, organic acid, and phenolic compounds formed during the hydrolysis of lignocellulosic biomass are considered as fermentative inhibitors that deter microbial metabolic growth and lead to the lowering of the fermentation efficiency.

The bioethanol production process from lignocellulosic biomass is completely different from that of food grade materials, such as corn and sugarcane. The bioethanol production process from sugarcane involves the simple extraction of sugarcane juice (which contains highly concentrated sucrose) and then fermentation (of hexose sugar using yeast).[59] In the case of starchy-type materials, cooking (85°C−105°C) is the primary step, in which the starch is gelatinized, thereafter it is hydrolyzed by amylase (enzyme) at 30°C−60°C for the production of glucose units that are further subjected to a fermentation process.[59] Therefore bioethanol production from food materials, such as corn (starch) and sugarcane (sucrose), is an industrially well-established process and the cost of ethanol production was found to be US$1.53 and US$1.14 per gallon, respectively.[60] However, the cost of bioethanol production from lignocellulosic biomass is found to be US$2.15 per gallon,[61] which is approximately 40%−88% higher than that of bioethanol produced from food grade materials. This could be due to the fact that lignocellulosic biomass require extreme conditions (like pretreatment at high temperatures and high enzyme loading) to convert the polymeric sugars into monomeric sugars.

Unlike starchy materials, the holocellulose fraction of lignocellulosic biomass cannot be gelatinized by hydrothermal processes (or simply cooking) and is not easily hydrolyzed by enzymes. This can be attributed to the fact that structural rigidity induced by lignin shows a recalcitrance effect on holocellulose hydrolysis.[62] Therefore a pretreatment process should be employed to hydrolyze most of the hemicellulosic fraction and disturb the complex network of lignin, which also enhances cellulose accessibility during enzymatic hydrolysis.[39] Pretreatment of lignocellulosic biomass is generally carried out at 121°C−180°C for 10−120 min using dilute acid or subcritical water as a catalyst.[1,39] Although performing the pretreatment process under high severity conditions cannot completely remove the lignin fraction, a portion of cellulose covered by lignin may not be accessed by the cellulase enzyme.[63] Moreover, during the pretreatment process, some sugar loss can also occur due to the combined action of temperature and catalyst strength.[41] Principally, during the pretreatment process, pentose and hexose sugars are dehydrated to form furfural and 5-HMF (also called sugar decomposition products), respectively. The decomposition of sugar into furans is generally dependent on the pretreatment severity.[27,64,65] These sugar decomposition products are toxic to microbial growth and deter the ethanol conversion efficiency during the fermentation process.[66]

Currently, the National Renewable Energy Laboratory (NREL, United States) is producing bioethanol from corn stover at a pilot-scale level.[61] According to a technical report (NREL/TP-5100-61563) by NREL in 2014, dilute sulfuric acid pretreatment (160°C−190°C for 1−10 min with 0.3−0.4 wt.%) and subsequent enzymatic hydrolysis (19−33 mg of protein/gram of glucan at 50°C for 1−4 days) of corn stover was carried to

TABLE 1.4 Bioethanol potential of corn stover according to the NREL report: NREL/TP-5100-61563.

Structural carbohydrates	I_{sc} (kg/t)	S_{uob} (%)	Monomeric sugar yield (kg) [Eq. (1.3)]	F_{efi} (%)	Ethanol (kg/t) [Eq. (1.4)]	Ethanol (L/t) [Eq. (1.2)]
Cellulose	374	14.8	318.40	95	154.26	195.51
Xylan	211	15	176.35	82	73.74	93.46
Arabinan	29	8	26.68	55	6.12	7.75
Total	614	NA	521.43	NA	234.12	296.72

NA, Data not available.

maximize the hydrolysis of hemicellulose and cellulose into monomeric sugars. Under these conditions, NREL conducted pilot-scale demonstration runs (five batches); the average values of hemicellulose and cellulose conversion are represented in Table 1.4. In this process, around 85.2% of cellulose, 85% of xylan, and 92% of arabinan were converted into monomeric sugars. Therefore at an average of 14.8% of cellulose, 15% of xylan, and 8% of arabinan are unavailable for the ethanol conversion process. This could be due to the loss of sugars and unconverted holocellulose (cellulose and hemicellulose) during pretreatment and enzymatic hydrolysis, respectively. Furthermore, the ethanol conversion efficiency of glucose, xylose, and arabinose was found to be 95%, 82%, and 55%, respectively (Table 1.4). This eventually produces between 266.47 and 276.33 L of bioethanol from 1 mt of dry corn stover [Eqs. (1.3) and (1.4); then (1.2)]. Dilute acid pretreatment followed by enzymatic hydrolysis of lignocellulosic biomass is the most preferred method globally. Therefore according to these calculations, the bioethanol potential of any lignocellulosic biomass can be predicted according to Eqs. (1.2)–(1.4):

$$\mathrm{MSY}(\mathrm{kg}/t) = \left[I_{\mathrm{sc}}\left(1 - \frac{S_{\mathrm{uob}}}{100}\right)\right] \tag{1.3}$$

$$\mathrm{EY}(\mathrm{kg}/t) = \left(\frac{T_{\mathrm{EY}} \times F_{\mathrm{efi}} \times \mathrm{Eq.\ (1.i)}}{100}\right) \tag{1.4}$$

$$V = \frac{\mathrm{Eq.\ (1.1)}}{\rho} \tag{1.2}$$

where MSY is the monomeric sugar (glucose, xylose, or arabinose) yield obtained after pretreatment and enzymatic hydrolysis, I_{sc} is the initial sugar content present in the biomass, S_{uob} is the percentage of unobtainable sugars which includes sugar loss (decomposition of glucose, xylose, or arabinose) during pretreatment, sugar oligomers formed during pretreatment and enzymatic hydrolysis, and holocellulose remaining after pretreatment and subsequent enzymatic hydrolysis. *EY*, Ethanol yield; T_{EY}, theoretical ethanol yield (which is constant 0.51 g_p/g_s); F_{efi}, fermentation efficiency (%); *V*, volume (L); *m*, mass (kg); and ρ, ethanol density (0.789 kg/L).

1.6 Conclusion

The potential availability of lignocellulosic materials, especially agricultural crop residues, has gained attention for bioethanol production. The utilization of bioethanol as a fuel source not only reduces the emission of greenhouse gasses, but also builds energy security and countries need not depend on the importation of crude oil. The Government of India, the United States, Brazil, China, and the EU-28 have already directed the use ethanol blends in transportation vehicles at a proportion of E5%–E15%. These ethanol blends can be used in regular motorcycles and cars without modifying their combustion engines. However, unlike edible materials (corn starch and sugarcane juice) bioethanol production from lignocellulosic materials is not well established yet and there should be more focus on conversion technologies where the economic cost will play a vital role.

Acknowledgment

Authors are thankful to Dr. Radha Krishna Gattu, Department of Chemistry, Indian Institute of Technology Guwahati for his valuable help in the representation of hemicellulosic structures of different lignocellulosic materials.

References

1. Behera S, Arora R, Nandhagopal N, Kumar S. Importance of chemical pretreatment for bioconversion of lignocellulosic biomass. *Renew Sustain Energy Rev* 2014;**36**:91–106.
2. Baeyens J, Kang Q, Appels L, Dewil R, Lv Y, Tan T. Challenges and opportunities in improving the production of bio-ethanol. *Prog Energy Combust Sci* 2015;**47**:60–88. Available from: https://doi.org/10.1016/j.pecs.2014.10.003.
3. Zabed H, Sahu JN, Suely A, Boyce AN, Faruq G. Bioethanol production from renewable sources: current perspectives and technological progress. *Renew Sustain Energy Rev* 2017;**71**:475–501. Available from: https://doi.org/10.1016/j.rser.2016.12.076.
4. Yang B, Systems B. Strengths, challenges, and opportunities for hydrothermal pretreatment in lignocellulosic biorefineries. *Biofuels Bioprod Biorefin* 2018;**12**:125–38.
5. Saini JK, Saini R, Tewari L. Lignocellulosic agriculture wastes as biomass feedstocks for second-generation bioethanol production: concepts and recent developments. *3 Biotech* 2015;**5**(4):337–53.
6. Gupta A, Verma JP. Sustainable bio-ethanol production from agro-residues: a review. *Renew Sustain Energy Rev* 2015;**41**:550–67.
7. Sarkar N, Ghosh SK, Bannerjee S, Aikat K. Bioethanol production from agricultural wastes: an overview. *Renew Energy* 2012;**37**(1):19–27.

8. Marriott PE, Gómez LD, McQueen-Mason SJ. Tansley review unlocking the potential of lignocellulosic biomass through plant science. *New Phytol* 2015;**209**(4):1366–81.
9. Zhou CH, Xia X, Lin CX, Tong DS, Beltramini J. Catalytic conversion of lignocellulosic biomass to fine chemicals and fuels. *Chem Soc Rev* 2011;**40**(11):5588–617.
10. Gírio FM, Fonseca C, Carvalheiro F, Duarte LC, Marques S, Bogel-Łukasik R. Hemicelluloses for fuel ethanol: a review. *Bioresour Technol* 2010;**101**(13):4775–800.
11. Tri CL, Khuong LD, Kamei I. The improvement of sodium hydroxide pretreatment in bioethanol production from Japanese bamboo *Phyllostachys edulis* using the white rot fungus Phlebia sp. MG-60. *Int Biodeterior Biodegrad* 2018;**133**:86–92.
12. Goshadrou A, Karimi K, Lefsrud M. Characterization of ionic liquid pretreated aspen wood using semi-quantitative methods for ethanol production. *Carbohydr Polym* 2013;**96**(2):440–9.
13. Frederick N, Zhang N, Ge X, Xu J, Pelkki M, Martin E, et al. Poplar (*Populus deltoides* L.): the effect of washing pretreated biomass on enzymatic hydrolysis and fermentation to ethanol. *ACS Sustain Chem Eng* 2014;**2**(7):1835–42.
14. Chen L, Fu S. Enhanced cellulase hydrolysis of eucalyptus waste fibers from pulp mill by tween80-assisted ferric chloride pretreatment. *J Agric Food Chem* 2013;**61**(13):3293–300.
15. Kim J, Lee J, Kim KH, Ok YS, Jeon YJ, Kwon EE. Pyrolysis of wastes generated through saccharification of oak tree by using CO_2 as reaction medium. *Appl Therm Eng* 2017;**110**:335–45.
16. Du X, Lucia LA, Ghiladi RA. Development of a highly efficient pretreatment sequence for the enzymatic saccharification of loblolly pine wood. *ACS Sustain Chem Eng* 2016;**4**(7):3669–78.
17. Cannella D, Sveding PV, Jørgensen H. PEI detoxification of pretreated spruce for high solids ethanol fermentation. *Appl Energy* 2014;**132**:394–403.
18. Zhu JY, Chandra MS, Gu F, Gleisnera R, Reinera R, Sessionsd J, et al. Using sulfite chemistry for robust bioconversion of Douglas-fir forest residue to bioethanol at high titer and lignosulfonate: a pilot-scale evaluation. *Bioresour Technol* 2015;**179**:390–7.
19. Ebrahimi M, Caparanga AR, Ordono EE, Villaflores OB, Pouriman M. Effect of ammonium carbonate pretreatment on the enzymatic digestibility, structural characteristics of rice husk and bioethanol production via simultaneous saccharification and fermentation process with *Saccharomyces cerevisiae* Hansen 2055. *Ind Crops Prod* 2017;**101**:84–91.
20. Zhang Q, Huang H, Han H, Qiu Z, Achal V. Stimulatory effect of in-situ detoxification on bioethanol production by rice straw. *Energy* 2017;**135**:32–9.
21. Tomás-Pejó E, Fermoso J, Herrador E, Hernando H, Jiménez-Sánchez S, Ballesteros M, et al. Valorization of steam-exploded wheat straw through a biorefinery approach: bioethanol and bio-oil co-production. *Fuel* 2017;**199**:403–12.
22. Yuan Z, Wen Y, Li G. Bioresource technology production of bioethanol and value added compounds from wheat straw through combined alkaline/alkaline-peroxide pretreatment. *Bioresour Technol* 2018;**259**:228–36.
23. Liu ZH, Qin L, Li BZ, Yuan YJ. Physical and chemical characterizations of corn stover from leading pretreatment methods and effects on enzymatic hydrolysis. *ACS Sustain Chem Eng* 2015;**3**(1):140–6.
24. Reddy N, Yang Y. Biofibers from agricultural byproducts for industrial applications. *Trends Biotechnol* 2005;**23**(1):22–7.
25. Qing Q, Guo Q, Zhou L, Wan Y, Xu Y, Ji H, et al. Catalytic conversion of corncob and corncob pretreatment hydrolysate to furfural in a biphasic system with addition of sodium chloride. *Bioresour Technol* 2017;**226**:247–54.

26. Gupta R, Hemansi, Gautam S, Shukla R, Kuhad RC. Study of charcoal detoxification of acid hydrolysate from corncob and its fermentation to xylitol. *J Environ Chem Eng* 2017;**5**(5):4573–82.
27. Deshavath NN, Mahanta S, Goud VV, Dasu VV, Rao PS. Chemical composition analysis of various genetically modified sorghum traits: pretreatment process optimization and bioethanol production from hemicellulosic hydrolyzates without detoxification. *J Environ Chem Eng* 2018;**6**(4):5625–34.
28. Kamireddy SR, Li J, Abbina S, Berti M, Tucker M, Ji Y. Converting forage sorghum and sunn hemp into biofuels through dilute acid pretreatment. *Ind Crops Prod* 2013;**49**:598–609.
29. Timung R, Mohan M, Chilukoti B, Sasmal S, Banerjee T, Goud VV. Optimization of dilute acid and hot water pretreatment of different lignocellulosic biomass: a comparative study. *Biomass Bioenergy* 2015;**81**:9–18.
30. Arni SAl. Extraction and isolation methods for lignin separation from sugarcane bagasse: a review. *Ind Crops Prod* 2018;**115**:330–9.
31. Chintagunta AD, Ray S, Banerjee R. An integrated bioprocess for bioethanol and biomanure production from pineapple leaf waste. *J Clean Prod* 2017;**165**:1508–16.
32. Patel VR. Cost-effective sequential biogas and bioethanol production from the cotton stem waste. *Process Saf Environ Prot* 2017;**111**:335–45.
33. Ebrahimi M, Caparanga AR, Ordono EE, Villaflores OB. Evaluation of organosolv pretreatment on the enzymatic digestibility of coconut coir fibers and bioethanol production via simultaneous saccharification and fermentation. *Renew Energy* 2017;**109**:41–8.
34. Eom IY, Yu JH. Structural characterization of the solid residue produced by hydrothermal treatment of sunflower stalks and subsequent enzymatic hydrolysis. *J Ind Eng Chem* 2015;**23**:72–8.
35. Papa G, Rodriguez S, George A, Schievano A, Orzi V, Sale KL, et al. Comparison of different pretreatments for the production of bioethanol and biomethane from corn stover and switch grass. *Bioresour Technol* 2015;**183**:101–10.
36. Eliana C, Jorge R, Juan P, Luis R. Effects of the pretreatment method on enzymatic hydrolysis and ethanol fermentability of the cellulosic fraction from elephant grass. *Fuel* 2014;**118**:41–7.
37. Timung R, Deshavath NN, Goud VV, Dasu VV. Effect of subsequent dilute acid and enzymatic hydrolysis on reducing sugar production from sugarcane bagasse and spent *Citronella* biomass. *J Energy*. 2016;**2016**:8506214. Available from: https://doi.org/10.1155/2016/8506214.
38. Światek K, Lewandowska M, Światek M, Bednarski W, Brzozowski B. The improvement of enzymatic hydrolysis efficiency of rape straw and *Miscanthus giganteus* polysaccharides. *Bioresour Technol* 2014;**151**:323–31.
39. Kumar P, Barrett DM, Delwiche MJ, Stroeve P. Methods for pretreatment of lignocellulosic biomass for efficient hydrolysis and biofuel production. *Ind Eng Chem Res* 2009;**48**(8):3713–29.
40. Lee HV, Hamid SBA, Zain SK. Conversion of lignocellulosic biomass to nanocellulose: structure and chemical process. *Sci World J* 2014;**2014**:20. Available from: https://doi.org/10.1155/2014/631013.
41. Deshavath NN, Mohan M, Veeranki VD, Goud VV, Pinnamaneni SR, Benarjee T. Dilute acid pretreatment of sorghum biomass to maximize the hemicellulose hydrolysis with minimized levels of fermentative inhibitors for bioethanol production. *3 Biotech* 2017;**7**(2):1–12.

42. Busse-Wicher M, Gomes TCF, Tryfona T, Nikolovski N, Stott K, Grantham NJ, et al. The pattern of xylan acetylation suggests xylan may interact with cellulose microfibrils as a twofold helical screw in the secondary plant cell wall of *Arabidopsis thaliana*. *Plant J* 2014;**79**(3):492–506.
43. Park YB, Cosgrove DJ. A revised architecture of primary cell walls based on biomechanical changes induced by substrate-specific endoglucanases. *Plant Physiol* 2012;**158**(4):1933–43.
44. Ebringerová A, Hromádková Z, Heinze T. Hemicellulose. *Adv Polym Sci* 2005;**186**:1–67.
45. Pereira H, Graca J, Rodrigues JC. *Wood chemistry in relation to quality*. 2003.
46. Alén R. *Structure and chemical composition of wood* 2000. Available from: https://doi.org/10.1002/prca.201200064.
47. Butardo VM, Sreenivasulu N. Tailoring grain storage reserves for a healthier rice diet and its comparative status with other cereals. *Int Rev Cell Mol Biol* 2016;**323**:31–70.
48. Mendis M, Simsek S. Arabinoxylans and human health. *Food Hydrocoll* 2014;**42**(P2):239–43.
49. Correia MAS, Mazumder K, Brás JLA, et al. Structure and function of an arabinoxylan-specific xylanase. *J Biol Chem* 2011;**286**(25):22510–20.
50. Sims IM, Craik DJ, Bacic A. Structural characterisation of galactoglucomannan secreted by suspension-cultured cells of *Nicotiana plumbaginifolia*. *Carbohydr Res* 1997;**303**(1):79–92.
51. Park YB, Cosgrove DJ. Xyloglucan and its interactions with other components of the growing cell wall. *Plant Cell Physiol* 2015;**56**(2):180–94.
52. de Vries RP, Visser J, Ronald P, de Vries R,P. *Aspergillus* enzymes involved in degradation of plant cell wall polysaccharides. *Microbiol Mol Biol Rev* 2001;**65**(4):497–522.
53. Sims IM, Munro SLA, Currie G, Craik D, Bacic A. Structural characterisation of xyloglucan secreted by suspension-cultured cells of *Nicotiana plumbaginifolia*. *Carbohydr Res* 1996;**293**(2):147–72.
54. Paulsen BS, Barsett H. Bioactive pectic polysaccharides. *Adv Polym Sci* 2005;**186**:69–101.
55. Antonova GF, Usov AI. Structure of an arabinogalactan from the wood of the S*iberian larch* (*Larix sibirica* Ledeb.). *Sov J Bioorganic Chem Dec* 1984;**10**(12):907–12.
56. Calvo-Flores FG, Dobado JA. Lignin as renewable raw material. *ChemSusChem* 2010;**3**(11):1227–35.
57. Zakzeski J, Bruijnincx PCA, Jongerius AL, Weckhuysen BM. The catalytic valorization of ligning for the production of renewable chemicals. *Chem Rev* 2010;**110**:3552–99.
58. Pandey KK. A study of chemical structure of soft and hardwood and wood polymers by FTIR spectroscopy. *J Appl Polym Sci* 1999;**71**(12):1969–75.
59. Vohra M, Manwar J, Manmode R, Padgilwar S, Patil S. Bioethanol production: feedstock and current technologies. *J Environ Chem Eng* 2014;**2**(1):573–84.
60. Humbird D, Davis RE, Tao L, Kinchin C, Hsu D, Aden A, et al. Process design and economics for biochemical conversion of lignocellulosic biomass to ethanol. *Renew Energy* 2011;**303**:147.
61. Tao L, Schell D, Davis R, Tan E, Elander R, Bratis A. NREL 2012 Achievement of ethanol cost targets: biochemical ethanol fermentation via dilute-acid pretreatment and enzymatic hydrolysis of corn stover; 2014 p. 81.
62. Jin M, Gunawan C, Uppugundla N, Balan V, Dale BE. A novel integrated biological process for cellulosic ethanol production featuring high ethanol productivity, enzyme recycling and yeast cells reuse. *Energy Environ Sci* 2012;**5**(5):7168–75.

63. Arora A, Carrier DJ. Understanding the pine dilute acid pretreatment system for enhanced enzymatic hydrolysis. *ACS Sustain Chem Eng* 2015;**3**(10):2423–8.
64. Deshavath NN, Dasu VV, Goud VV, Rao PS. Development of dilute sulfuric acid pretreatment method for the enhancement of xylose fermentability. *Biocatal Agric Biotechnol* 2017;**11**:224–30.
65. Mohan M, Timung R, Deshavath NN, Banerjee T, Goud VV, Dasu VV. Optimization and hydrolysis of cellulose under subcritical water treatment for the production of total reducing sugars. *RSC Adv* 2015;**5**:103265–75.
66. Palmqvist E, Hahn-Hägerdal B. Fermentation of lignocellulosic hydrolysates I: inhibitors and mechanism of inhibition. *Bioresour Technol* 2000;**74**(1):25–33.

Chapter 2

Rural biorefinery: A viable solution for production of fuel and chemicals in rural India

Nilutpal Bhuyan[1], Debashis Sut[1], Lina Gogoi[1], Rupam Kataki[1] and Vaibhav V. Goud[2]

[1]*Department of Energy, Tezpur University, Tezpur, India,* [2]*Department of Chemical Engineering, Indian Institute of Technology, Guwahati, India*

2.1 Introduction

Over the past few centuries, conventional fossil fuels have been one of the most significant factors in the rapid advancement of civilization, contributing about 88% of the world's energy requirements.[1] The global demand for petroleum oil is around 84 million barrels per day and by 2030 it is expected to increase to about 116 million barrels per day. In addition to its use as a primary energy source for the transport sector (about 60%), fossil fuels are also being used as a feedstock in oil refineries for the production of chemicals and other related products.[2,3] It is observed that almost 4% of oil is used globally for the production of chemicals and essential commodities including plastic materials.[2,4,5] However, with the increasing population and industrialization, the consumption of conventional fuel has also increased manifold, which has an adverse impact on the environment like greenhouse gas (GHG) emissions and global warming.[6]

In India, more than 80% of the energy consumed in rural households comes from noncommercial or traditional sources, such as animal dung, agricultural wastes, and fuelwood. However, the use of traditional biomass as a fuel is hindered by its low combustion efficiency, resulting in higher levels of indoor air pollution and an increase in respiratory infections both in adults and children. Thus growing environmental concerns as well as the declining petroleum reserves have contributed to a growing interest in the research and development of renewable energy as an alternative to fossil fuel.[6,7] Among the renewables, biomass energy appears to have maximum potential as these can be used to produce energy as well as chemicals and materials.[8]

Sustainable Bioenergy. DOI: https://doi.org/10.1016/B978-0-12-817654-2.00002-2

According to a 2011 census, 68.84% of the total population of India lives in nearly 6 lakh villages.[9] The main livelihood option in rural India is agriculture and its allied occupations, and therefore, these areas also produce various waste biomasses, and the utilization of these can be the key to a number of inadequacies in rural India. The use of appropriate technology for waste biomass utilization in a biorefinery approach can be helpful to meet some of the needs of the rural areas to a certain extent, and thereby can contribute to the sustainable development of such areas as well.[10] As such, it can open up many possibilities like generating an additional income for farmers, creating new jobs for rural youths, providing clean fuel, hygienic and smoke-free kitchens, by-products for the management of soil, water purification, and others.

There are various reports available on the biorefinery concept using biomass such as wheat straw,[11] corn,[12] forest residues,[7] among others. In this chapter, different biomass feedstocks with region-wise potential for agricultural waste generation, the conversion of biomass to biofuel and chemicals, and the potential of the implementation of biorefineries in rural India are discussed.

2.2 Biomass resources in rural India

Biomass is defined as bioresidue obtained from living or dead plant wastes from the forestry, agriculture, or food industries as well as water-based vegetation.[13] Biomass resources are basically composed of a mixture of carbon (C), hydrogen (H), nitrogen (N), and oxygen (O). Approximately 100 billion tons of organic dry matter of land biomass and 50 billion tons of aquatic biomass are found on Earth per annum.[6] Only 1.25% of the total land biomass is used as food while some fraction of it is used as energy sources and feedstock for industry. The rest of the biomass remains unused, which can be used as a basic material for the production of chemicals. Currently, various biomass-based products, such as starch, sugar, oil and fat, cellulose, and rubber, have been used industrially.[6,14] Thus biomass-based biorefinery has received strong scientific emphasis.

India is prosperous in various biomass resources such as wood and its waste, agro-crop residues, agro-industrial processing residues, animal residues, and aquatic plants, which could be utilized for its energy requirements. There are 150 million hectares (Mha) of land available in India, including land used for dumping, degraded crops, mixed crops, and vegetation.[15]Approximately 50 million metric tons (MMT) of liquid fuels are consumed per annum in India, but with the complete utilization of the available biomass waste, India can generate almost twice of that amount per annum.[16] The major biomass sources found in rural India are discussed next.

2.2.1 Agricultural waste (or residues)

The agriculture sector is considered to be the pillar of India's financial system. Although agriculture contributes only 17% to India's GDP, nearly 60% of the population depend on agricultural activity.[17] Different agricultural practices are carried out in about 60% of the country's land area.[18]

India is the largest producer of jute in the world and the second largest producer of rice, wheat, sugarcane, cotton, and groundnuts. In India during the period of 2011–12, the total production of wheat, rice, maize, millets, sugarcane, fiber crops (jute, mesta, cotton), pulses, and oilseed crops were 93.9, 104.6, 21.6, 20.7, 357.7, 8.1, 17.2, and 30.0 Mt, respectively.[19] It is estimated that about 500–550 Mt of crop residues are generated from these crops both on-field and off-field annually in India.[20] However, various factors, such as diverse cropping practices, different soil types, fertilizers, and various agroclimatic environments, affect the distribution, amount, shape, composition, and availability of residues. The obtained crop residues are used as feed for animals, in soil mulching, manure making, thatching for rural homes, and fuel for domestic and industrial use.[3,6] These residues have the potential to be used as raw material for the production of energy.[13,21]

Agricultural residues or wastes are organic materials generated as derivatives in agricultural production processes or in agricultural harvesting. These can be categorized as primary and secondary residues. Primary residues are field-based residues, obtainable during harvest. On the other hand, secondary residues are known as processing-based residues that are gathered during processing.[13,22] Rice straw, sugarcane tops, maize stalks, coconut empty bunches and fronds, arhar stalks, jute sticks, etc., are examples of primary residues, whereas rice husk, coffee husk, coconut shells, coconut husk, groundnut shells, and bagasse are secondary residues. Primary residues are also used as animal feed, fertilizers, and others, which minimize their availability for biorefinery application, while secondary residues are obtained in large quantities at the yielding site and can also be considered as a feedstock. These residues, comprising of stalks, husks, stems, leaves, shells, cores, and other by-products of harvested crops are important biomass resources as they represent a potential feedstock for biorefinery operations.

In a developing country like India, biomass such as fuelwood, animal dung, and agricultural residues are usually used as energy sources. These biomass residues are also utilized for fulfilling other traditional needs, like feed for livestock, roof thatching, composting, etc. Their availability may also vary in different states of India. For example, the rural people of Uttarakhand rely mostly on fuelwood for cooking and heating purposes while crop residues are used as animal feed. Sugarcane tops in most areas are either used for feeding dairy animals or burnt in the field for crop regrowth. The rural population of Assam and West Bengal commonly use rice straws as animal feed. Similarly, wheat straws are commonly used as animal feed from the Trans-Gangetic plains (Punjab, Haryana, Chandigarh, and Delhi) to Bihar.[17,23] In a state like Bihar, crop residues such as wheat straw, rice straw, and gram have limited use as fuels. Due to the comparatively sturdy nature of wheat straw, these can be used as feed for livestock only after the mechanical threshing of wheat straw into *bhusa*.[17,23] Similarly, the availability and use of maize residue and other crops also vary between regions and provide important feed sources. On the other hand, people from other regions of India with limited availability to forest resources use easily available crop residues and dung cake as energy sources for cooking. The residues of cotton, pulses, chilli, oilseed crops, coconut shells, rapeseed and mustard stalks, jute and mesta stalks, sunflower, and others are also used to meet domestic energy needs. Groundnut residues are burnt as fuel in brick and lime kilns.

In Haryana, Punjab, and Uttar Pradesh, rice straw is not favored as cattle feed.[17,24] These residues are either left in the field or burnt after harvesting. In northwest and central Uttar Pradesh, the burning of rice straw and wheat straw is rarely observed or is limited, while cereal residues are burnt in these regions. It is estimated that 22% of rice straw and 10% of wheat straw are burned in the fields in Uttar Pradesh,[25] while 80% of rice residues are burnt by farmers in Punjab.[26] Thus a huge amount of residues, such as rice straw, coconut shell and husk, rice husk, oilseed cakes, sugarcane bagasse, and pods, are typically burnt on-farm to clear the field for the sowing of the succeeding crop. About 35 million tons of rice straw is burned in Punjab and Haryana each season.[27] This extensive crop residue burning, lasting for more than 3 weeks every year, during October and November, has been contributing to atmospheric pollution across the entirety of the Indo-Gangetic plains and affecting the health of people in nearby regions.

Instead of the direct burning of these agro-residues, they could be valuable feedstocks for biorefinery operations that can yield probable outputs like fuel and chemicals. Region-wide availability of various crop residues in India is shown in Table 2.1.

2.2.2 Animal waste

Animal wastes can also be classified as biomass due to their biodegradable nature. India has a large livestock population of 512.05 million, consisting mainly of cattle, buffalo, yaks, goat, sheep, horses and ponies, pigs, donkeys, and camels.[30] Dung production depends on the cattle population and dung yield per animal. Plenty of animal wastes are generated from these animals, although all of these wastes cannot be collected. The primary products of animal wastes are organic material, moisture, and ash, which can decompose into carbon dioxide (CO_2) and stabilized organic materials under aerobic conditions. In addition, methane gas (CH_4) is produced under anaerobic conditions, which has a high energy potential.[13,31]

TABLE 2.1 Region-wise availability of various crop residues in India.[28,29]

Region	Crop residues available
North (Haryana, Punjab, Uttar Pradesh, Himachal Pradesh, Uttaranchal)	Sawdust, rice husk and straw, wheat husk, mustard stalk and straw, juliflora, and cane trash
Northeast (Assam, Jharkhand, West Bengal)	Rice husk, wood chips, jute and mesta sticks, and sugarcane
West (Gujarat, Rajasthan)	Maize stalks, cotton stalks, arhar stalks, castor waste, groundnut shell, maize cobs
Central (Maharashtra)	Sawdust, groundnut shell, arhar stalks, cotton stalks, sugarcane waste, and maize waste (stover, cob, and silk)
South (Tamil Nadu, Andhra Pradesh, Telangana, Karnataka, Kerala)	Rice husk, sawdust, groundnut shell, tamarind shell, coir pith, sugarcane waste, coffee husk, coconut shell, bengal gram, chilly stalk, maize waste (stover, cob, and silk)

The pyrolysis of animal bone is already performed for bone char and tar production and it has been found that bone char is a cheap and effective filtering and adsorbent material after its activation.[32]Also, a new curing agent for epoxide resin has been prepared by mixing bone tar with diethylenetri-amine.[33]

2.2.3 Household wastes

Solid wastes generated in rural areas are predominantly organic and biodegradable. The kitchen wastes produced by rural people include vegetable flay, leftover food (e.g., uneaten rice, pulses, vegetables, etc.), fruit and vegetable rejects, rotten fruit, egg shells, etc. All of these can be treated in a biorefinery after drying.

2.2.4 Aquatic biomass

Aquatic weeds are undesirable plants, growing and completing their life cycles in water. They negatively affect water bodies and form noxious aquatic vegetation. When these plants die and decay, they may pollute aquatic environments as well as the drinking water quality of water bodies. Moreover, they provide a convenient breeding site for mosquitoes and snails, invade large areas, and impede the free flow of natural water bodies and canals as well as affecting irrigation systems and fish cultures.

India has the largest canal network in the world with more than 1.2 Mha of irrigation canals and 4 Mha of water in reservoirs, ponds, and tanks, etc. Aquatic weeds reduce the velocity of flowing water by about 30%–40%. Moreover, in some states of India, weeds occurring in submerged rice fields and other arable lands constitute a serious threat to agriculture, For example, water hyacinth has become a problem in some states, including Assam, West Bengal, Kerala, and Madhya Pradesh.[34] Algae, water hyacinth, Ipomoea (*Ipomoea carnea*), alligator weeds, Salvinia, and some other submerged and floating aquatic weeds are available in rural India in huge amounts and can be used as feedstock for biorefinery. It is estimated that 160 tons of woody stems could be harvested from a 1 ha area of land covered by *Ipomoea* feedstock in India.[35]

2.2.5 Forest wood and sawdust

Woody biomass is the largest biomass energy source for most of the world, especially for developing countries. India is a large developing country known for its diverse forest ecosystems and mega biodiversity. In India, natural forests constitute the second largest land use after agriculture, covering 21.54% of the country's total land. It ranks 10th among the most forested nations of the world with 708,273 km^2 of forest area.[36,37] With nearly 173,000 villages classified as forest fringe villages, there is evidently a large dependence of communities on forest resources.[38] As these areas are rich in forest wood, wastes such as tree branches, chips, and leaves are found there. Also, many wood-based industries, such as sawmills, are found in rural India where a lot of sawdust is produced during the processing of wood into primary wood products.

Moreover, a lot of solid wastes in the form of seed cover and deoiled cake are obtained from several nonedible oilseed varieties (*Jatropha, Mesua ferrea*, castor, *Pongamia*, etc.) found in forests. These wastes (i.e., seed cover and deoiled cake) have become available due to the Government of India's biofuel policy, which stipulates a 10% replacement of fossil fuel with biobased fuel, like biodiesel and bioethanol. These wastes cannot be utilized as food or feed for animals due to the presence of different antinutritional and toxic (potentially fatal) components, such as karanjin, pongamol (*Pongamia*), ricin (castor), and phorbol ester (*Jatropha*).[39,40] Furthermore, the utilization of these biowastes (generated during the conversion of nonedible oilseeds to biodiesel) is also necessary to improve the economic and environmental scenario for downstream application. It is seen that the chemical composition of solid residues obtained from oilseeds are comparable to various biomass feedstocks. Seed covers contain a high amount of cellulose and hemicelluloses, hence this feedstock can be converted to syngas, pyrolysis oil, and other biofuels and chemicals by different conversion processes.

In the agricultural field, weeds cause 10%–80% crop yield losses with degraded product quality as well as posing health and environmental hazards.[41] Invasive weeds act as a chief inhibitor to agriculture, forestry, and aquatic environments. For example, crop-specific problematic weeds, like weedy rice in rice, are appearing as a menace to agriculture, affecting crop production and product quality. Lantana (*Lantana camara* L.), Black Wattle (*Acacia mearnsii*, De Wild), Mikania (*Mikania micrantha*, Kunth), Royle's Spurge (*Euphorbia royleana*, Boiss), and Parthenium (*Parthenium hysterophorus*) are some examples of these; they are found in different wastelands, forestlands, grazing lands, and the roadside in various regions and they spread extremely rapidly, covering thousands of hectares of land. Therefore these weeds become a major threat to natural forests as well as agricultural systems in rural areas.[41] These species, after appropriate processing, can be converted in a thermochemical biorefinery into valuable chemicals and fuels, which could also be an option for weed management in those areas.

2.2.6 Energy crops

India has more than 55 Mha of wasteland, of which nearly 20 Mha are considered barren and uncultivatable lands.[42,43] These lands are usually unsuitable for agricultural practice due to various reasons including the low content of nutrients and moisture in the soil and landscape factors like steep land or rugged terrain, among others and, hence, would involve a higher investment with a low return. Wastelands also include degraded forest areas, but exclude protected lands, such as water and soil conservation areas, natural forests, national parks, wildlife sanctuaries, watersheds, and other protected forest regions. Many afforestation, reforestation, and tree planting programs related to energy crops have been developed in these degraded lands to meet renewable energy requirements due to their ability to be used in creating alternative fuels. Energy crops are characterized by their high photosynthetic ability and high output. They have the ability to survive and flourish in deprived environments, like under drought or in salinized soils, which makes them ideal for marginal and unused land. As liquid biofuel from energy crops do not compete with the food industry, they strictly follow the planning for biofuel production. Presently some well-known biomasses, such as *Melia dubia*, kadam, *Prosopis juliflora*, bamboo, babul, and others, are used in energy crop plantation.[13]

Thus with all the mentioned biomass resources, rural India has huge potential for biofuel and chemical production from biomass. To understand the potential of biomass resources as useful energy, knowledge of some properties like moisture content, ash content, calorific value, carbon ratio, among others are significant.[13] Literature surveys reveal that biomass wastes, such as rice husk and straw, cotton stalk, sugarcane bagasse, coconut shells, jute waste, soya husk, groundnut shells, sawdust, and coffee waste are already used in pyrolytic conversion processes in India.[17,44,45]

2.3 Biorefinery concept

The National Renewable Energy Laboratory introduced the term "biorefinery" for the first time in 1990, for the utilization of biomass for the production of fuels and other biobased products. According to the International Energy Agency, a biorefinery can be defined as a facility that optimizes the integrated production of an array of products, including materials, fuels, energy, and chemicals, through the utilization of various biomass feedstocks and conversion technologies.[46] Thus the term "biorefinery" refers to a facility, method, plant, or group of facilities which, in a sustainable way, combine the manufacture of a spectrum of viable products, including both intermediate products (platforms) and final products; for instance, materials, food, feed, and chemicals; or fuel products with energy production in the form of power and/or heat.[6,47] The biorefinery system composed of the production of biomass, the conversion or processing of it, and its application with minimal waste emissions. Thus their sustainability feature is the main driving force in the establishment of biorefineries. The biorefinery concept is similar to that of current fossil fuel-based (petroleum, coal, and natural gas) oil refinery processes, which can produce a wide range of biobased products in an efficient production unit. A biorefinery can use all the various kinds of biomass that were discussed earlier, such as residues from industry and households, wood, agricultural crops, plant- and animal-derived organic residues, forest residues, and different aquatic biomass.

Products from biorefineries, like biofuels, are commonly known for their contribution to sustainability, reducing GHG emissions, promoting regional growth, improving socioeconomic structures, and improving the efficiency of the agricultural sector.[48–52] The main reason that biofuels are gaining so much attention is that these represent a substitute for fossil fuels. The government has also taken initiative through suitable policies or directives that promote the growth of the biofuel-based industry. This helps in the development of rural and economic prospects of industrialized as well as developing countries by attracting significant capital investment and creating employment related to biorefineries in rural areas.[10,53]

On the basis of various characteristics, namely the utilization of feedstock, use of different technological platforms or processes (chemical, biochemical, thermochemical, etc.), and the generation of products, biorefineries can be classified into different types.[12,54,55]In a typical biorefinery, various feedstocks are first transformed into multiple biorefinery streams and then a spectrum of marketable products is generated using a single or a combination of various processes, namely thermal, chemical, and biochemical.

However, in this chapter emphasis is given only to the thermochemical biorefinery system.

A thermochemical or lignocellulosic biorefinery is based on the utilization of dry biomass, woody energy crops, and various waste biomasses for the production of biobased fuels and chemicals via thermochemical conversion. Different conversion processes are discussed here.

2.4 Conversion processes

Biomass can be converted using a number of technologies to generate the necessary hydrocarbons in the form of fuel, chemicals, or petrochemicals; each of these technologies has specific requirements, advantages, and disadvantages. India's waste biomass potential is very high and these can be converted into different biofuels, such as biodiesel, bio-oil, bioethanol, and chemicals. Various factors influence the choice of conversion processes, which include: (1) the type and quantity of biomass feedstock; (2) end-use requirements; (3) environmental standards; and (4) techno-economic feasibility. The conversion of biomass to energy is commenced using two main process technologies: thermochemical and biochemical/biological. Physicochemical technology is used for producing biodiesel from plant and vegetable oils by transesterification and other processes. The thermal conversion processes consist of pyrolysis, biomass gasification, combustion, and liquefaction, and these are discussed next.

2.4.1 Biochemical conversion

There are mainly two processes, fermentation and anaerobic digestion, which are used in this conversion process.

2.4.1.1 Fermentation

Fermentation is used in various countries for wide-scale production of ethanol from sugarcane, bagasse, sugar beet, maize, wheat, sweet sorghum, and others. The biomass is first ground and then enzymes are used to convert the starch into sugars, followed by the conversion of the sugars into ethanol using yeast. In this process, some solid residues or sugarcane bagasse are obtained, which can be used as animal feed, in the gasification/pyrolysis process, or as a fuel for boilers.[13] Lignocellulosic biomass consists of some longer-chain polysaccharide molecules, so the conversion of lignocellulosic biomass, like wood, grass, or other biomass, is more complex and, hence, acid or enzymatic hydrolysis is necessary prior to the fermentation of sugars into ethanol.

2.4.1.2 *Anaerobic digestion*

In this process, organic material is directly converted to a gas termed as biogas. It is a mixture of mainly methane and carbon dioxide with small quantities of other gases, such as hydrogen sulfide. In an anaerobic environment, bacteria converts the biomass to a gas with an energy content of about 20%–40% of the lower heating value of the feedstock.[56] Anaerobic digestion is a commercially proven technology and is widely used for treating high moisture content (80%–90%) organic waste. Biogas can be directly used in gas engines and gas turbines as well as a fuel for natural gas vehicles. Biogas quality can be upgraded by removing CO_2 with a conversion efficiency of 21%.[57]

2.4.2 Thermochemical conversion

Thermochemical conversion technologies can be considered more advantageous than biochemical conversion technologies in terms of reaction time, flexibility, and also the potential to utilize all types of biomass. Thermochemical biorefineries are composed of one or more thermochemical conversion method merged to produce valuable hydrocarbons or energy. Such methods are especially important in reducing reliance on fossil resources by efficiently using the available biomass. As a result, high energy efficient, value-added products can be obtained. The primary aim of thermochemical biorefineries is the production of transportation biofuels as their main product. Various processes are used for the thermochemical conversion of biomass such as combustion, gasification, torrefaction, and pyrolysis; which are shown in Fig. 2.1.[58–60]

2.4.2.1 *Combustion*

The combustion process involves the burning of biomass in the presence of air. Biomass can store chemical energy which, through the combustion process, is converted into heat energy, shaft power, and electricity by employing different methods as well as devices like furnaces, stoves, steam turbines, boilers, and others. Although any type of biomass can be burnt, biomass with a low moisture content is preferred for combustion. Biomass with a higher amount of moisture is suitable for biological conversion processes or hydrothermal liquefaction.[61] Combustion plants from small-scale (e.g., for domestic heating) to large-scale industrial plants are in the range of 100–3000 MW. Biomass combustion power plants have a bioenergy conversion efficiency of 20%–40%. The Stirling cycle uses combustion to provide shaft power directly, but the development of the cycle is presently limited to small power outputs. The co-combustion of biomass with coal can also be carried out in existing coal-based power plants, which may be an attractive alternative due to its high conversion efficiency.

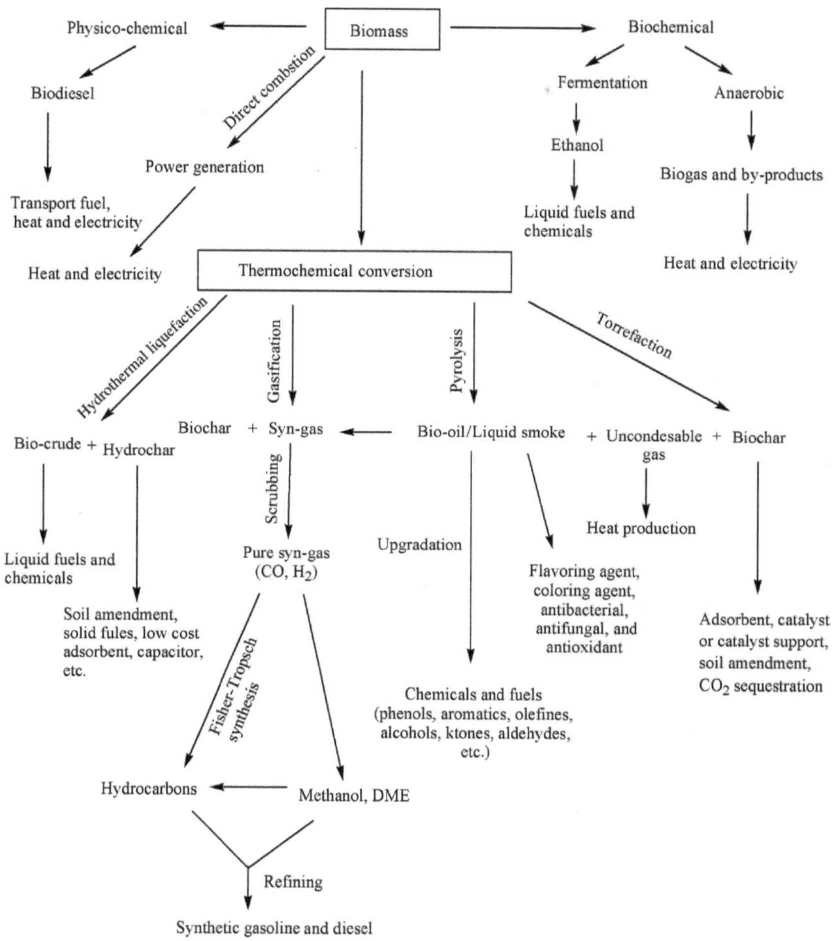

FIGURE 2.1 Various conversion routes and utilization of lignocellulosic biomass.

2.4.2.2 Gasification

Gasification is the conversion of biomass into synthesis gas (a mixture of CO and H_2) by the partial oxidation of biomass (in the presence of limited air or O_2) at a high-temperature range (800°C–900°C). This lower heating value gas can be burnt directly or applied in gas engines/gas turbines as fuel. Also, the produced gas can be utilized as a chemical platform for the production of some chemicals, such as methanol, dimethyl ether, and olefins.[62–64]

Biomass-derived syngas is also utilized in the renowned Fisher–Tropsch synthesis-based biomass to liquid (BTL) technology for the production of liquid fuels, such as gasoline, diesel, and jet fuel. However, there are some flaws of BTL technology: it may produce poor quality biomass-derived syngas with a low calorific value, low H_2 to CO ratio, low selectivity to desirable olefins, catalyst poisoning, and the formation of tar and methane which can harm the process equipment with long-term operation leading to complicated cleaning steps. These drawbacks of BTL technology draw attention toward the production of syngas from economical natural gas or coal as a feed instead of biomass.[65] Moreover, the major techno-economic barrier for the large-scale implementation of BTL is that it requires large gasification systems, involving high investment and running costs as well as uncertainty in terms of the availability and quality of feedstock.[66]

2.4.2.3 Pyrolysis

Pyrolysis can be considered as a viable option, in which product selectivity can be controlled and tuned to suit end-use interests by regulating the various process parameters.[67] Pyrolysis lies at the heart of all thermochemical fuel conversion processes and is assumed to become an avenue to petroleum-type products from biomass resources. In pyrolysis, lignocellulosic biomass is converted into liquid (bio-oil, pyrolysis oil, or biocrude), solid carbonaceous residue (biochar), and gaseous fractions at high temperatures in the absence of air.[68,69] So, biorefineries based on pyrolysis involve less complicated process equipment and lower temperatures which considerably lowers the capital investments compared to BTL.

Fast pyrolysis can be used to obtain a higher yield of bio-oil, enabling the conversion of biomass into biocrude with an efficiency of up to 80%. There are more than 300 chemical compounds identified in bio-oil. Also, the properties of pyrolysis oil may vary depending on the type of biomass used and the pyrolysis conditions. Pyrolysis oil exhibits characteristics such as high acidity ($pH \sim 2-3$), high viscosity, high moisture content, low thermal stability, and the presence of oxygenated compounds, which may create problems like corrosivity in engines and turbines. To overcome such problems, it is necessary to upgrade such oils by lowering the oxygen content and removing alkalis by means of hydrogenation and cracking with suitable catalysts.[70]

Hence, pyrolysis can be considered as one of the most important conversion technologies for biorefineries to convert biomass into bio-oil can be further used as a source of platform chemicals. Solid biochar and noncondensable gases that are coproduced along with the liquid fraction during pyrolysis also show versatile applications and are discussed further in the next section. Since the biorefinery concept covers the complete utilization of biomass with zero waste and leads to an array of valuable products similar to petroleum oil refineries, pyrolysis has the ability to treat the side streams that originate from other conversion processes. Considering all aspects, it is expected that pyrolysis can be placed at the center of a biorefinery for the supply of fuel, chemicals, and other ingredients.

Pyrolysis ensures adequate solid waste management, which is considered to be one of the key challenges in sustainable development. Pyrolysis also provides an opportunity for the upliftment of rural and urban areas where solid waste management is characterized by low waste collection coverage, a lack of treatment, and inadequate disposal. Pyrolysis offers a viable option for the treatment of biowaste and agricultural waste by converting them into solid, liquid, and gaseous fuel, which reduces the waste substances from disposal sites and could also become a probable source of income. Thus there remains an immense scope for establishing either a central processing unit or a peripheral unit of pyrolysis-based biorefineries in areas with abundantly available bioresources as feedstock. Furthermore, pyrolysis-based biorefineries are comparatively easier to build and operate.

2.5 Utilization of the product formed in the rural biorefinery

Biorefineries utilize cheaply available feedstock for the production of various commodities, such as fuel, chemicals, pharmaceuticals, detergents, synthetic fiber, plastics, pesticides, fertilizers, lubricants, solvent, waxes, and coke.[71,72] These are quite similar to crude oil refineries. But as the use of fossil resources is debatable from an economic, ecological, and environmental perspective due to nonsustainability,[73] there is renewed interest world in the production and utilization of renewable liquid biofuels from plants or organic waste due to their renewability, sustainability, common availability, regional development, capacity to manufacture rural jobs, reduction of GHG emissions, and biodegradability.[10,74,75] The term biofuel or renewable fuel is referred to as solid, liquid, or gaseous fuels that are predominantly produced from biomass and which could help to minimize the burning of fossil fuel and CO_2 production as well as reduce the world's dependence on oil.[76–83] Therefore the main focus of biorefinery systems is the production of biofuels, such as ethanol, biodiesel, pyrolysis liquids, green gasoline, butanol, methanol, syngas liquids, biohydrogen, algae jet fuel, and hydrocarbons.

2.5.1 Bio-oil as a chemical platform

The thermochemical routes generally have the ability to process the entire biomass at elevated temperatures.[84] In these routes, biomass constituents are thermally decomposed to form a complex chemical composition of oxygenated compounds known as bio-oil or pyrolytic oil. The bio-oil produced in pyrolysis processes cannot be used without upgradation either as fuel or as chemicals due to many limitations, such as high oxygen content (typically 10%–44%), high acidity, high moisture content (~15–30 wt.%), low calorific value (17–25 MJ/kg), low oxidative stability or storage stability, and inorganic impurities. Moreover, pyrolysis oil derived from proteinaceous biomass contains heteroatoms such as nitrogen.[85,86] As bio-oils are a complex mixture of various chemical constituents, it is not possible to separate pure compounds using conventional methods, like solvent extraction or distillation. So, these properties make the direct utilization of bio-oils difficult. However, it has great prospects as platform chemicals to generate high value-added chemicals as well as hydrocarbons.[69]

In some cases, pyrolysis oil can also be added to oil refinery feedstock or catalytically upgraded to obtain transport grade fuels.[50] Thus a systematic discussion is required on the diverse upgrading techniques of bio-oil, including physical, chemical refining, co-pyrolysis refining, and physical–chemical refining routes.[87] These methods include emulsion, hot-gas vapor filtration, solvent addition, supercritical CO_2 extraction, catalytic hydrogenation, hydrodeoxygenation, steam reforming, microwave pyrolysis, fluidized catalyzed cracking, catalytic esterification, among others.[88–90] In general, various chemical reactions, like dehydration, hydrogenation, and others, are used to enhance the specificity of certain chemical substances in bio-oil as well as to improve the stability of it. Although pyrolysis oils exhibit a lower calorific value than petroleum fuels, it is higher than that of the parent biomass. These oils can be directly used in some diesel engines or low rpm machinery utilized in the agricultural sector for operating turbines and firing boilers.[91,92]

Among the different upgrading techniques mentioned earlier, catalytic upgrading is possibly the most researched process. Usually in this process, bio-oils containing more aromatic and olefinic compounds are produced with a lower amount of oxygenated compounds.[93] Bio-oil quality is also dependent on many factors, namely the catalyst used, types of biomass feedstock, and different reaction parameters, such as temperature, heating rate, residence time, the catalyst to biomass ratio, and pyrolysis environment as well as the upgradation type (i.e., in situ or ex situ). In the in situ upgrading process, biomass is uniformly mixed with a catalyst and then allowed to thermally decompose under high temperature in an inert environment to deoxygenate the produced pyrolysis vapors. On the other hand, in the ex situ process the pyrolysis vapor streams are passed through a fixed catalyst bed and in that intermediate stage, deoxygenation occurs.[94] In both of these processes, acidic zeolites, such as HZSM-5, β-zeolite, faujasite, Y-zeolite, mordenite, ferrierite, modified zeolites, and metal impregnated catalysts, are usually investigated.[95–98]

The in situ process has some drawbacks, including the deposition of coke, catalyst recovery, regeneration or reuse of catalyst, while the ex situ processes can eliminate such limitations. However, both in situ and ex situ processes result in a stable upgraded bio-oil having a low oxygen content with a number of aromatics.[96] Hence, these oils can be used as a chemical intermediate or precursor to other useful chemical compounds and hydrocarbons. Various reactions such as cracking, hydrocracking, and hydrogenation can be employed on pyrolysis oils to produce alkenes which on subsequent hydrogenation yield alkanes. Similarly, hydroformylation reactions yield alcohols that are used to produce various other important products.[99] Pyrolysis parameters can affect the product composition, hence, by altering the chemical environment, catalyst, or other parameters, various chemical substances (phenols, levoglucosenone, aromatic aldehydes, etc.) can be produced.[100,101] The upgrading of pyrolysis oil can also be carried out using domino reactions with the help of industrially useful hydrodeoxygenation catalysts to produce crude oil like fuel and furan-based hydrocarbon building blocks[90,102] (Fig. 2.2).

The pyrolysis oils derived from deoiled cake and seed cover of several nonedible oilseed varieties (*Jatropha, M. ferrea, Cascabela thevetia, Pongamia*, etc.) are identical to that of lignocellulosic biomass in terms of properties, like energy content, antimicrobial activity, and others.[103–105] Another exceptionality of these bio-oils is that they are widely used in the pharmaceutical, agrochemical, and dye industries due to the presence of nitrogen-containing heterocycles (pyridine, pyrroles, indoles, 2-nitropyrrole, etc.).[85,106,107] The potential usefulness of nonedible oilseeds in the biorefinery approach has been represented in Fig. 2.3.

Another major advantage associated with pyrolysis oil is the possibility for storage and transport. Thus small, low-cost decentralized pyrolysis units

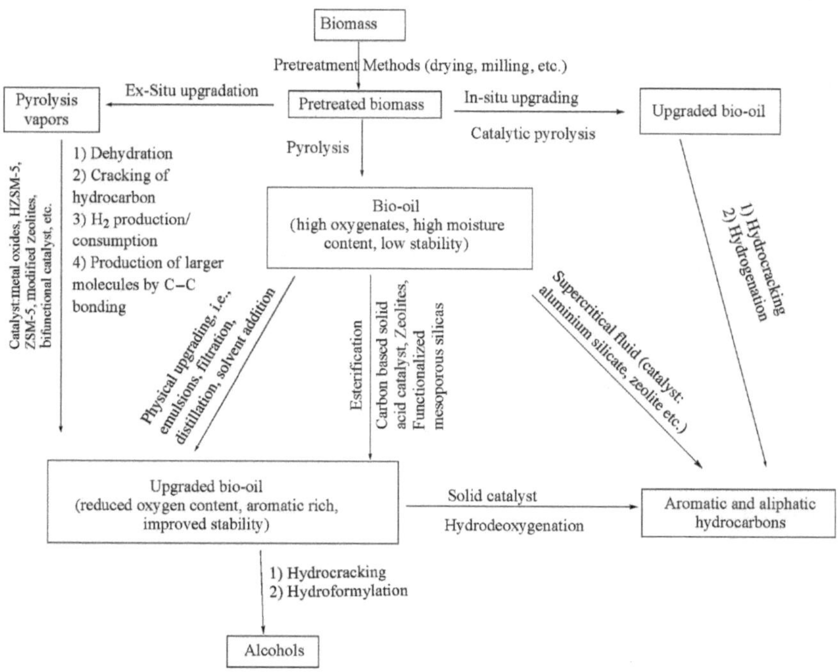

FIGURE 2.2 Routes for the upgradation of bio-oil or pyrolysis oil.

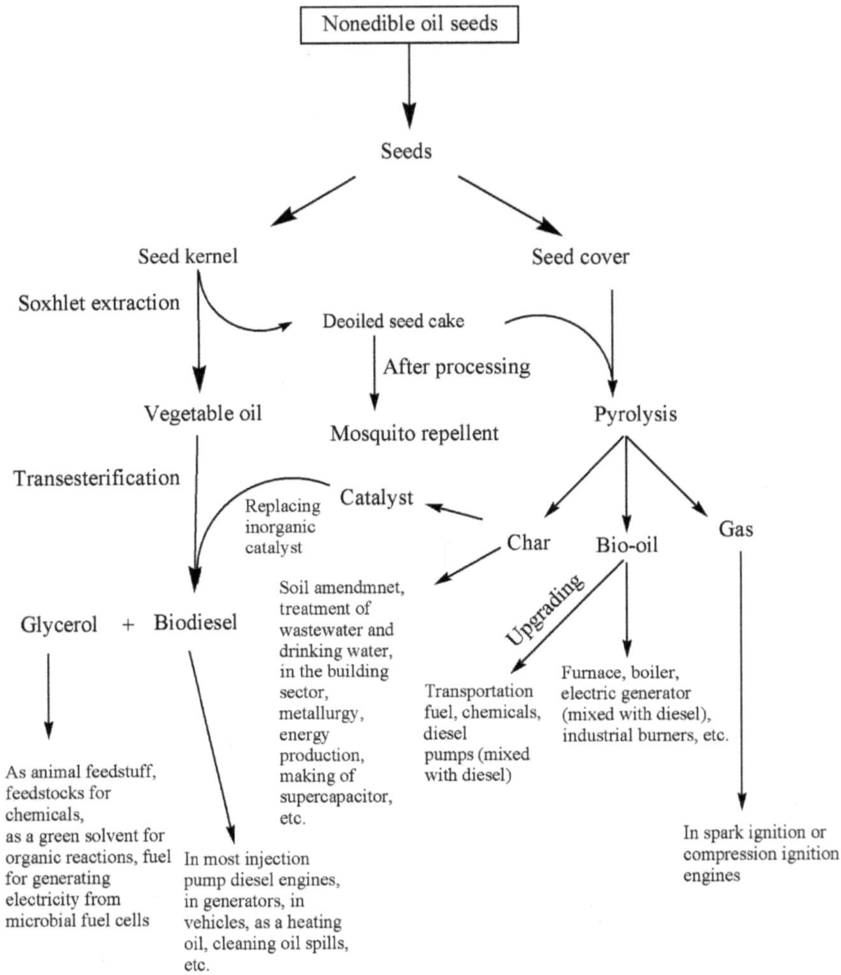

FIGURE 2.3 Nonedible oilseed-based biorefinery.

can be established in the rural areas of India for pyrolysis oil production, which could be easily collected at centralized infrastructures (e.g., existing refinery units) for their upgradation to high-value products.[93,108]

2.5.2 Application of biochar

Biochar produced in pyrolysis conversion processes is carbon-enriched material anticipated to be used as a soil amendment agent to sequester carbon and boost soil quality. The addition of biochar to soil can improve soil quality by raising soil pH, increasing moisture holding capacity, attracting more beneficial fungi and microbes, improving cation exchange capacity, and retaining nutrients in the soil.[37] Biochar may persist in soil for a long period of time since it is highly resistant to microbial decomposition and mineralization. Therefore the use of sustainable biochar as a soil amendment is an innovative and highly promising practice for sustainable agriculture.[15] Biochar has the capacity to remediate contaminated water and provides additional benefits to the environment. Biochar, which requires low investment as compared to activated carbon, absorb hydrocarbons, other organics, and some inorganic metal ions, is emerging as a promising wastewater treatment technology.[41] Biochar has numerous applications; as heating fuel and energy fuel, as a biomass-based energy carrier (replacing fossil energy carriers), as raw material for the preparation of high-quality adsorbents, supercapacitors, catalysts, and catalyst supports.[92,109–111]

2.6 India's need for rural biorefineries

India has the sixth highest energy demand in the world accounting for 3.5% of commercial energy consumption globally.[46] Oil demand in India is expected to grow by a factor of 2.2 by 2030, thus increasing the oil import dependency from 69% at present to 91%.[46] Therefore it is high time to think of other sources of energy to supplement petro-oil. India is also the third largest GHG emitter after China and the United States. According to the Ministry of New and Renewable Energy, the total annual production of agro-processing and household wastes in India is 200 Mt. Generally, the disposal of these wastes is carried out by poor farmers, rural labors, unorganized or small agro-based industries in an unscientific and dispersed way, and as a result, major amounts of wastes end up being burned.

There are many adverse consequences associated with the burning of crop residues in agricultural fields or the incomplete combustion of fuelwood. For instance, crop residue burning causes air pollution due to the emission of particulate matter (PM), smoke, dioxins, furans, and mercury as well as some chemically important trace gases, such as carbon monoxide (CO), oxides of nitrogen (NO_x), sulfur (SO_x), and other hydrocarbons into the atmosphere which adversely affects human and animal health. In fact, many pollutants found in biomass smoke are regarded as carcinogens and these could also cause different respiratory diseases. Moreover, the release of GHGs is a major cause of global warming and it decreases the nitrogen (N), phosphorus (P), potassium (K), and sulfur (S) concentrations in plants. For example, the burning of 1000 kg of rice straw emits about 3 kg of PM, 60 kg of CO, 1460 kg of CO_2, 199 kg of ash, and 2 kg of SO_2.[112] Moreover, the burning of agricultural residues can cause some important resources to be wasted, which could otherwise be a source of carbon, bioactive compounds, feed, and energy for rural households and small industries. It also reduces beneficial microbial populations due to the elevation of soil temperature in agricultural fields. Although this effect is transitory, repeated burning can permanently damage the soil. The burning of crop residues instantly increases the exchangeable $NH_4^+ - N$ content and the bicarbonate P content, but no nutrients are developed in the soil. So, if the burning of residues continues for a long time, it diminishes the total nitrogen and carbon content and potentially the mineralizable nitrogen in the upper soil layer.

On a per capita basis, India emits about 1.6 tons of CO_2 per person, in comparison to the 15.5 tons per person of the United States and the world average of 4.4 tons per person.[113,114] The severity of air pollution is so high that the life expectancy among Indians on average is reduced by 3.4 years while among residents of Delhi it is reduced by almost 6.3 years. According to the 2016 EPI, more than 3.5 billion people—half of the world's population—are exposed to unsafe air quality, which also includes 75% of India's population. Some reports, including one by the WHO, claim that 300,000–400,000 people die of indoor air pollution and carbon monoxide poisoning in India due to biomass burning.[115]

India is also facing a major problem due to its ever increasing population. With the increasing population, the unemployment rate is also increasing day by day, which has hindered the development of the country. Since the advancement of rural areas is essential for the development of the country, the government should encourage and develop agricultural waste-based biorefineries in rural areas so that rural candidates can get employment opportunities in these refineries. The advancement of society with the improvement of environmental issues is another key sustainability factor of the establishment of biorefineries. The presence of biorefineries in rural communities could potentially reduce the rate of migration away from rural areas. Local application of biorefinery products (like biochar) will benefit the local farmers of an agriculture-based country like India. Biorefinery plants will also create jobs indirectly because more chemical industries will be established to transform the produced bio-oil into various chemicals. Also, for the maintenance of biorefineries or to make them more efficient for continuous operation, new jobs will be created. On the other hand, if biorefinery plants are located near to agricultural fields, it could influence the profitability of farmers meeting the plant demand for agricultural wastes by lowering transportation costs. Moreover, these biorefineries will contribute to an increased competitiveness and wealth of the country by responding to the need to supply a wide range of biobased products and energy sources in an economically, socially, and environmentally sustainable manner.

Liquid biofuels for use in transportation and produced in biorefineries have attracted huge attention in different countries all over the world for many obvious reasons. Rural biorefineries offer an alternative to conventional fuel production by providing the opportunity to produce fuel, such as bio-oil, onsite to reduce costs associated with biomass transportation thus making the fuel more economically viable. These biofuels have another advantage as they emit no net CO_2; the CO_2 released in burning is balanced by the CO_2 captured by the plant during photosynthesis.[6] Thus biofuels and bioproducts produced in biorefineries from biomass would also diminish global warming.

Thus it can be summarized that the establishment of biorefineries in the countryside, especially in rural India, is favorable because: (1) India has very high potential to produce biofuel and chemicals due to its abundant biomass resources; (2) the distributed operation of thermochemical biorefineries in rural areas will reach a near commercialization state by generating various rural earnings and employment options; and (3) these medium- or small-scale rural biorefineries will, in some way, reduce the GHG emissions caused by the burning of agricultural wastes so that it will eliminate air pollution to some extent.

2.7 Challenges of biorefineries

Biorefineries are considered sustainable from an environmental, economic, and social point of view. However, there are some major issues associated with the use of some biomasses in biorefineries. As reported, 64% of households in rural India are dependent on firewood for cooking, while others are dependent on crop residue or animal wastes for the same purpose.[116] According to Census of India 2011, only 26.3% of urban households depend on solid biomass as a primary energy source for cooking. So, each year 220 Mt of firewood is used for cooking in the rural areas of India.[117] Depending on the income level of families, firewood contributes a key portion (65%–80%) of the final energy consumption of communities.[118] Apart from domestic utilization, biomass is also used in various rural enterprises and small-scale industries such as brick making, charcoal making, tobacco and tea curing, rice parboiling, hotels, restaurants, and potteries.[117] Also, some biomasses such as rice straw and corn straw are used for animal feed, soil mulching, manure making, and thatching for rural homes.

The incorporation of crop residue results in the higher activity and greater population of microbes in soil and the subsequent transformation of nutrients.[119] Thus the establishment of biorefineries may result in the unsystematic elimination of crop residues, which has adverse impacts on soil properties and their productivity, soil organic matter (SOM) dynamics, water and wind erosion and, hence, on the production of crops.[83,120] Although the complete or over elimination of residues has detrimental effects on soil, the removal of a portion of waste would have no severe harmful impact.[121–124] However, where, when, and how much waste can be eliminated sustainably is still uncertain.[17] Another problem associated with the use of low energy density biomass in biorefineries is the bulky nature of feedstock, which causes problems with the collection, handling, storage, transportation, processing, and conversion of these biomasses and these may increase the overall cost of the feedstock. Therefore biorefinery plants should be located in the vicinity of where an adequate amount of biomass resources are available. On the other hand, the thermal conversion of biomass generally requires low moisture content. As biomass generally has a high moisture content (about 9%–35% depending on the plant species for herbaceous and agricultural biomass), it would adversely impact on the overall energy stability for the conversion process.[56,125]

Another major obstacle to deal with in the overall development of rural biorefineries is the financial barrier which can be overcome through the adoption of people-friendly banking policies by the government. Government agencies that could be directly involved in the establishment of biorefineries include the Ministries of New and Renewable Energy, Rural Development, Agriculture and Farmers' Welfare, Science and Technology, Panchayati Raj, state government departments, and local governments. These agencies can help potential participants in the establishment of biorefineries by financing them with a loan guarantee and can then attract investors, credit institutes, policy makers, suppliers of raw materials, and end users toward this new business venture. Also, participants can gather a portion of the funding in the form of a low-cost bank loan. As there are many considerable restrictions and approvals associated with a conventional loan approval, a facile approach from bank authorities to the first few biorefinery start-ups shall create a conducive atmosphere for others to follow. On the other hand, if government ministries or departments decide to support the loan by acting as a loan guarantor, it shall encourage educated and skilled youths to accept these kinds of unconventional business practices. Thus policy makers need to come forward with a stable policy toward the establishment of biorefineries.

2.8 Conclusion

In this chapter, the resources and potential of biomass have been presented along with the feasibility and importance of the establishment of biorefineries and their advantages and disadvantages in the rural areas of India. It is found that biorefineries show promise both for industrialized and developing countries. The establishment of biorefineries could enrich rural areas by supplying a wide range of biobased products and energy on demand in an economically, socially, and environmentally sustainable manner. At the regional, national, and global levels, there are three main factors that act as a driving force for the use of biomass residues to generate bioenergy, biofuels, and biochemicals. These are (1) climate change; (2) energy security; and (3) rural development. These factors independently, or in combination, have also influenced the planning of energy policies. Besides, biomass is probably the only viable alternative to fossil resources for the production of transportation fuels and chemicals, while other renewable sources are used to generate heat and electricity.

It can be concluded that there is vast potential for the conversion of available biomass into energy in India. As various biomass resources are available in different forms and large volumes of waste biomass can be obtained from these resources (e.g., agricultural waste, food wastes, etc.), this can be taken as an opportunity to switch over from conventional sources of energy to renewables. Already various agencies and industries are practicing the conversion of different waste biomass into energy in India and reporting on the benefits of these. As thermochemical conversion is one of the major technologies used to convert biomass into energy, various pyrolysis-based biorefinery plants can be installed in rural India for fulfilling the country's energy and other requirements. These plants will not only have the potential to supply much-needed energy and biofuels to people in remote areas, but the by-product (biochar) obtained from these plants can also be utilized in the agricultural field for better crop yields.

However, the success of biorefineries depends on the development of other possible biomass integrated conversion technologies for the production of goods and services. Also, the government needs to provide financial assistance to encourage the use of nonconventional sources of energy by establishing various biorefineries in rural India, which will help in the sustainable development of the nation. It should encourage companies to upgrade the bio-oil produced in these biorefineries for producing transportation fuel and chemicals from it. For that, the extensive installation of rural biorefineries will surely help in the future.

References

1. Weiland P. Biogas production: current state and perspectives. *Appl Microbiol Biotechnol* 2009;**85**(4):849–60.
2. Cherubini F. The biorefinery concept: using biomass instead of oil for producing energy and chemicals. *Energy Convers Manage* 2010;**51**(7):1412–21.
3. Iea.org. <https://www.iea.org/publications/freepublications/publication/WEO_2007.pdf>; 2018 [accessed 01.12.17].
4. Lucia L, Argyropoulos D, Adamopoulos L, Gaspar A. Chemicals and energy from biomass. *Can J Chem* 2006;**84**(7):960–70.
5. Nossin PMM. White biotechnology: replacing black gold? In: *Fifth international conference on renewable resources and biorefineries*, Ghent, Belgium, 10–12 June 2009.
6. Naik SN, Goud VV, Rout PK, Dalai AK. Production of first- and second-generation biofuels: a comprehensive review. *Renew Sustain Energy Rev* 2010;**14**(2):578–97.
7. Mabee W, Gregg D, Saddler J. Assessing the emerging biorefinery sector in Canada. *Appl Biochem Biotechnol* 2005;**123**(1–3):0765–78.
8. Clark J, Deswarte F. The biorefinery concept—an integrated approach. In: Clark J, Deswarte F, editors. *Introduction to chemicals from biomass*. Chichester: John Wiley & Sons, Ltd; 2008.
9. Censusindia.gov.in. *Census of India Website: Office of the Registrar General & Census Commissioner, India*. <http://www.censusindia.gov.in/>; 2018 [accessed 20.06.18].

10. Demirbas A. Biorefineries: current activities and future developments. *Energy Convers Manage* 2009;**50**:2782–801.
11. Deswarte FEI, Clark JH, Hardy JJE, Rose PM. The fractionation of valuable wax products from wheat straw using CO_2. *Green Chem* 2008;**8**:39–42.
12. Haung HJ, Ramaswamy S, Tschirner UW, Ramarao BV. A review of separation technologies in current and future biorefineries. *Sep Purif Technol* 2008;**62**:1–21.
13. Kumar A, Kumar N, Baredar P, Shukla A. A review on biomass energy resources, potential, conversion and policy in India. *Renew Sustain Energy Rev* 2015;**45**:530–9.
14. Davis SC, Hay W, Pierce J. *Biomass in the energy industry: an introduction.* London: BP; 2014978-0-9928387-1-3. p.l.c.
15. Kataki R, Goswami K, Bordoloi N, Narzari R, Saikia R, Sut D, et al. Biomass resources for biofuel production in Northeast India. In: Purkayastha J, editor. *Bioprospecting of indigenous bioresources of North-East India.* Singapore: Springer; 2016.
16. Goyal S. *Biorefinery prospects in India. BioEnergy Consult.* <https://www.bioenergyconsult.com/biorefinery-india/>; 2018 [accessed 01.07.18].
17. Hiloidhari M, Das D, Baruah DC. Bioenergy potential from crop residue biomass in India. *Renew Sustain Energy Rev* 2014;**32**:504–12.
18. Data.worldbank.org. *Agricultural land (% of land area) | Data.* <https://data.worldbank.org/indicator/AG.LND.AGRI.ZS>; 2018 [accessed 03.05.18].
19. Agarwal A, Dintwa E, Joshi P. Analysis of agro residue burning and present scenario in key areas of Northern Plains in India. *Int J Adv Res* 2016;**4**(3):1499–509.
20. Iari.res.in. <http://www.iari.res.in/files/Important_Publications-2012-13.pdf>; 2018 [accessed 27.12.18].
21. Prasad S, Singh A, Joshi HC. Ethanol as an alternative fuel from agricultural, industrial and urban residues. *Resour Conserv Recycl* 2007;**50**:1–39.
22. Murali S, Shrivastava R, Saxena M. Quantification of agricultural residues for energy generation—a case study. *J Inst Public Health Eng* 2008;**3**:27–31.
23. Saud T, Singh DP, Mandal TK, Ranu G, Pathak H, Saxena M, et al. Spatial distribution of biomass consumption as energy in rural areas of the Indo-Gangetic plain. *Biomass Bioenergy* 2011;**35**:932–41.
24. Asadullah M, Rahman MA, Ali MM, Motin MA, Sultan MB, Alam MR. Jute stick pyrolysis for bio-oil production in fluidized bed reactor. *Bioresour Technol* 2008;**99**:44–50.
25. Beri V, Sidhu BS, Gupta AP, Tiwari RC, Pareek RP, Rupela OP. *Organic resources of a part of Indo-Gangetic Plain and their utilization.* Ludhiana: Department of Soils, Punjab Crop University; 2003. National Crop Technology Project (NATP).
26. NAAS. Innovative viable solution to rice residue burning in rice-wheat cropping system through concurrent use of super straw management system-fitted combines and turbo happy seeder. In: *Policy brief no. 2.* National Academy of Agricultural Sciences, New Delhi; 2017.
27. The Economic Times. *Remedy in sight to end straw burning in northern India?* <https://economictimes.indiatimes.com/news/science/remedy-in-sight-to-end-straw-burning-in-northern-india/articleshow/61064753.cms>; 2018 [accessed 28.12.17].
28. Agarwal AK. Part II Bio-energy options and rural India, electrification and bio-energy options in rural India. In: *India infrastructure report-2007*, pp. 165–177.
29. Devi S, Gupta C, Jat SL, Parmar MS. Crop residue recycling for economic and environmental sustainability: the case of India. *Open Agric* 2017;**2**:486–94.
30. <http://pib.nic.in/newsite/PrintRelease.aspx?relid = 109280> [accessed 25.07.18].

31. Varshney R, Bhagoria JL, Mehta CR. Small scale biomass gasification technology in India-an overview. *J Eng Sci Manag Edu* 2010;**3**:33–40.
32. Cecen F, Aktas O. *Activated carbon for water and wastewater treatment.* Wiley-VCH Verlag GmbH & Co. KGaA; 2012.
33. Purevsuren B, Davaajav Y, Namkhainorov J, Glavcheva-Laleva ZI, Genadiev V, Glavchev IK. Pyrolysis of animal bone, characterization of the obtained char and tar and application of bone tar for crosslinking of epoxy resin. *Bulg Chem Commun* 2017;**49**(1):34–9.
34. Varshney JG, Kumar S, Mishra JS. Current status of aquatic weeds and their management in India. In: Sengupta M, Dalwani R, editors. *Proceeding of Taal 2007—the 12th world lake conference, Jaipur*; 2008, pp. 1039–1045.
35. Saikia P, Gupta UN, Barman RS, Kataki R, Chutia RS, Baruah BP. Production and characterization of bio-oil produced from *Ipomoea carnea* bio-weed. *Bioenergy Res* 2015;**8**(3):1212–23.
36. Fao.org. *India at a glance | FAO in India | Food and Agriculture Organization of the United Nations.* <http://www.fao.org/india/fao-in-india/india-at-a-glance/en>; 2018 [accessed 30.05.18].
37. Fsi.nic.in. <http://fsi.nic.in/details.php?pgID = sb_64>; 2018 [accessed 25.05.18].
38. Kishwan J, Pandey R, Dadhwal VK. Emission removal capability of India's Forest and Tree Cover. *Small-Scale For* 2012;**11**:61–72.
39. Regina CAL. *Castor* and *jatropha* oils: production strategies – a review. *OCL* 2009;**16**(4):241–7.
40. Rao SBN, Prasad KS, Rajendran D. Recent advances in amelioration of anti nutritional factors in livestock feedstuffsJanuary In: Sampath KT, Ghosh J, Bhatta R, editors. *Animal nutrition & reproductive physiology (recent concepts).* Delhi: Satish Serial Publishing House; 2013.
41. Rao AN, Chauhan BS. *Weeds and weed management in India – a review.* Hyderabad: Weed-Science in the Asian-Pacific Region. Indian Society of Weed Science; 20159788193197806. p. 87–118.
42. Hassan M, Natarajan K, Pelkonen P, Zyadin A, Pappinen A. Perspectives of feedstock supply for biomass-based energy plant development in India: views from an expert survey. *Challenges* 2015;**6**(1):71–87.
43. Ravindranath NH, Sita Lakshmi C, Manuvie R, Balachandra P. Biofuel production and implications for land use, food production and environment in India. *Energy Policy* 2011;**39**:5737–45.
44. Jorapur R, Rajvanshi AK. Sugarcane leaf bagasse gasifiers for industrial heating applications. *Biomass Bioenergy* 1995;**8**:91–8.
45. Kapur T, Kandpal TC, Garg HP. Electricity generation from rice husk in Indian rice mills: potential and financial viability. *Biomass Bioenergy* 1998;**14**:573–83.
46. Sankar PD, Saleh MA, Selvaraj CI, Palanichamy V, Mathew R. Progress of biorefinery in India. *Res Biotechnol* 2013;**4**(2):26–35.
47. Fernando S, Adhikari S, Chandrapal C, Murali N. Biorefineries: current status challenges and future direction. *Energy Fuels* 2006;**20**(4):1727–37.
48. Demirbas A. Combustion systems for biomass fuels. *Energy Source Part A* 2007;**29**:303–12.
49. Demirbas A. Combustion of biomass. *Energy Source Part A* 2007;**29**:549–61.
50. Demirbas A. Present and future transportation fuels. *Energy Source Part A* 2008;**30**:1473–83.

51. Ilkilic C, Yucesu HS. The use of cottonseed oil methyl ester on a diesel engine. *Energy Sources Part A* 2008;**30**:742–53.
52. Nas B, Berktay A. Energy potential of biodiesel generated from waste cooking oil: an environmental approach. *Energy Source Part B* 2007;**2**:63–71.
53. Keskin A. Biodiesel production from free fatty acids obtained with neutralization of the crude glycerin. *Energy Source Part A* 2009;**31**:17–24.
54. Chew TL, Bhatia S. Catalytic processes towards the production of biofuels in a palm oil and oil palm biomass-based biorefinery. *Bioresour Technol* 2008;**99**:7911–22.
55. Koutinas AA, Wang RH, Webb C. The biochemurgist—bioconversion of agricultural raw materials for chemical production. *Biofuel Bioprod. Biorefinery* 2007;**1**:24–38.
56. McKendry P. Energy production from biomass (part 1): overview of biomass. *Bioresour Technol* 2002;**83**:37–46.
57. Nelson SSM, Ivan W, Aimoni MC, Reinaldo B, Reginaldo FS. Electric energy microproduction in a rural property using biogas as primary source. *Renew Sustain Energy Rev* 2013;**28**:385–91.
58. Cho DW, Kwon EE, Song H. Use of carbon dioxide as a reaction medium in the thermochemical process for the enhanced generation of syngas and tuning absorption ability of biochar. *Energy Convers Manage* 2016;**117**:106–14.
59. Cherubini F, Jungmeier G, Wellisch M, Willke T, Skiadas I, Ree RV, et al. Toward a common classification approach for biorefinery systems. *Biofuel Bioprod Biorefinery* 2009;**3**:534–46.
60. Katyal S. Effect of carbonization temperature on combustion reactivity of bagasse char. *Energy Source Part A* 2007;**29**:1477–85.
61. Sharma A, Unni BG, Singh HD. A novel fed batch system for bio methanation of plant biomasses. *J Biosci Bioeng* 1999;**87**(5):678–82.
62. Bridgwater AV. The technical and economic feasibility of biomass gasification for power generation. *Fuel* 1995;**74**(5):631–53.
63. Ganesh A, Banerjee R. Biomass pyrolysis for power generation: a potential technology. *Renew Energy* 2001;**22**:9–14.
64. Maity SK. Opportunities, recent trends and challenges of integrated biorefinery: Part I. *Renew Sustain Energy Rev* 2015;**43**:1427–45.
65. Huber GW, Iborra S, Corma A. Biomass to liquids (BtL) – European biofuels technology platform. *Chem Rev* 2006;**106**:4044–98.
66. Mikkola JP, Sklavounos E, King AWT, Virtanen P. The biorefinery and green chemistry. In: Lukasik RB, editor. *Book series RSC green chemistry*. 2015. https://pubs.rsc.org. doi:10.1039/9781782622598-00001.
67. Bordoloi N, Narzari R, Sut D, Saikia R, Chutia RS, Kataki R. Characterization of bio-oil and its sub-fractions from pyrolysis of *Scenedesmus dimorphus*. *Renew Energy* 2016;**98**:245–53.
68. Laird DA, Brown RC, Amonette JE, Lehmann J. Review of the pyrolysis platform for coproducing bio-oil and biochar. *Biofuel Bioprod Biorefinery* 2009;**3**(5):547–62.
69. Qi Z, Jie C, Tiejun W, Ying X. Review of biomass pyrolysis oil properties and upgrading research. *Energy Convers Manage* 2007;**48**:87–92.
70. Mohan D, Pittman CU, Steele PH. Pyrolysis of wood/biomass for bio-oil: a critical review. *Energy Fuels* 2006;**20**:848–89.
71. Bender M. Potential conservation of biomass in the production of synthetic organics. *Resour Conserv Recycl* 2000;**30**:49–58.

72. Demirbas MF. Current technologies for biomass conversion into chemicals and fuels. *Energy Source Part A* 2006;**28**:1181–8.
73. Kamm B, Gruber PR, Kamm M. *Biorefinery industrial processes and products. Status and future direction*, vols. 1 and 2. Weinheim: Wiley-Verlay Gmbtt and Co KGaA; 2006.
74. Demirbas A. New liquid biofuels from vegetable oils via catalytic pyrolysis. *Energy Edu SciTech* 2008;**21**:1–59.
75. Demirbas A. Production of biofuels from macroalgae and microalgae. *Energy Edu SciTech* 2007;**18**:59–65.
76. Balat M. Progress in biogas production processes. *Energy Edu SciTech* 2008;**22**:15–36.
77. Berktay A, Nas B. Biogas production and utilization potential of wastewater treatment sludge. *Energy Source Part A* 2008;**30**:179–88.
78. Chhetri AB, Islam MR. Towards producing a truly green biodiesel. *Energy Source Part A* 2008;**30**:754–64.
79. Demirbas A. Bio-fuels from agricultural residues. *Energy Source Part A* 2008;**30**:101–9.
80. Konwer D, Kataki R, Saikia P. Pyrolysis of some indigenous tree species of northeast India: effect of pyrolysis temperature and heating rate on the products yield and char quality. *Energy Source Part A* 2007;**29**:1433–42.
81. Demirbas A. Converting biomass derived synthetic gas to fuels via Fisher–Tropsch synthesis. *Energy Source Part A* 2007;**29**:507–12.
82. Demirbas A. Biofuels sources, biofuel policy, biofuel economy and global biofuel projections. *Energy Convers Manage* 2008;**49**:2106–16.
83. Demirbas A. Producing and using bioethanol as an automotive fuel. *Energy Source Part B* 2007;**2**:391–401.
84. Wood C, Muthukumarappan K, Rosentrater KA. Optimization of the pyrolysis of distillers dried grains with solubles. In: *Presented at 2013 ASABE annual international meeting*, Kansas City, Missouri.
85. Chutia RS, Kataki R, Bhaskar T. Characterization of liquid and solid product from pyrolysis of *Pongamia glabra* deoiled cake. *Bioresour Technol* 2014;**165**:336–42.
86. Oasmaa A, Czernik S. Fuel oil quality of biomass pyrolysis oils: state of the art for the end users. *EnergyFuels* 1999;**13**(4):914–21.
87. Yang H, Yao J, Chen G, Maa W, Yan B, Qi Y. Overview of upgrading of pyrolysis oil of biomass. *Energy Proc* 2014;**61**:1306–9.
88. Budarin VL, Shuttleworth PS, Dodson JR, Hunt AJ, Lanigan B, Marriott R, et al. Use of green chemical technologies in an integrated biorefinery. *Energy Environ Sci* 2011;**4**:471–9.
89. Hoekstra E, Hogendoorn KJA, Wang X, Westerhof RJM, Kersten SRA, van Swaaij WPM, et al. Fast pyrolysis of biomass in a fluidized bed reactor: in situ filtering of the vapors. *Ind Eng Chem Res* 2009;**48**(10):4744–56.
90. Zhao C, Lercher JA. Upgrading pyrolysis oil over Ni/HZSM-5 by cascade reactions. *Angew Chem* 2012;**124**:6037–42.
91. Czernik S, Bridgwater AV. Overview of applications of biomass fast pyrolysis oil. *Energy Fuels* 2004;**18**:590–8.
92. Tang Z, Lu Q, Zhang Y, Zhu X, Guo Q. One step bio-oil upgrading through hydrotreatment, esterification, and cracking. *Ind Eng Chem Res* 2009;**48**:6923–9.
93. Sharma RK, Bakhshi NN. Catalytic upgrading of pyrolysis oil. *Energy Fuel* 1993;**7**:306–14.
94. Ma R, Xu Y, Zhang X. Catalytic oxidation of biorefinery lignin to value added chemicals to support sustainable biofuel production. *ChemSusChem* 2015;**8**:24–51.

95. Budhi S, Mukarakate C, Iisa K, Pylypenko S, Ciesielski PN, Yung MM, et al. Molybdenum incorporated mesoporous silica catalyst for production of biofuels and value-added chemicals via catalytic fast pyrolysis. *Green Chem* 2015;**17**:3035–46.
96. Mukarakate C, Watson MJ, Dam JT, Baucherel X, Budhi S, Yung MM, et al. Upgrading biomass pyrolysis vapors over β-zeolites: role of silica-to-alumina ratio. *Green Chem* 2014;**16**:4891–905.
97. Wan S, Pham T, Zhang S, Lobban L, Resasco D, Mallinson R. Direct catalytic upgrading of biomass pyrolysis vapors by a dual function Ru/TiO_2 catalyst. *AIChE J* 2013;**59**(7):2275–85.
98. Yung MM, Starace AK, Mukarakate C, Crow AM, Leshnov MA, Magrini KA. Biomass catalytic pyrolysis on Ni/ZSM-5: effects of nickel pretreatment and loading. *Energy Fuels* 2016;**30**(7):5259–68.
99. Mante OD, Rodriguez JA, Senanayake SD, Babu SP. Catalytic conversion of biomass pyrolysis vapors into hydrocarbon fuel precursors. *Green Chem* 2015;**17**:2362–8.
100. Mukarakate C, McBrayer JD, Evans TJ, Budhi S, Robichaud DJ, Iisa K. Catalytic fast pyrolysis of biomass: the reactions of water and aromatic intermediates produce phenols. *Green Chem* 2015;**17**:4217–27.
101. Zhang Z, Lu Q, Ye X, Wang T, Wang X, Dong C. Selective production of levoglucosenone from catalytic fast pyrolysis of biomass mechanically mixed with solid phosphoric acid catalysts. *BioEnergy Res* 2015;**8**:1263–74.
102. Mercader FM, Groeneveld MJ, Kersten SRA, Way NWJ, Schaverien CJ, Hogendoorn JA. Production of advanced biofuels: co-processing of upgraded pyrolysis oil in standard refinery units. *Appl Catal B* 2010;**96**:57–66.
103. Patra JK, Hwang H, Choi JW, Baek KH. Bactericidal mechanism of bio-oil obtained from fast pyrolysis of *Pinus densiflora* against two foodborne pathogens, *Bacillus cereus* and *Listeria monocytogenes*. *Foodborne Pathog Dis* 2015;**12**(6):529–35.
104. Phukan MM, Chutia RS, Kumar R, Kalita D, Konwar BK, Kataki R. Assessment of antimicrobial activity of bio-oil from *Pongamia glabra, Mesua ferrea* and *Parachlorella* spp. deoiled cake. *Int J Pharma Bio Sci* 2013;**4**(4):910–18.
105. Sut D, Chutia RS, Bordoloi N, Narzari R, Kataki R. Complete utilization of non-edible oil seeds of *Cascabela thevetia* through a cascade of approaches for biofuel and by-products. *Bioresour Technol* 2016;**213**:111–20.
106. Kataki R, Konwar LK, Mikkola JP, Bordoloi N, Saikia R, Chutia RS. Side-streams from bioenergy and biorefinery complexes as a resource for circular bio-economy. In: Bhaskar T, Pandey A, Mohan SV, Lee DJ, Khanal SK, editors. *Waste biorefinery: potential and perspectives*. Elsevier; 2018. p. 85–125 Available from: https://doi.org/10.1016/C2016-0-02259-3.
107. Xu L, Yao Q, Deng J, Han Z, Zhang Y, Fu Y, et al. *Renewable N heterocycles production by thermocatalytic conversion and ammonization of biomass over ZSM 5. ACS Sustain Chem Eng*, 3. 2015. p. 2890–9.
108. Ruddy DA, Schaidle JA, Ferrell III JR, Wang J, Moens L, Hensley JE. Recent advances in heterogeneous catalysts for bio-oil upgrading via "ex situ catalytic fast pyrolysis": catalyst development through the study of model compounds. *Green Chem* 2014;**16**:454–90.
109. Dehkhoda AM, West AH, Ellis N. Biochar based solid acid catalyst for biodiesel production. *Appl Catal A* 2010;**382**:197–204.
110. Lehmann J, Joseph S, editors. *Biochar for environmental management science, technology and implementation*. Routledge, NY: Oxon; 2015.

111. Vardon DR, Moser BR, Zheng W, Witkin K, Evangelista RL, Strathmann TJ, et al. Complete utilization of spent coffee grounds to produce biodiesel, bio-oil and biochar. *ACS Sustain Chem Eng* 2013;**1**(10):1286–96.
112. Gadde B, Bonnet S, Menke C, Garivait S. Air pollutant emissions from rice straw open field burning in India, Thailand and the Philippines. *Environ Pollut* 2009;**157**(5):1554–8.
113. U.S. Energy Information Administration. <http://www.ieee.es/Galerias/fichero/OtrasPublicaciones/Internacional/2016/EIA_Country_Analysis_Brief_India_14jun2016> [accessed 10.09.17].
114. International Energy Agency. <http://www.indiaenvironmentportal.org.in/files/file/CO2EmissionsfromFuelCombustionHighlights2017.pdf>; 2017 [accessed 20.07.18].
115. *Green stoves to replace chullahs - Times of India.* The Times of India. <https://timesofindia.indiatimes.com/india/Green-stoves-to-replace-chullahs/articleshow/5293563.cms> [accessed 25.09.17].
116. Bhattacharyya SC. Energy access problem of the poor in India: is rural electrification a remedy? *Energy Policy* 2006;**34**:3387–97.
117. Kishore VVN, Bhandari PM, Gupta P. Biomass energy technologies for rural infrastructure and village power—opportunities and challenges in the context of global climate change concerns. *Energy Policy* 2004;**32**:801–10.
118. Ekholm T, Krey V, Pachauri S, Riahi K. Determinants of household energy consumption in India. *Energy Policy* 2010;**38**:5696–707.
119. Blanco-Canqui H, Lal R. Crop residue removal impacts on soil productivity and environmental quality. *CRC Crit Rev Plant Sci* 2009;**28**:139–63.
120. Kludze H, Deena B, Weersink A, van Acker R, Janovicek K, Laporte AD. Estimating sustainable crop residue removal rates and costs based on soil organic matter dynamics and rotational complexity. *Biomass Bioenergy* 2013;**56**:607–18.
121. Graham RL, Nelson R, Sheehan J, Perlack RD, Wright LL. Current and potential U.S. corn stover supplies. *Agronomy J* 2007;**99**:1–11.
122. Kim S, Dale BE. Global potential bioethanol production from wasted crops and crop residues. *Biomass Bioenergy* 2004;**26**:361–75.
123. Lindstrom MJ, Skidmore EL, Gupta SC, Onstad A. Soil conservation limitations on removal of crop residues for energy production. *J Environ Qual* 1979;**8**:533–7.
124. Nelson RG. Resource assessment and removal analysis for corn stover and wheat straw in the Eastern and Midwestern United States-rainfall and wind-induced soil erosion methodology. *Biomass Bioenergy* 2002;**22**:349–63.
125. Vassilev SV, Baxter D, Andersen LK, Vassileva CG. An overview of the chemical composition of biomass. *Fuel* 2010;**89**:913–33.

Chapter 3

Sustainable bioenergy development in Africa: issues, challenges, and the way forward

Saidu Muhammad Maishanu[1], Abubakar Sani Sambo[2] and Mahmoud Muhammad Garba[1]

[1]Sokoto Energy Research Centre, Usmanu Danfodiyo University, Sokoto, Nigeria, [2]Faculty of Engineering and Environmental Sciences, Usmanu Danfodiyo University, Sokoto, Nigeria

3.1 Introduction

Energy is commonly associated with the capacity to do work and it plays a pivotal role in the technological and economic advancement of all nations. Moreover, it provides vital services and means to improve the quality of life in today's societies.[1] Africa is a continent made up of 55 countries, covering a land area of about 30.37 million km^2 constituting about 20.3% of the total land area on Earth, and with a population of about 0.9334 billion, but 2016 estimates put the continent's total population at 1.216 billion.[2] Fig. 3.1 shows a map of Africa showing the 55 countries and other vital statistics. Improving the quality of energy supply in Africa is central to sustainable development and poverty reduction efforts. Bioenergy is renewable energy derived from biomass and various biological sources, such as fuelwood, charcoal, livestock manure, microbial biomass, agricultural waste and its by-products, energy crops, and others.[3,4] Biofuels might be solid, such as fuel-wood, charcoal, and wood pellets; liquid, such as ethanol, biodiesel, and pyrolysis oils; or gaseous, such as biogas. The estimated contribution of bioenergy to the global energy supply has remained relatively stable for more than a decade, at around 10%–11% of the total global primary energy supply. Hence, the development of modern bioenergy, from sustainably derived biomass resources, is seen as an alternative energy option with the potential to alleviate poverty and to contribute to rural development in many African countries where there is a heavy reliance on traditional bioenergy systems.[2,5,6]

Sustainable Bioenergy. DOI: https://doi.org/10.1016/B978-0-12-817654-2.00003-4

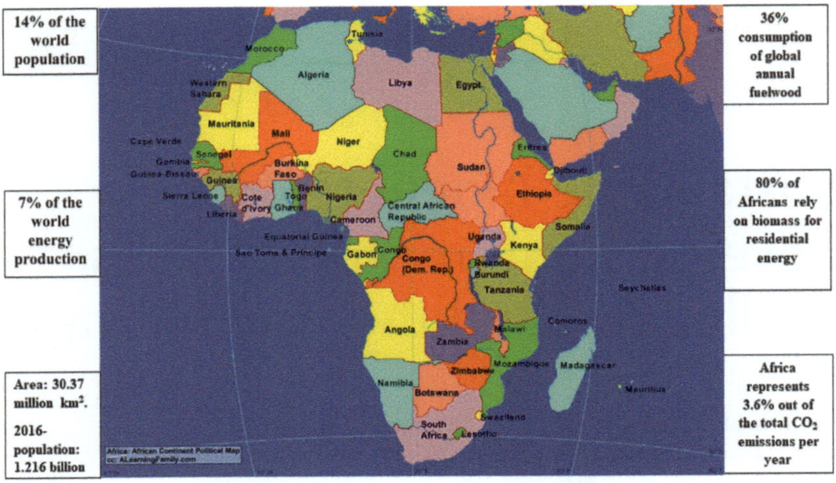

FIGURE 3.1 Political map of Africa.

Bioenergy production from wood is one of the oldest forms of energy and it was considered as a primitive energy source for a long time in many countries. However, it is currently experiencing an increase in attention worldwide. Globally, wood is the most important source of renewable energy and is used to produce more energy than all other renewable energy sources combined.[7,8] According to the Food and Agriculture Organization (FAO),[7] the global annual wood fuel consumption, which comprises of fuelwood, charcoal, and other wood-based energy sources, sums up to 1.86 billion m^3. Of this amount, 17% is consumed in the tropical and subtropical regions of America and in the Caribbean, 36% in Africa, 40% in Asia and the Pacific region, and 8% in Europe. In total, almost three-quarters of global fuelwood consumption occurs in tropical countries. Hence, energy demand in Africa as a whole is expected to double, from 500 Mtoe (million tonnes oil equivalent) in the year 2000 to 1000 Mtoe in 2030, with serious implications for energy security. It is important to note that this projection does not take into account the development needs of Africa to enhance energy security, where 70% of rural Africa does not have access to clean energy.[9] Table 3.1 shows the regional distribution of the population, gross domestic product (GDP), percentage access to electricity, and power generation in Africa signifying that different regions are facing different challenges in getting access to energy, and therefore, each region has to look inwards to harness its wealth of energy resources and that each country should strive to meet its own energy needs from multiple sources.

In sub-Saharan Africa, traditional biomass still dominates the energy mix and the number of people cooking with traditional biomass is expected to increase through to 2030.[10,11] Figs. 3.2 and 3.3 and Table 3.2 are examples

TABLE 3.1 Regional energy landscape in Africa.[9]

Region	Population (million people)	GDP billion/ year (current international US$)	Access to electricity (% population)	Electricity (kW h per capita)
North Africa	175	1936	98	1574
West Africa	327	1310	47	188
East Africa	303	646	23	91
Central Africa	115	227	25	167
Southern Africa	177	1100	43	2061

GDP, Gross domestic product.

FIGURE 3.2 Trucks of fuelwood heading for the big markets in the city in Nigeria.

of how fuelwood has continued to be the primary energy source in many African countries. A report by the OECD/IEA (International Energy Agency) indicates that "it is difficult to quantify the traditional use of biomass precisely given the unregulated nature of its use and a lack of detailed and coordinated efforts necessary to more accurately gauge consumption levels." However, according to the report, the current estimates indicate that over 2.5 billion people still rely on the traditional use of biomass as their principal

FIGURE 3.3 Piles of fuelwood in a local African fuelwood market.

TABLE 3.2 Total and percentage population relying on traditional use of biomass in Africa and globally.

Region	Population relying on traditional use of biomass (millions)	Percentage of population relying on traditional use of biomass
World	2722	38
Africa	754	68
North Africa	1	0
Sub-Saharan Africa	753	80

source of energy, which translates to 28 EJ of solid biomass resource and around ~10% of the total global energy demand.[9]

In many African countries today, the main focus of bioenergy development is placed on biofuels, such as biodiesel, bioethanol, and biogas, to replace fossil fuels. With this in view, the chapter looks into the issues surrounding the sustainable use of bioenergy and its prospects in Africa, such as poverty reduction, economic development, health benefits, energy security, and environmental benefits, such as alleviating desertification and reducing greenhouse gas (GHG) emissions as well as slowing land degradation. It also examines the challenges, such as impacts on biodiversity, insecurity of food supply, deforestation, land degradation, and water supply. It then proposes a way forward for sustainable bioenergy development in Africa.

3.2 Sustainable bioenergy production

The widespread cognizance of wood as an important energy source in modern times has been mainly driven by the pressure of diminishing fossil fuel resources in industrialized countries as well as the wish to become more independent from nuclear power and its risks in some developed countries. In addition, bioenergy was identified as an alternative to fossil fuels, which could also help to prevent the worsening of anthropogenic climate change by attempting to reduce GHG emissions. The global production of biomass and biofuel is growing rapidly due to the increasing price of fossil fuels, growing environmental concerns, and considerations regarding the security and diversification of energy supply.[12,13] There are many scenarios that predict a high potential for biomass in the future. There have also been many studies performed in the past few decades to estimate the future demand and supply of bioenergy. Overall, the world's bioenergy potential seems to be large enough to meet the global energy demand in 2050.[14,15] The current stock of standing forests is a large reservoir of bioenergy and in line with the theoretical potential of biomass energy. However, most of the research studies on biomass potentials ignore existing studies on the demand and supply of wood, despite the extensive literature and data on the subject. Taking into account data from a variety of international sources, rough estimates of the energy production potential of woody biomass from forests show that, in theory, the demand for wood fuel and industrial roundwood in 2050 can be met, without further deforestation, although regional shortages may occur. Africa's biomass energy resources vary geographically and are not uniformly distributed.[15–17]

Biomass energy use depends on a number of issues, including geographical location, land-use patterns, preferences, and cultural and social factors. Income distribution patterns also contribute to variations in biomass energy use; with poorer African countries relying on traditional forms of biomass and wealthier African countries using more modern biomass energy technologies.[15,18,19] Low per capita national incomes as well as the slow growth in conventional energy use influences the heavy reliance on biomass energy in Africa and is considered unlikely to change in the near future. Estimates indicate that by 2020, traditional biomass energy use is expected to increase roughly at the same rate as population growth rates, resulting in modest changes in the share of biomass in total energy supply[18] (see Table 3.1).

The sustainable use of biomass as an energy source requires the comprehensive management of natural resources, such as land and water. Unsustainable biomass production would erode the climate-related environmental advantages of bioenergy.[17,19–21] It is more important now than ever before to reliably demonstrate that the advantages of biofuels exceed the cost of the potential environmental damage caused by their production. Therefore a major challenge in increasing the production of biomass and biofuels is their production in a sustainable way. Many African countries have huge potential to generate and consume energy from both traditional and modern bioenergy resources, yet the continent is starved of energy. Some examples of these include the renewable form of municipal solid waste; industrial waste of nonrenewable origin (solids or liquids) burnt directly for the production of electricity or heat. Others include organic, nonfossil materials of biological origin, such as charcoal and wood, wood waste, bio-briquettes, and other solid waste that may be used as fuel for heat production or electricity generation. Gaseous biofuels, such as biogas; and liquid biofuels, like biodiesel and bioethanol; and the use of modern technologies, such as gasification and fast and slow pyrolysis are all additional energy resources or technologies available in Africa. In a situation where many African countries continue to struggle to provide sufficient energy to address the basic needs of all their citizens, bioenergy, especially biogas, may be a viable and attractive option for supplying cheap and consistent energy to rural populations.

3.3 Bioenergy for sustainable development in Africa

Bioenergy is seen as a solution to the increasing energy costs in Africa.[21] Much literature shows that the production of biofuels represents, especially for developing economies, both a great opportunity and a great threat.[22–25] In many sub-Saharan African countries, the ease of access to biomass energy in the form of fuelwood has led to increased dependence on biomass as the primary source of energy in both rural and urban areas and also a source of gainful employment. Much work has been done on the topic of energy accessibility in this region, through the use of improved stoves[26,27] and fuelwood plantations. The unsustainable use of fuelwood and charcoal and ineffective forest and land management have led to large-scale deforestation and degradation. To halt this trend, African countries must increase the use of sustainable practices in the effective management of forests and land, improve biomass yield for food and fuel, as well as improve the biomass supply chain, and the use of bioenergy equipment with high efficiency.[21,22] An FAO report (2018) has indicated that, for example, in Burkina Faso and Tanzania, fuelwood and charcoal meet 80% of household energy needs. Strategies in these countries have supported fuelwood plantations and many African countries have programmes for improved stoves, yet biomass is generally regarded as an affordable but unclean source of energy. Despite the fact that the demand for fuelwood has contributed to forest degradation, woody biomass, if well managed, can be affordable, clean, and renewable.[26]

3.4 Advantages linked to bioenergy sector implementation or development

It is important to note that sub-Saharan Africa has limited case-specific data to determine the sustainability of bioenergy systems. However, part of the existing literature on the subject suggests that bioenergy holds significant promise for technical improvement, particularly in the Sahel countries. Presently biomass energy (electricity, heat, and ethanol) has a capacity factor of 85%–95%, but in practice, most plants do not regularly operate at these levels. Plants in India, China, and the rest of the world, including Africa operate above 60%, while in Europe and North America, it can reach up to 80%.[28] In addition, bioenergy can play a dual role in GHG mitigation related to the objectives of the Kyoto protocol and other climate change initiatives. It can act as a source of sustainable energy to substitute fossil fuels and as a carbon store.[12]

3.4.1 Poverty reduction

The lack of access to modern energy constrains economic and social development, contributes to poverty, and reduces the chances of achieving the millennium development goals (MDGs). Modern energy, provided at affordable prices in sustainable and appropriate ways to people living in rural and semi-urban areas, is key to poverty reduction, increasing food production and consumption, and improvements in the environment, education, and health.[28,29] Bioenergy can play a major role in helping to achieve broader energy access, agriculture, poverty reduction, and national development strategies. In order to end energy poverty, it is necessary to have access to modern and cleaner energy carriers (e.g., biogas, electricity) or better conversion technologies and equipment.[30] Diaz-Chavez et al.[30] also noted that bioenergy crops provide an opportunity for diversifying rural activities and creating jobs and infrastructure. Although there is yet little evidence that biofuel production can reduce poverty directly, it has been demonstrated that it plays an important role in creating jobs and local markets.[30–32]

3.4.2 Bioenergy and economic development

By providing energy to rural areas and improving agricultural productivity, bioenergy has the potential to bring significant economic and social improvements in living conditions. In particular, the income-generating potential of bioenergy is linked to the production, transformation, and commercialization of biomass resources.[33,34] Local production and consumption of biomass can play an important role in sustainable livelihood strategies. First, biofuel production can be used to improve agricultural employment and livelihoods, especially when the cultivation involves small-scale farmers and the conversion facilities are located near crop sources in rural areas. Many biofuel production processes are labor-intensive and most bioethanol-related jobs involve low-skilled and poor workers in rural areas, thus, bioenergy programs can generate employment opportunities and contribute to the overall development of these regions.[35]

Biofuel production is expected to generate more employment per unit of energy than conventional fuels and more employment per unit investment than in the industrial, petrochemical, or hydropower sectors.[17] For instance, the World Bank reports that biofuel industries require about 100 times more workers per unit of energy produced than the fossil fuel industry. In the case of the Brazilian bioethanol industry, it has directly provided more than half a million jobs. In sub-Saharan Africa, the job creation potential of bioenergy is estimated to be between 0.7 and 1.1 million through the regional adoption of E10 for petrol and E5 for diesel.[36] For example, in Senegal, intercropping of Jatropha with peanut plants and bananas has enabled the creation of an inter-seasonal vegetable crop. Agricultural workers have been able to stay employed throughout the year instead of having to stop work during the rainy season.[37,38]

Biofuels can also provide opportunities for agricultural diversification. For instance, the reform of the EU sugar regime in 2006 implies that biofuel production could provide an opportunity for the diversification for many traditional sugar-producing countries in Africa whose sugar exports enjoy preferential access to this market, especially the ACP countries (African, Caribbean, and Pacific sugar exporters).[39,40] Biofuels are also being promoted as production alternatives to less desirable crops, such as tobacco in Malawi. All of these aspects enhance rural development, especially in African countries. In the determination of biofuel economic impacts, the entire value chain comprising of the production of feedstock, their processing, blending, distribution, and marketing will have to be taken into account. In Africa, there is a need to implement policy measures to motivate the private sector to invest in the value chain, ranging from producers to consumers of bioenergy.[41]

3.4.3 Improving health benefits and enhancing the role of women

Women and children spend several hours of the day collecting fuelwood and other forms of biomass. This is often detrimental to productive activities and education. In rural sub-Saharan Africa, many women carry about 20 kg of fuelwood for an average of 5 km every day. Beyond the considerable loss of time to collect them, the use of traditional fuels through traditional stoves or open fireplaces is not only inefficient but also damaging to health. Combustion is incomplete and results in substantial emissions that, due to poor ventilation, produce high indoor pollution. Exposure to this polluted air leads to, among others, respiratory illness, cancer, tuberculosis, low birth weight, and eye disease.[26,27,42,43] Women and children are the first to be affected. The WHO estimates that 4 million women and young children in developing countries die prematurely each year from breathing smoke from indoor biomass stoves.[44] According to FAO studies,[45,46] women in most developing countries are the mainstay of the agricultural sector, the farm labor force, food systems, and day-to-day family subsistence, but they are the last to benefit.

Throughout sub-Saharan Africa, women are responsible for between 70% and 80% of household food production. They tend to livestock and the land, grow vegetable gardens, and cultivate subsistence crops, such as rice. Women are also responsible for gathering wood for home cooking and heating and carrying water for household needs and farm irrigation. This consumes their days and leaves little time for studying or income earning. Reducing the exposure of women and children to these risks should be a priority that will ensure universal access to modern energy services. This can be achieved by promoting the use of advanced cooking stoves that burn biomass more cleanly and efficiently. This can contribute substantially to improving the air quality, health, and nutrition of rural populations. The use of processed wood pellets or ethanol gel in stoves reduces time spent collecting firewood and frees up time to spend on other activities particularly on new income-generating activities.[42–44] By growing biofuel feedstock alongside food crops, women can gain access to additional energy sources. Women organizations can be empowered to implement a sustainable food-and-fuel system, allowing local energy needs to be met and providing income opportunities from the sale of biofuels and by-products to broader markets. Moreover, using bioenergy permits the wider use of refrigeration to keep vaccines and other heat-sensitive medical products safe from deterioration, thus improving health conditions for children in rural communities.[45,46]

3.4.4 Bioenergy and improvement in energy security

Closely linked to the policy goal of energy security is the goal of reducing the oil import bill. As noted previously, most sub-Saharan countries are dependent on petroleum imports. Near record oil prices are increasing oil import bills, consuming the budgets of cash-strapped countries, hampering economic growth, and threatening the gains made in economic development. For instance, the 2005 oil price surge reduced the GDP growth of net oil-importing countries from 6.4% to 3.7%, and, as a consequence, the number of people in poverty rose by as much as 4%–6%, with nearly 20 countries experiencing increases of more than 2%.[47] Also, the 2008 record high oil prices made many countries reconsider their outlook of future energy supplies and energy security.[48]

As projected in Table 3.3, the primary source of energy in Africa, mainly fuelwood, will continue to play a major role in the energy supply and demand markets in Africa for some time in the future. Even though the production of liquid biofuels was reported to be more than the consumption in Africa, these countries must brace themselves for the accelerated production of these fuels for both domestic and international markets. Domestic biofuel production offers a promising alternative to traditional biomass fuels and fossil fuels for domestic uses and small-scale industries. Bioenergy can assist in advancing energy security in the region by diversifying the regions energy mix, reducing demand for petroleum imports, and decreasing the impact of fossil fuel price uncertainty on the balance of payments. Biofuel production

TABLE 3.3 Projected annual consumption and production of bioenergy in Africa, 2005–30 (Mtoe).[48]

Item	Consumption				Production			
Energy type	2005	2010	2020	2030	2005	2010	2020	2030
Primary solid biomass	295.2	314.1	350.8	377.4	295.2	314.1	350.8	377.4
Biogas	0	0	0	0	0	0	0	0
Ethanol	0	0	0.8	1.1	0.3	0.7	2.4	3.2
Biodiesel	0	0	0.9	1.3	0	0	1.2	3.5
Total bioenergy	295.2	314.1	352.5	379.8	295.4	314.9	354.4	384.1
Total primary energy supply (TPES)	466.1	517.1	625.8	744.7	n/a	n/a	n/a	n/a
Bioenergy share of TPES (%)	63.3	60.7	56.3	51	n/a	n/a	n/a	n/a
Transport fuels	35.3	40.7	55	75.8	n/a	n/a	n/a	n/a
Bioenergy share of transport fuels (%)	0	0.1	3.1	3.1	n/a	n/a	n/a	n/a

offers poor oil-importing countries an opportunity to replace oil imports and improve their trade balance. Energy diversification, such as the exploitation of renewables, makes countries less vulnerable to oil price shocks, which can compromise macro-stability and affect variables such as the exchange rate, inflation, and debt levels.[49]

In 2015, the bioenergy markets were reported to be dominated by bioethanol and biodiesel.[50,51] Biofuels are a rational choice, especially in those countries where feedstock can be produced at a reasonable cost without adverse social and environmental impacts. For example, in Brazil, biofuels that are locally produced can offer a highly competitive alternative to other fuels. The Brazilian ethanol industry is estimated to have the lowest ethanol production costs in the world.[52] The average production cost of hydrous ethanol in Brazil in 2013 was approximately US$0.60–0.71 per liter, which steadily rose to US$1.65 in the first week of 2018/2019 cane processing season.[53,54] The experience in Brazil, for instance, suggests that replacing imported gasoline with bioethanol saved the country US$43.5 billion between 1976 and 2000 (US$1.8 billion/year).[55]

3.4.5 Environment and greenhouse gases

Reducing GHG emissions is one of the main reasons behind the policy for promoting biofuels. According to the IEA, biofuels used in transport are expected to enable a 6% global GHG reduction. However, there are mixed views among the scientific community regarding the actual GHG reductions associated with biofuels.[56] There are two ways in which biofuels can reduce carbon emissions.[57] First, over their life cycle, biofuels absorb and release carbon from the atmospheric pool without adding to the overall pool (in contrast to fossil fuels). Second, they displace the use of fossil fuels. However, the production of biofuels does in most cases involve, to some extent, the consumption of nonrenewable fuels. While sugarcane-based ethanol and second-generation biodiesels may achieve 70%–100% reduction in GHG emissions relative to gasoline, cereal-based ethanol shows lower reductions or even gains in the case of inefficient early experiences with bioethanol production from maize.[55] Higher use of mechanization and fertilizers during production is associated with higher emissions.[20]

The 2007 UN energy report suggests that the value of biofuels in combating climate change through reduced GHG emissions will come from fuel switching from coal in the combined heat and power sector, rather than the current focus on the transportation sector.[17] However, a decade after the report, several newer reports[58–63] have highlighted the significant role of biofuels and other forms of bioenergy in the 2°C mitigation scenario pathways to 2050 and beyond, as reported by STAP.[64] In developing countries, the prospect of bilateral or multilateral aid transfers for climate change mitigation is also generating significant interest in biofuels. African countries like other developing countries do not currently have binding GHG reduction targets under the Kyoto Protocol and are typically more concerned with potential climate change impacts than with reducing their own GHG emissions.[60] The Clean Development Mechanism (CDM), however, offers an opportunity to developing countries to benefit directly from investment in mitigation technologies that generate GHG reduction credits for industrialized countries. The CDM allows developed countries to implement project activities that reduce emissions in developing countries in return for certified emission reductions (CERs). However, the existing experience with CDM projects shows that approved projects are strongly concentrated in a handful of large developing countries, with over 60% of all CDM projects distributed across China, India, and Brazil alone. Figures from the United Nations Framework Convention on Climate Change have indicated that of the total 7804 registered projects in 2018, African countries only had about 218 projects, which are mostly located in South Africa, while Asia and Pacific countries, mainly China and India, received 6532 projects and Latin America and Caribbean countries, mainly dominated by Brazil, received about 1007 projects.[65] Hence, for African countries to benefit fully from the CDM projects, bioenergy projects specifically, should include all land-use activities that will boost bioenergy generation, consumption, and exportation in addition to afforestation and reforestation, which have been the focal point of CDM approved projects in the past.[66]

3.4.6 Slowing land degradation and improved waste management

The United Nations Environment Program report "Global Environment Outlook"[67] singled out land degradation as the biggest threat to Africa's achievement of the MDGs. According to a 2013 report, about 500,000 km^2 of land in Africa is being degraded due to soil erosion, salinization, pollution, and deforestation, which can adversely affect agricultural productivity, nutrition, and human health.[68] In the arid and semiarid regions of Africa, land degradation, deforestation, and desertification, compounded by global warming, are extremely alarming issues. The development of agroforestry, reforestation, and afforestation through the generation of bioenergy can play an important role in addressing these key environmental challenges.

For degraded areas, the planting of perennial biofuel crops, such as Jatropha, has the potential to regenerate agricultural potential, providing shade and nutrients for other crops.[69–71] Second-generation biofuels will further reduce environmental impacts relative to first-generation plants used as feedstock for biofuels because of their production from waste (weeds), by-products of agriculture and forestry, and indigenous perennial plants.[17] Many African countries have a high percentage of unused arable lands. The cultivation of certain bioenergy crops, such as Jatropha, with the potential to grow on marginal lands, could help revive certain areas. However, more agronomic research is needed to determine the best species and conditions for these lands as poor yields have been recorded in many African countries.[70,71] In addition, modern biomass technologies can enhance waste management practices in Africa. For example, biomass power plants can process a range of agricultural, forestry, and crop residues as a feedstock for energy production.

3.4.7 Export development

Many countries see in biofuels an opportunity to develop a new export market for their agricultural produce and to increase export revenues. This stems from the fact that the main international consumers (northern countries) will not have the domestic capacity to supply their entire domestic demands, while many developing countries located in tropical and subtropical areas have, or may develop, advantages in biofuel production.[72] However, the assessment of the biomass potential of African countries requires reliable, up-to-date data. According to the available results, the tremendous potential of energy crops in Africa can only be realized through an enormous increase in agricultural production efficiency comparable to that obtained in developed countries. Therefore the expansion and improvement of the current agricultural production systems in Africa will play a crucial role in the sustainable development of modern bioenergy systems.[73] On average, biomass in tropical and subtropical areas is five times more productive, in terms of photosynthetic efficiency, than biomass produced in temperate regions.[74] Countries with large land endowments, significant cost advantages in agricultural commodity production, and long experience in feedstock production and trade are seizing opportunities from supplying biofuels to international markets. Likewise, those countries enjoying preferential access to key consumer countries are already expanding production to access the market. Targets may not only involve traditional north–south trade but also regional south–south trade. Even though there are several prospective benefits to bioenergy development, a number of barriers exist.[74]

The broader picture is not so attractive. A number of concerns are raised by these developments. Without subsidies, most biofuels cannot compete in price with petroleum products in most regions of the world, especially in Africa. The surface area of cultivable land that they require is significant and has put pressure on food and water prices. A report by the OECD/FAO[75] expected food prices to rise by between 20% and 50% in 2016. The growing use of cereals, sugar, oilseeds, and vegetable oils to satisfy the needs of a rapidly increasing biofuels industry is one of the main drivers, according to the report.

3.5 Barriers to biofuel commercialization in Africa

There exist many problems that hinder the commercialization of bioenergy in both developed countries and developing countries. However, the magnitude and characteristics are more pronounced in developing countries. The main problems of biofuel commercialization in Africa are related to economic and political issues.

3.5.1 Economic issues

Probably the most important challenge to the widespread development of the biofuel industry is its cost. Biomass feedstock from agricultural and forest resources have low mass and energy density with the current harvest and collection technologies. The low density of these feedstocks makes them cost-prohibitive to transport, handle, and store. Other problems that have been identified include: the equipment for biomass feedstock logistics systems is inefficient, the quality characteristics of biomass feedstocks are variable and inconsistent, and lastly the transportation of biomass is costly and can strain transportation networks.[76] Some estimates show biofuels to be twice as costly as conventional fuels.[55] However, the current oil prices make the production from the most efficiently producing countries more competitive. The economics of biofuel production and consumption will depend on a number of factors specific to the local situation. These factors include: the cost of biomass materials, which varies among countries depending on land availability, agricultural productivity, labor costs, and others[77–79]; biofuel production costs, which depends on the plant location, size, and technology, all of which vary a great deal among countries; the cost of corresponding fossil fuel (e.g., gasoline, diesel) in individual countries, which depends on fluctuating petroleum prices and domestic refining characteristics; and the strategic benefit of substituting imported petroleum with domestic resources.[77,80]

3.5.2 Economic costs of biofuels (financial barriers)

The economics of biofuel production and use, therefore, will depend on the specific country and project situation. The variation of cost with location affects the cost of the construction plant directly and indirectly. The cost of biofuels tends to differ widely depending on the type of biofuel, the price of fossil fuels, the price of feedstock, the country of provenance, the technology used, the cost of conversion (including investment needs), and the revenues generated by the coproducts.[79–81] For African countries to maximize on the economic benefits of bioenergy, all value-adding should, therefore, take place within the country to produce finished products rather than exporting raw feedstock. This will increase both investment and job opportunities within the country as well as increasing the value of the finished product.[79]

3.5.2.1 Cost of feedstock

This accounts for a major portion of the total cost of biofuel production, varying between 50% and 75%.[49] Several studies on biofuel production have shown that the cost of raw feedstock is considerably high, comprising of 75%–80% of the total operating costs.[82–85] The highest share of the cost is for those energy crops used for biodiesel production. This implies that any changes in feedstock production costs or in their international prices can have a huge impact on the overall cost of biofuel production. Brazil is by far the most efficient biofuel-producing country, with production costs starting from US$23, while in temperate countries, such as the United States and the European Union, these are not lower than US$40 and 50, respectively.[73,74] Low cost sugar-producing countries, such as Thailand and Australia, also show relatively lower costs. For biodiesel, costs tend to be higher than those for bioethanol. The lowest cost of production seems to come from palm oil-based biodiesel. For coconut oil, the estimated cost of conversion in Pacific Island countries is US$0.30–0.60 per liter, which is currently perceived to be too high to make it viable as a feedstock. In India, the production costs of Jatropha-based biodiesel are estimated to be between US$0.40 and 0.53 per liter, but with strong dependence on the price of by-products.[74] Many studies have reported that today the most economically viable option for biofuel is sugarcane-based bioethanol. In the long term, it is argued that prospects for bioethanol are better than those for biodiesel as the costs for the former are expected to fall faster, especially once cellulosic bioethanol becomes commercially competitive.[56,76]

3.5.2.2 Transportation, distribution costs, and logistics

The cost of biofuels is also affected by the available infrastructure for the transport and distribution of both the feedstock and the biofuel. For instance, the sale price of Brazilian bioethanol in the European Union varies in the range of €200–300/toe (tonnes of oil equivalent) and transport and distribution costs can add an extra €150–200/t.[77] Infrastructure involves suitable roads, waterways, and pipelines to transport products to the markets, but also communication infrastructure, which is essential for notifying producers, processors, and traders on weather and market conditions. Poor telecommunications infrastructure and high cost of services also represent a barrier to biofuel commercialization.[86,87] Biofuels can generally be accommodated and distributed via the existing fossil fuel distribution infrastructure. However, in their pure forms, some special measures to upgrade the existing infrastructure must be taken. The experience in Sweden and Switzerland suggest that these costs are relatively low.[17] There are also differences in infrastructure needs depending on whether the product is destined for local use or for national and international consumption. For the former, distribution and logistics are relatively straightforward, but for the latter they could be a major challenge for market development.

Pipelines are the safest and most cost-efficient method of transporting liquid fuels. Bioethanol absorbs water and impurities that normally reside in fuel pipelines, which can cause the phase separation of bioethanol–gasoline blends thereby reducing engine performance.[88,89] An additional problem is stress corrosion cracking (SCC)—which is the formation of brittle cracks caused by corrosion processes and stress—associated with the transport of high bioethanol blends. The need to avoid contamination from water and SCC explains why bioethanol is not distributed through existing fuel pipelines.[90]

Biodiesels based on tropical oils (e.g., palm or coconut oil) typically have higher viscosities than biodiesel based on temperate oils (rapeseed or soy) and therefore require thinning agents to be exported to temperate climates (which involves additional costs). Inadequate or lack of infrastructure is a major constraint in Africa, and therefore it can constitute an important factor undermining the commercial viability of biofuel production.[91] Attempts to import biofuel technology from developed countries (technology transfer) to Africa may likely fail due to the lack of proper understanding of African features (the technology being transferred is not appropriate to the local context and demands or is not adapted to the local environment).[92]

3.5.2.3 *Biofuel economics are sensitive to fossil fuel prices*

The price of oil is one the most important determinants in the economics of biofuel. If world oil prices remain high for a prolonged period of time, biofuel programs have a better chance of becoming financially viable without sustained government support in a larger number of countries. At present, the costs involved in the production of biofuels in the form of alcohol or biodiesel are higher than the costs of extracting and refining petroleum and transporting the fuels produced to where they will be used. However, with oil prices above US$45–50 a barrel (and other variables, such as feedstock prices, remaining constant), biofuels produced in the most efficient countries become commercially competitive without public subsidies.[80] The IEA does not expect oil prices to fall below US$48 until 2030. In addition, a 5-year oil market forecast expects oil demand to grow strongly at least until 2022 with the main developing economies leading the way. However, there is a risk of prices rising more sharply by 2022 if the spare production cushion is eroded. It should be noted, however, that biofuels are price-takers in the energy market, representing a tiny fraction of the total value of the market.[56]

3.5.2.4 *Costs of biofuels are sensitive to by-products*

The economics of biofuels are affected by the revenues generated from the associated by-products. For bioethanol, the main by-products are soluble from maize and wheat (a high protein animal feed), for sugarcane it is maize oil and bagasse (used for electricity generation), from cellulosic bioethanol it is lignin (used for wood products). For biodiesel, the key by-products are glycerin (used as a degreaser, hand soap, or for general use in the cosmetic industry, and others) and oil seed meal (animal feed).[93] The sale of by-products may offset the high cost of biofuels production. In the United States, for instance, by-product sales amounted to an estimated 24% of the total revenue generated from maize-based bioethanol in 2004.[47] On the other hand, increases in biofuel production can decrease the price of by-products (e.g., as is happening with glycerin), negatively affecting they biofuel economy.

3.5.2.5 *Investment needs*

The production of biofuels requires channeling investment for feedstock production, processing, and distribution. For example, the establishment in Southern African countries of a new estate and sugar factory processing 2 Mt of cane per year would cost US$200–300 million; the capital costs for an ethanol distillery are in the range of US$50–80 million; and the costs of a cogeneration plant are about US$1000–1200 per installed kilowatt.[18] Further costs for transport, infrastructure, and storage would also be incurred. Nevertheless, it will be difficult to attract such investments to regions with poor infrastructure and high investment rates.[91] So for these regions, governments will play a crucial role in creating the appropriate environment for investment. The fundamentals for attracting investment include the establishment of a clear, stable, and transparent legal and fiscal framework and an efficient administration, the creation of capital markets to favor small-scale investments as normally required for some biomass energy, and they must also facilitate low-interest rate credit.

3.6 Policy issues: biofuels public policies, their costs, and benefits

Policy and regulatory support are necessary for the successful implementation of improved and modern bioenergy projects as has been recommended by some international initiatives, such as the Global Network on Energy for Sustainable Development. In contrast to the development of bioenergy policies in other regions of the world, Africa does not have a comprehensive regional policy on biofuels to regulate the growing industry.[79] Many African countries are only now introducing biofuel policies. Mali, Nigeria, Senegal, Tanzania, Ethiopia, Angola, Mozambique, South Africa, and Swaziland have implemented specific biofuel policies. Burkina Faso, Côte d'Ivoire, Ghana, Guinea-Bissau, Senegal, Sierra Leone, DRC, Equatorial Guinea, Kenya, Rwanda, Madagascar, Mauritius, Zambia, and Botswana are also developing explicit policies.[94] This lack of a regional policy and strategy has led to an underinvestment in biofuels research and development (R&D) in Africa. Regional economic communities in Africa, such as ECOWAS, SADC, AU/NEPD, and EAC are playing and must play an important role in supporting the development of the biofuels industry in Africa. A number of international aid organizations are collaborating with different countries in Africa toward the generation of policies; such as GTZ in Mozambique, Swedish International Development Cooperation Agency in Tanzania, and Centre de Coopération Internationale en Recherche Agronomique pour le Dévelopment in Burkina Faso. There are a number of international, national, and regional initiatives in Africa regarding policies and plans.[95–98] For instance, the 2007 Addis Abba declaration that emanated from the first high level meeting of African bioenergy stakeholders, committed the continent to sustainable bioenergy development. The seminar was organized by the African Union Commission along with the United Nations Industrial Development Organization and the Brazilian government.[95–98]

The political declaration put out will, among other things, facilitate the development of not only the biofuel sector, but also the bioenergy sector as a whole because available experience suggests that the biofuel sector is not commercially viable on its own, at least in the initial phases of sector development due to several mitigating factors, such as the food versus energy problem, land tenure systems, and social and environmental consequences, such as the depletion of biodiversity, water resources, soil fertility, and landscapes, among others.[99] The sector requires some form of policy support that will define and adopt sustainability criteria that will eventually enhance food security, rural development, poverty alleviation, land rights and tenure, environmental protection, social equity and well-being, cultural heritage, and macroeconomic impacts, at least in the initial phases of sector development. Also, these policies need to respond to specific goals associated with biofuel production. These policy interventions are expected to put costs and benefits to the public that need to be weighed against the equivalent costs of the energy being replaced.[99]

There are many types of policy tools needed to support the sector:

- The setting of national targets for biofuels (voluntary or mandatory).
- The use of financial incentives to support biofuels production and consumption (e.g., subsidies for feedstock and biofuel production, fuel tax exemptions).
- Price and production controls.
- Trade policy measures (e.g., tariffs and quotas).
- Investment incentives, such as soft loans, grants, tax reductions, accelerated depreciation.
- Policy incentives targeting the inclusion of small-scale producers.
- Support of demonstration projects and pilot projects.
- Support of R&D for technology development.
- Support of the introduction of flexible-fuel vehicles.

Such policies differ in terms of their trade impact (e.g., producer subsidies and trade tariffs vs support of R&D), social returns (e.g., policies to promote small-scale producer inclusion vs policies based on large-scale production), GHG reductions (e.g., environmental taxes on biofuels vs other forms of carbon reduction), and economic cost involved.[97,99] In many countries, the main rationale policy behind biofuels production is to decrease the costs associated with imported fossil fuels. One of the costs of such a policy is the foregone duty on fuel imports, which results in a decline in government revenues. For instance, in Brazil, the forgone tax revenue in the state of São Paulo, which accounts for more than one-half of the total hydrous ethanol consumption in the country, was about US$0.6 billion in 2005.[47] In the United Kingdom, the government estimates that under present arrangements, annual foregone fuel duty revenue will total US$136 million if biofuels achieve a 1% market share.[55] This poses a significant challenge for developing countries, especially in Africa where there is a multitude of urgent needs competing for scarce fiscal resources. Moreover, in some cases, the diversion of feedstock exports for biofuel production into domestic and local markets means that countries may suffer reductions in their export earnings.[72] Another issue is that once subsidies are granted and the biofuel industry has been launched, it will difficult to withdraw them. Every country with a biofuel program has provided subsidies to the industry, and none of them has yet removed those subsidies entirely. In Brazil, the most competitive country globally, success has come through 20 years of government support. According to ESMAP,[47] promoting biofuels for energy diversification makes sense if large government subsidies are not required. So, the promotion of the bioenergy sector should not be done solely by government grants. Other alternatives must be found. The first alternative policy can be the liberalization of the bioenergy sector (benefits of diversification can also be greatly enhanced if trade of biofuels is liberalized). Government can also provide basic rural and road infrastructure. The less favorable risk rating compared to better-established energy technologies implies that small-scale biofuels projects could face challenges obtaining finance from traditional financing institutions.[17] In those cases, governments can help by providing market infrastructure, including microcredit or other alternative credit delivery systems and/or acting as a guarantor to assist them in all stages of the process.

Foreign investment can also constitute an important source of capital but governments need to make sure, however, that the investment involves important domestic benefits (in terms of technology transfer, for example). Investments can also be obtained from international financial institutions that play a key role in reducing political risk in developing countries. For instance, the Inter-American Development Bank (IDB) approved a US $120 million biofuels project in Brazil in 2006. The same bank, in collaboration with Brazil's development bank BNDES, approved a US$750 million conditional credit facility for the country's Sustainable Energy Financing Program in 2016. The IDB is also analyzing the possibilities of facilitating technology transfer and technical assistance from Brazil so that other countries in the region can benefit from Brazilian know-how.[100] The BNDES bank in Brazil provides specific credit lines for the sugar, ethanol, and bioenergy industries to fund investments in sugarcane production, expansion of the industrial capacity for sugar and ethanol, cogeneration, logistics, and multimodal transportation.[99,100] In Africa, similar financial institutions, such as the African Development Bank and regional, such as ECOWAS–Ecowas Bank for Investment and Development, and national initiatives or foundations, can provide credit facilities and other financial support to national bioenergy programmes.[101] Developing countries can also make use of carbon finance markets for attracting investment in biofuel projects using the market value of the expected GHG emission reductions. Developed countries can use the CERs generated by such project activities to help meet their emission targets under the Kyoto Protocol.[65,101]

Although the CDM is a potential source of financing for biofuels projects, taking advantage of it can present a number of challenges for the developing country hosts. For example, so far there is no liquid-biofuels baseline and monitoring methodology approved. The calculation of GHG emissions is not straightforward and for many countries, biofuels are still a relatively expensive means of reducing GHG emissions relative to other mitigation measures. In addition to these barriers that must be overcome to develop the production and use of bioenergy in Africa in a sustainable way, many poor African countries do not show the inertia and flexibility to change. There are rigidities in structures and behaviors, hence, change takes a long time and faces great resistance. Furthermore, serious drawbacks associated with the development of this sector are observed and some of them are enumerated here.

3.7 Challenges of bioenergy in Africa

Biofuel production will increase the demand for agricultural land at the expense of natural ecosystems. Perhaps more critically, it will also require large quantities of water, which is already a major constraint to agriculture in many parts of the world.[21,102] Pursuing biofuel production in water-short countries will put pressure on an already stretched resource and will turn green energy into a major threat to resources. Indeed, there are other cases where major switches have occurred in response to the volatility of certain commodities. For example, a major switch was observed in Latin America and the Horn of Africa, from traditional food crops and coffee to cocoa production as a response to diminishing returns in the commodities sector.[103,104] This concern over food security is highly sensitive in sub-Saharan Africa given the large number of famine cases; therefore assigning land for fuel will consist of significant policy challenges.

3.7.1 Food or fuel

There are potential dangers associated with the emergence of agriculture as a producer of transport fuels for which the market appears to be without limits. The temptation to increase the production of biofuels will inevitably increase the pressure to clear land for activities such as expanding sugarcane production (in the case of ethanol) and palm oil plantations, thereby posing new threats to animal and plant diversity.[105] Furthermore, with the price of crude oil expected to continue rising, investments in fuel crop production and competition for land will create increasing pressure on traditionally farm-based economies. This soaring demand for agro-based fuel is coming at a time when world food stocks are at their lowest level in almost 40 years and there are 83 million more people to feed each year.[53,57,78]

Policies over the past two decades in this region have favored export crops over basic food crops, promoting declines in domestic per capita food production, though some African countries are reported to give equal investment incentives to both export and basic food crops.[106] For countries dependent on fuel and food imports, such as several in Africa, cost increases in both fuel and staple food imports threaten the survival of the rural poor in Africa. A clear pathway for the whole world to make biofuel sustainable is by emulating the way Brazil achieved prominence in its ethanol program, which did not show any competition between food and fuel and that was due to the thoroughness of the program and its association with the political leadership of that country which ensured the existence of political will at the highest level.[53,99,100,105]

3.7.2 Deforestation

Large plans to produce biodiesel could lead to bad environmental impacts. Brazil, being the largest biofuel producer among developing countries, has been at the center of the biofuel–deforestation debate. The expansion of feedstock production capacity, such as sugarcane or palm oil plantations, can impose severe environmental damages. In Brazil, for example, large areas of forest have been destroyed for the production of soya, and there are concerns that the new plans of the Brazilian government to produce biodiesel will cause even more deforestation in the Amazon. In addition, the expansion of bioenergy production in Brazil could indirectly lead to increased deforestation as cattle ranching displaced from pasturelands by sugarcane and soya production could encroach on Amazon forest areas.[107] Biodiesel from soybean in the Brazilian state of Mato Grosso was reported to be responsible for deforestation[108] and another report claimed that over the past decade, up to 5.9% of the direct annual deforestation was attributed to land clearing for soybean cultivation.[109] Other sources argue that the Brazilian soya industry has little to do with forest clearing and plays an important role in promoting regional economic development.[110–112] In Indonesia and Malaysia, biodiesel from oil palm may have been responsible for up to 2.8% and 6.5% of direct deforestation, respectively.[113] The more palm oil is used in the biofuel industry, the more forests are cleared for its cultivation. At present, 1 ha of oil palm may yield 5000–6000 L of crude biodiesel, less than that of sugar cane which is 6000 L/ha, but palm oil is superior as it has 27% higher energy content (30.53 MJ/L) than ethanol from sugarcane (24 MJ/L).[114] In comparison, soybean and maize currently generate less than 1500 and 1500–4000 L/ha, respectively. Presently, much of the world's palm oil comes out of the forests of Southeast Asia, mainly in the biodiverse rainforests of Indonesia and in Malaysia. Due to the increasing demand for biofuels, oil palm cultivation has expanded in Indonesia from 8.5 Mha in 2005 to more than 12.5 Mha in 2015 and is expected to double by 2025.[109,113]

In the case of Africa, and in the specific case of sub-Saharan Africa, the major issue associated with deforestation is that in addition to it paving the way for agricultural land and development projects, it is also being exploited to provide fuelwood and charcoal, which are major cooking fuels in the households and service sectors of the overwhelming majority of African countries. Charcoal production for urban energy consumption has been reported to be the main driver of forest degradation in sub-Saharan Africa. In their work, Sedano et al.[114] reported that charcoal production is a major contributor of forest degradation and can be as high as 50% of deforested areas. In addition, many developing countries use simple charcoal-making technologies with low conversion efficiency of only 10%–22%, as compared to advanced technologies with over 30% efficiency.[44] Charcoal is the fuel of choice mainly in urban areas due to its higher energy density, low transport costs, and relative cleanness,[115–117] although it emits carbon monoxide.[118] A practical solution in this regard, is the development of plantations of quick-growing tree species to provide fuelwood, while natural forests are left intact.[114,119] Large-scale cultivation of fast-growing trees, like poplar, willow, and eucalyptus has the potential for bioenergy production.[120] In the rural areas of many African countries where biomass is freely available and distribution channels for alternative fuels or appropriate smoke-free stoves are not in place, only an estimated one-fifth of the population have access to clean cooking facilities.[121] Hence, there is the compelling need to also strongly advocate for the use of efficient wood-burning stoves.[26,27,115]

3.7.3 Land ownership

Sub-Saharan Africa has more than 1 billion hectares of land with the potential for rain-fed crop production according to the FAO, of which less than one-quarter is being cultivated.[122] The need to reduce the production costs of biofuels offers considerable incentives for large-scale, mechanized agribusiness and concentrated land ownership. Displaced small-scale farmers can either be absorbed into agricultural businesses or migrate elsewhere.[123] The primary threat associated with biofuels is landlessness and the resultant deprivation and social upheaval. Given the importance of access to land for growing agricultural or forestry-based feedstock, some experts fear that small producers may be forced off of their land in favor of large or foreign producers. To the majority of Africans, land is the primary source of livelihoods and a measure of social class and economic well-being. In many African countries, population growth and slow technological progress have forced people to rely on extensive agricultural practices that have exhausted fertile land and become increasingly dependent on degraded and marginal land. In the recent past, large land acquisitions for the production of biofuel feedstocks in several African countries have attracted global media attention. A World Bank report[124] showed that before 2008, an annual average of less than 4 Mha of land was being sourced, as opposed to 45 Mha in 2009. At least 70% of this land (31.5 Mha) was sourced in Africa mainly for biofuels production.

3.7.4 Water availability

Water availability is essential for food security and sustainable agriculture. The water requirements of feedstock, such as sugarcane and cotton, for example, need to be carefully considered given the groundwater shortages and limited rainfall in many countries of sub-Saharan Africa. The continued effect of the recent drought and consequent water shortages in the Western Cape, South Africa, have resulted in significant damage to orchards, such that recovery to full production levels will be prolonged and water allocation toward agriculture could remain constrained.[125] The large-scale cultivation of biomass for fossil fuel substitution influences future water demand.[126] The water footprint of a product, such as a biofuel, is defined as the volume of freshwater used for its production at the place where it was actually produced.[36] It shows large variations for similar crop types, dependent on agricultural production systems used and climate conditions.[127] According to Gerbens-Leenes et al.,[126] the water footprint of a product is described using green, blue, and gray. The green refers to rainwater that evaporated during the growth of the feedstock, blue refers to the amount of surface and groundwater used for irrigation that evaporated during crop growth, and gray refers to the volume of water that becomes polluted during production. The latter is also defined as the amount of water needed to dilute pollutants discharged into natural water systems to the extent that the quality of the ambient water remains above the agreed water quality standards.[127] As with any other crop, the impacts of biofuels on the environment depend very much on farming techniques. The use of heavy machinery compacts the soil, thereby reducing oxygen and water availability and in turn biodiversity. Agrochemicals leave residues in runoff with onsite and downstream impacts.[128]

The expansion of irrigation systems and greater access to untapped surface and groundwater resources through the use of bioenergy powered pumps may be an important option in some regions. Moreover, certain bioenergy feedstocks, such as Jatropha, can improve water retention in fragile soils, thus improving water access for neighboring plants and countering desertification.[129] The Comprehensive Assessment of Water Management in Agriculture estimated that 1.2 billion people live in areas affected by water scarcity, where water resources are insufficient to meet growing needs.[130] In these water-short countries, increasing the agricultural production of biofuels will simply add more pressure on the stressed water resources. Almost all of India's sugarcane (the country's major ethanol crop) is irrigated, as is about 45% of China's top biofuel crop, maize. The water needed to process crops into biofuel is negligible compared with the amounts that go into growing them. Research at the International Water Management Institute (IWMI) in Sri Lanka has shown that at a global average, the biomass needed to produce 1 L of biofuel evaporates between 1000 and 4000 L of water, depending on the type of feedstock and conversion techniques used. For example, when data for the two main ethanol producing countries, Brazil and the United States, were compared, Brazilian ethanol from sugar cane was found to be more efficient than that from maize (99 against 140 m^3/GJ ethanol); however, in the United States, maize is more attractive than sugar cane (78 against 104 m^3/GJ ethanol).[127,130] Also, the amount of water needed to produce 1 L of biodiesel from soybean or rapeseed, on average, is about 14,000 L, while Jatropha will require about 20,000 L of water.[126] In many African countries, irrigation must make up the shortfall, especially in the drier areas of Africa.[131] Table 3.4 shows suitable and available areas for bioenergy crops in sub-Saharan arid/semiarid regions.[132,133]

Because of their limited water resources, such countries will face serious challenges to meet the predicted increase in demand for food production, let alone sustain any further growth prompted by expanding biofuel production. The example of China and India in this case is highly descriptive. By 2030, India's demand for cereal is set to rise by 60%, and more than double for sugar.[130] An analysis by the IWMI indicates that even under the most optimistic scenario, the demand for irrigation water will increase by 13%. Growing sugarcane to produce the 9 billion liters of bioethanol needed to meet just 10% of India's petrol demand by 2030 could add another 22,000 billion liters of irrigation water to the figure.[134] In China, cereal demand is expected to grow by 45%, mainly to feed animals. Growing maize to produce enough ethanol to meet 9% of China's predicted demand for gasoline by 2030 could add another 26,000 billion liters.[135] The problem is aggravated by the drying up of rivers, such as the Yellow River, which no longer reaches the sea during dry months; as well as the falling water tables in the North China Plain and in India's breadbasket region, Punjab; and pollution and intense competition over water.[134]

TABLE 3.4 Suitable and available areas for bioenergy crops in sub-Saharan arid/semiarid regions.[132,133]

Country	Senegal	Burkina Faso	Mali	Kenya	Tanzania	Zambia	Botswana	South Africa
Total area (km^2)	196,013	272,339	1,252,281	581,871	941,375	751,920	587,337	1,221,361
Arid and semiarid (km^2)	111,147	149,973	673,960	457,908	316,738	160,281	58,160	901,345
Arid and semiarid available and suitable (km^2)	15,783	22,576	192,438	379,698	147,252	67,383	291,860	722,874
% Arid available and suitable	6	15	29	75	46	42	42	70
% Semiarid available and suitable	72	0	31	91	–	–	80	94

3.7.5 Biosafety and biodiversity

One of the most cost-effective means of enhancing biofuel feedstock yields is through the genetic modification of planting material.[47] A preference for genetically modified crops will create risks mainly associated with the increased use of herbicides, affecting soil microorganisms, birds, and so on. The invasion of vegetation by biofuel crops is a major threat to biodiversity. For example, castor oil (which has seeds that are highly toxic to humans), livestock and wild animals, and Jatropha, may threaten vegetation. The contamination of genetically modified material from fuel to food is also a fear, as is the loss of agricultural biodiversity associated with biofuel monocultures.[136] Second-generation biofuels generated through enzymatic breakdown are likely to be highly dependent on genetic engineering technology to produce the necessary enzymes. Several short-cycle woody plants that hold promise for second-generation biofuels are also known as invasive (e.g., *Acacia* and *Pinus* species in South Africa and Uruguay).[57] Political support for biofuels in Africa is strong and growing, but development models need to be catalyzed. Some believe that the public–private partnership approach that is succeeding in India can also succeed in Africa with appropriate adjustments to local conditions. The agribusiness-incubator approach can also be implemented in Africa and investors must seek to help accelerate this pro-poor strategic development intervention.

3.8 Guidelines for affordable biofuels

African countries are looking for affordable and employment-generating ways for biofuel production to maximize gains at a national level. The challenges for pro-poor biofuel development relate to carrying out policy research for bioenergy, including appropriate forms of government support, identification of barriers to uptake, and policy response to these barriers.[137] Pro-poor bioenergy development can be made possible by:

- Undertaking R&D in the bioenergy sector in order to identify bioenergy needs in the context of specific African countries. For developing countries with a large number of poor people reliant on agriculture, the first priority should be given to the effective use of existing agricultural wastes for energy generation. This option has the least adverse impacts on the poor and could provide additional revenue for poor rural communities.[138]
- Financial support for bioenergy; many developing countries encounter difficulties in financing their bioenergy projects and programmes. The Global Bioenergy Partnership (GBEP), Competence Platform on Energy Crop and Agroforestry Systems for Arid and Semiarid Ecosystems (COMPETE), and similar institutions are trying to bring together public, private, and civil society stakeholders in a joint commitment to promote

bioenergy for sustainable development, especially in African countries. In its 2010 report, GBEP[139] gave some useful insights into available options for financing bioenergy projects at global and regional levels. Although some external funding sources exist for financing bioenergy development in Africa, African countries should first mobilize internal sources of funding for bioenergy development. Financial assistance can take many forms: subsidies, tax reductions, tax credits, loan guarantees, subsidized credits, income tax reductions, tax holidays, and cash subsidies linked to production levels and import tariffs. Financial schemes must be established at various levels, including those for small-scale producers. Microfinance and other innovative mechanisms can be utilized. Public sector loan guarantees and other risk-mitigation mechanisms should be provided to enable more private investment in new technologies. Financial support can also be achieved by enabling public–private partnerships. In a nutshell, governments should provide market infrastructure, robust micro-credit system, support public-private partnerships and if need be, act as guarantor to assist in all the stages of the process.

- Facilitating collaborations among researchers nationally and internationally with academic and research communities, entrepreneurs, investors, and financiers.[2] In addition, maximizing smallholder participation by establishing effective revenue-sharing mechanisms that ensure higher revenues from the exploitation of agricultural wastes are shared in an equitable fashion and flow to all stakeholders, including low-income farmers.[48] This will be enhanced by reducing border barriers to imports of technologies and facilitate the transfer of materials needed, technologies, and the sharing of information.[48]
- Developing intersectoral plans and programmes and intersectoral cooperation among all sectors involved and affected by bioenergy.[74]
- Enacting a legal and regulatory framework that allows for the development of bioenergy and that provides, among other incentives, access to the power grid and transport fuel market. In some cases, mechanisms for the efficient centralization of bioenergy production would need to be in place.[140–142]
- Building the capacity, technical skills, and managerial skills of producers, researchers, and entrepreneurs, and educating all participating decision-makers (the public and consumers, including nongovernmental organizations).[143]
- Ensuring flexibility in food and energy production technologies in order to minimize risk while carefully balancing any associated trade-offs between food security and energy generation.[48]
- Taking policy measures for environmental protection by setting up an appropriate system of tax-based and regulatory incentives to counter environmentally perverse incentives to convert natural forests (and other landscapes) into short-term biofuel stocks or plantations.[17,140]

- Careful appraisal of the potential for agricultural expansion into unfarmed areas, such as marginal lands and indigenous vegetation. In addition, developing best-practice regulations or incentives for on-farm production in order to minimize water depletion, soil erosion, agrochemical runoff, and GHG emissions and avoiding land-use changes from natural vegetation to crop monocultures.[31]

3.9 The way forward

The African bioenergy policy document (Africa Bioenergy Policy Framework and Guidelines: Towards Harmonizing Sustainable Bioenergy Development in Africa)[143,144] has clearly enumerated the way forward for Africa to achieve sustainable bioenergy development. The document has noted as follows:

Africa is home to substantial bioenergy resources and potential as indicated in Table 3.5, though these resources are mostly underdeveloped (processing, farming, and household wastes) or poorly used (inefficient energy conversion processes and poor cooking devices). There is an urgent need to formulate policies that can mobilize resources and stakeholders to make proper use of the resources to the benefit of humans and the ecosystem. The development of zero emission, clean, alternative energy systems through the utilization of compressed waste materials in the form of briquettes and the promotion of fast-growing trees to reduce unsustainable means regarding the cutting of fuelwood in the developing world are some necessary steps that must be taken to achieve sustainability in bioenergy production and utilization.[30]

These key messages provide the way forward for the development of the bioenergy sector in Africa:

- A sustainable energy transition will not succeed without paying attention to modern bioenergy.
- Bioenergy is already an important source of energy for households, but the mode of production, transformation, and consumption is inefficient.
- To reach a sustainable energy mix, it is crucial to modernize the traditional biomass sector and develop modern bioenergy while encompassing other sectors. There is a need to adopt a range of appropriate technologies, such as low emission cooking stoves and efficient charcoal kilns, and to ensure that these are affordable and meet the needs of the end-users, particularly women.
- A holistic approach to bioenergy development is essential, including a broad developmental agenda that takes bioenergy beyond the transport sector aiming at improving access to energy at the household level (rural and urban) for cooking and lighting, as well as at the commercial or industrial levels; focusing on nonfood feedstock; and evaluating each

TABLE 3.5 Potential rainfall, land, and cultivation practices for bioenergy crops in African countries.[132]

Bioenergy crops	Rainfall (cm)	Land and climate	Cultivation practice	Suitable countries for large-scale production
Sugarcane	150–250	Tropical and subtropical roughly between the latitudes 37°N and 31°S. Well-drained, deep, loamy soil with high water-holding capacity	Intensive irrigation and fertilization required at all times	Sudan, Kenya, Swaziland, Zimbabwe, Mauritius, Malawi, parts of Cameroon, Nigeria, and Ghana
Oil palm	180–500	Humid tropical, moderate to high quality land. An annual rainfall of 120–170 cm is the optimum, but 70–300 cm is acceptable	No irrigation required in humid areas; with fertilization for high yields	Angola, Democratic Republic of Congo, Nigeria, Ghana, Tanzania, and Malawi
Maize (grain)	70–150	Semiarid, moderate to high quality land. An annual rainfall of 60–110 cm is optimum, but rainfall of about 40 cm is acceptable	Intensive irrigation and fertilization for high yields	Sudan, Chad, Nigeria, Ghana, Angola, Tanzania, and South Africa
Jatropha (seeds)	60–120	Tropical, subtropical, and semiarid poor to moderate land. Can thrive in rainfall of 25–30 cm	No irrigation and fertilizer application, drought resistant	Sahel, Togo Ghana, Niger, Nigeria, Tanzania, Mozambique
Woody biomass (ligneous biomass)	50–500	Arid, semiarid, humid tropical to temperate poor quality land, drought-resistant	Less intensive fertilization and no irrigation required	Many African countries in the Sahel, Sudan, Guinea, and rain forest zones
Sweet sorghum	50–100	Arid and semiarid. Best adapted to areas having an average annual rainfall between 45 and 65 cm	More drought tolerant than sugarcane, suitable for wider areas whilst still achieving high yields	Nigeria, Congo, South Africa, Egypt, Sudan
Cassava	100–150	Semiarid, moderate to high quality land. Precipitations should be evenly distributed	Well-drained light, sandy loam soils with medium soil fertility and with good drainage	Nigeria, Chad, Sudan Democratic Republic of Congo, and South Africa

bioenergy feedstock for its economic, social, and environmental benefits as well as costs prior to issuing investment contracts.

- The process of policy development is as important as the policy itself. Assessing global and regional dynamics and opportunities; identifying the needs and societal concerns; putting in place the necessary legal and institutional frameworks for coordinating and integrating economic, social, and environmental objectives; mobilizing and building capacities (human and institutional); consulting and engaging stakeholders; and setting up monitoring mechanisms are all critical to the success of sustainable bioenergy policies.
- Ensuring that the reviewed bioenergy policies are passed into law by parliaments. This is necessary for the majority of African countries in view of the frequent and unnecessary policy changes from one elected administration to the next one.
- Developing bioenergy is not without risks and sustainable criteria are paramount to ensure sound scaling up of bioenergy.
- Using technology in the production of modern bioenergy is an important component of the sustainable energy solution and of lowering carbon emissions.
- Financing, regulatory frameworks, R&D, and capacity building are crucial to bioenergy deployment on a sustainable basis.
- Policies and regulatory frameworks should be harmonized across countries to facilitate regional cooperation and trade.
- The "Africa Bioenergy Policy Framework and Guidelines" is an important tool to overcoming the barriers in the deployment of sustainable and modern bioenergy and ensuring that bioenergy contributes to economic growth, rural development, energy and food security, poverty reduction, and the protection of natural resources.
- A more sustainable approach to fuelwood is its provision from plantations of quick-growing tree species and its utilization with efficient wood-burning stoves.
- Success in using bioresources for energy can only be realized with sound policies driven by political will at the highest level.
- The establishment of information and modeling systems at a national and provincial level to determine the bioenergy potential available, demand, and projections in each country so as to ensure that the production and exploitation of bioenergy is achieved in an environmentally sustainable way and with due consideration of the sustainability of soil and water resources.

3.10 Conclusion

Currently, energy use in Africa is mainly based on the use of traditional bioenergy, where fuelwood is used in simple inefficient stoves. Low per capita national incomes, slow growth in conventional energy use, and easy access to biomass, influence the heavy reliance on biomass energy in Africa and the trend is expected to continue for some time in the future. The low efficiency levels of just 10%–20% for burning biomass in Africa have led to environmental degradation, particularly deforestation. Many African countries have suitable conditions for bioenergy development, such as abundant labor and sufficient available arable land and water resources, and yet these countries have low access to energy. Bioenergy can play a major role in helping to achieve broader energy access and security, agricultural development, poverty reduction, and national development strategies. To achieve these, African countries need to improve their agricultural infrastructure to improve in food and feedstock production. They must also cooperate in order to promote knowledge, technology, and experience transfer on modern bioenergy at national, regional, and international levels. The formulation and harmonization of policies and developmental plans to guide the development and sustainability of the bioenergy sector; environmental protection; the development of technical and managerial skills; flexibility in food and energy production technologies; and bioenergy markets and the financing of the sector must all be put into place. Finally, there is a need for the establishment of information and modeling systems at national and provincial levels to assess the bioenergy potential available and meet the demand and projections in each country, so as to ensure that the production and exploitation of bioenergy is achieved in a sustainable way.

References

1. Ambali A, Paxie WC, Owen C, Willem H, van Zyl. A review of sustainable development of bioenergy in Africa: an outlook for the future bioenergy industry. *Sci Res Essays* 2011;**6** (8):1697–708.
2. Sambo AS. Renewable energy development in Africa: issues, challenges and prospects. In: Sayigh Ali, editor. *Renewable energy in the service of mankind*, vol. II. Cham, Switzerland: Springer International Publishing; 2016.
3. Emad AS. Biofuel: sources, extraction and determination, liquid, gaseous and solid biofuels by Zhen Fang. *IntechOpen* 2013. Available from: https://doi.org/10.5772/51943.
4. Ubaid R, Hemalatha S. A review on bioenergy and biofuels: sources and their production. *Braz J Biol Sci* 2016;**3**(5):3–21.
5. Garba MM. *Thermal energy systems materials optimisation for solar drying and cooking applications* [Ph.D. thesis]. United Kingdom: The University of Nottingham; 2009.
6. Garba MM, Atiku AT, Sambo AS. Comparative studies of some passive solar dryers. Proceedings of the 1st world renewable energy congress, Reading, UK, 23–28 September 1990 In: Sayigh AAM, editor. *Energy and the environment into the 1990s*. Oxford: Pergamon Press; 1990. p. 927–31.
7. FAO. *FAOSTAT*. Rome: FAO. Available from: <http://www.fao.org/faostat/en/#home>; 2017.
8. Ren21. *Renewables 2013. Global status report*. Paris: ren21 Secretariat; 2013.
9. IEA/OECD. *Technology roadmap: delivering sustainable bioenergy*; 2017.

10. World Bank. *Atlas of sustainable development goals 2018: world development indicators*. Washington, DC: World Bank; 2018 Available from: https://doi.org/10.1596/978-1-4648-1250-7. License: Creative Commons Attribution CC BY 3.0 IGO.
11. Morrissey J. *The energy challenge in sub-Saharan Africa: a guide for advocates and policy makers: Part 2: Addressing energy poverty. Oxfam research backgrounder series*. <https://www.oxfamamerica.org/static/media/files/oxfam-RAEL-energySSA-pt2.pdf>; 2017.
12. IEA. *Sustainable production of woody biomass for energy. A position paper prepared by IEA Bioenergy*; 2002.
13. Kapasi ZA, Nair AR, Sonawane S, Satpute SK. Biofuel—an alternative source of energy for present and future. *J Adv Sci Technol* 2010;**13**(3):105–8.
14. IEA. *IEA bioenergy. Annual report*; 2009.
15. Karakezi S, Kithyoma W, Muzee K, Oruta A. Scaling up bioenergy in Africa, In: *Presented during the international conference on renewable energy in Africa, Dakar 16 – 18 April, 2008*; 2008.
16. Halsnaes K, Shukla PR, Christensen JM. The global energy scene in 2050. In: Larsen HH, editor. *Non-fossil energy technologies in 2050 and beyond*. Roskilde, Denmark: Denmark's Tekniske Universitet, Risø Nationallaboratoriet for Bæredygtig Energi; 2010. pp. 9–13. Risø Energy Report 9; Forskningscenter Risoe. Risoe-R, No. 1729 (EN).
17. UN Energy. *Sustainable bioenergy: a framework for decision makers*. New York: United Nations. <http://www.fao.org/docrep/010/a1094e/a1094e00.htm>; 2007.
18. IEA. *Energy balances for Africa*. <http://www.iea.org/Textbase/stats/balancetable.asp>; 2003.
19. IEA. Bioenergy—a sustainable and reliable energy source. A review of status and prospects. In: *IEA Bioenergy annual report*; 2010.
20. World watch Institute. *Biofuels for transportation: global potential and implications for agriculture and sustainable energy in the 21st century. Final report prepared by Worldwatch for GTZ*. Washington, DC: World Watch Institute; 2006.
21. NEA. *Sustainable biomass production and use: lessons learned from the Netherlands Programme Sustainable Biomass (NPSB) 2009–2013*. A report published by Netherlands Enterprise Agency; 2014.
22. Janssen R, Rutz D. *Bioenergy for sustainable development in Africa*; 2012.
23. World Energy Council. *World energy resources: bioenergy 2016*, 2016.
24. Reijnders L. Sustainability of bioenergy systems. In: Wang Lijun, editor. *Sustainable bioenergy production*. Boca Raton: CRC Press; 2014. p. 115–32.
25. Barbanti L. Novel cropping technologies and management applied to energy crops. In: Wang Lijun, editor. *Sustainable bioenergy production*. Boca Raton: CRC Press; 2014. p. 19–65.
26. FAO. *The State of the World's Forests 2018—forest pathways to sustainable development*. Rome; 2018. Licence: CC BY-NC-SA 3.0 IGO.
27. Maishanu SM, Danshehu BG, Sambo AS. Saving fuelwood with improved wood-burning stoves. Proceedings of the 1st world renewable energy congress, Reading, UK, 23–28 September 1990 In: Sayigh AAM, editor. *Energy and the environment into the 1990s*. Oxford: Pergamon Press; 1990. p. 1923–7.
28. IRENA. *Renewable power generation costs in 2017*. Abu Dhabi: International Renewable Energy Agency; 2018.
29. GNESD. Bioenergy: the potential for rural development and poverty alleviation. Global Network on Energy for Sustainable Development (GNESD). Summary for policy-makers. In: *GNESD-SPM-BET-11/2011.2011*.

30. Diaz-Chavez R, Johnsonb FX, Richardc TL, Charakyad H. Biomass resources, energy access and poverty reduction. In: Souza GM, Reynaldo LV, Carlos AJ, Luciano MV, editors. *Bioenergy and sustainability: bridging the gaps*. São Paulo, Brazil: Scientific Committee on Problems of the Environment (SCOPE); 2015. p. 710–28.
31. IRENA. *Renewable energy and jobs—annual review 2017*. Abu Dhabi: International Renewable Energy Agency; 2017.
32. IRENA. *Renewable energy and jobs—annual review 2018*. Abu Dhabi: International Renewable Energy Agency; 2018.
33. Clancy J. *Biofuels and rural poverty*. London: Earthscan; 2013.
34. Hans MH, Edward SE, David Z. Bioenergy economics and policies. In: Souza GM, Reynaldo LV, Carlos AJ, Luciano MV, editors. *Bioenergy and sustainability: bridging the gaps*. São Paulo, Brazil: Scientific Committee on Problems of the Environment (SCOPE); 2015. p. 682–708.
35. AFREA. *Wood-based biomass energy development for sub-Saharan Africa: issues and approaches*. Washington, DC: The International Bank for Reconstruction and Development/The World Bank Group; 2011. p. 1–45.
36. Worldwatch Institute. *Biofuels for transportation: global potential and implications for agriculture*. London, UK: Earthscan; 2007.
37. Campbell D. *Hope or hype? A critical assessment of Jatropha curcas for domestic biofuel production in Senegal*. A thesis submitted to Victoria University of Wellington in Fulfilment of the Requirements for the Degree of Masters in Environmental Studies, Victoria University of Wellington; 2014.
38. Dafrallah T, Ackom E. Analysis of national Jatropha biodiesel programme in Senegal. *AIMS Energy* 2016;**4**(4):589–605.
39. Renwick A, Reverode-Giha C, Philippidis G, Bourne M, Lang B, Reader M. *Assessment of the impact of the 2006 EU sugar regime reforms*. Final report prepared for Department for Environment, Food and Rural Affairs (DEFRA). Land Economy and Environment Research SAC; 2011.
40. Goodison P. *The impact of EU Sugar reforms on traditional African Caribbean and Pacific Sugar Exporters*. Initiativet Handel og Udvikling; 2015.
41. Diop D, Blanco M, Flammini A, Schlaifer M, Kropiwnicka MA, Markhof MM. Assessing the impact of biofuels production on developing countries from the point of view of policy coherence for development. In: *Contract no. 2012/299193, FWC COM 2011—Lot 1—studies and technical assistance in all sectors final report February, 2013*, 2013.
42. Bruce N, Perez-Padilla R, Albalak R. Indoor air pollution in developing countries: a major environmental and public health challenge. *Bull World Health Organ* 2000;**78**(9):1078–92.
43. Kilabuko JH, Nakai S. Effects of cooking fuels on acute respiratory infections in children in Tanzania. *Int J Environ Res Public Health* 2007;**4**(4):283–8.
44. WHO. *Burning opportunity: clean household energy for health, sustainable development, and wellbeing of women and children*. Geneva: World Health Organization; 2016.
45. Balakrishnan R. *Rural women and food security in Asia and the Pacific: prospects and paradoxes*. Bangkok, Thailand: FAO Regional Office for Asia and the Pacific, Food and Agriculture Organization of the United Nations, Regional Office for Asia and the Pacific; 2005.
46. SOFA Team and Doss C. The role of women in agriculture. In: *ESA working paper no. 11-02 March 2011 Agricultural Development Economics Division*. The Food and Agriculture Organization of the United Nations (FOA). <www.fao.org/economic/esa> 2011.

47. ESMAP. *Potential for biofuels for transport in developing countries. Energy Sector Management Assistance Programme.* Washington, DC: World Bank; 2005.
48. Cushion E, Whiteman A, Dieterle G. *Bioenergy development: issues and impacts for poverty and natural resource management.* Washington, DC: TBRD/The World Bank; 2010.
49. OECD. *Agricultural market impacts of future growth in the production of biofuels AGR/CA/APM(2005)24/FINAL.* Paris: Directorate for Food, Agriculture and Fisheries, Committee for Agriculture. Organisation for Economic Co-operation and Development; 2006.
50. WIREs Energy Environ. Definitions and dimensions of energy security: a literature review. *Adv Rev* 2018;**7**:e268 Available from: https://doi.org/10.1002/wene.268.
51. IEA. *Medium-term renewable energy market report 2015*, 2015.
52. Martins-Filho J, Burnquist HL, Vian CEF. Bioenergy and the rise of sugarcane-based ethanol in Brazil. *Choices* 2006;**21**:91–6.
53. Bentivoglio D, Finco A, Bacchi MRP. Interdependencies between biofuel, fuel and food prices: the case of the brazilian ethanol market. *Energies* 2016;**464**:1–16.
54. REUTERS. *Brazil ethanol prices plunge in the first week of new cane crop.* <https://www.reuters.com/article/us-brazil-ethanol/brazil-ethanol-prices-plunge-in-the-first-week-of-new-cane-crop-idUSKBN1HG32R>; 2018 [accessed 28.10.18].
55. Dufey A. Biofuels production, trade and sustainable development: emerging issues. In: *Environmental economics programme/sustainable markets group discussion paper 2.* London, UK: International Institute for Environment and Development; 2006.
56. IEA. *World energy outlook.* Paris: International Energy Agency; 2006.
57. Kartha S. *Environmental effects of bioenergy. Brief 4. Bioenergy and agriculture: promises and challenges. IFPRI 2020 vision, Focus 14.* Washington, DC: International Food Policy Research Institute; 2006.
58. Fatih B. *World energy outlook 2015.* International Energy Agency; 2015.
59. IPCC. In: Edenhofer O, Pichs-Madruga R, Sokona Y, et al., editors. *Special report on renewable energy sources and climate change mitigation. Prepared by Working Group III of the Intergovernmental Panel on Climate Change.* Cambridge, UK: Cambridge University Press; 2011.
60. IPCC. Mitigation of *climate change: technical summary in IPCC Fifth Assessment Report (AR5).* Geneva: Intergovernmental Panel on Climate Change; 2014. <http://www.ipcc-wg3.de/>. GEA. Austria: The International Institute for Applied Systems Analysis (IIASA) *Global energy assessment—toward a sustainable future.* Cambridge/New York: Cambridge University Press and Luxemburg; 2012.
61. GEA. The International Institute for Applied Systems Analysis (IIASA), Luxemburg, Austria *Global energy assessment—toward a sustainable future.* Cambridge/New York: Cambridge University Press; 2012.
62. IEA. *Energy technology perspectives.* Paris: International Energy Agency, Organisation for Economic Co-operation and Development; 2012 Available from: http://www.iea.org/etp/.
63. IEA. *Energy technology perspectives.* Paris: International Energy Agency, Organisation for Economic Co-operation and Development; 2014 Available from: http://www.iea.org/etp/.
64. STAP. *Optimizing the global environmental benefits of transport biofuels.* Washington, DC: Scientific and Technical Advisory Panel of the Global Environment Facility; 2014. Authored and edited by Bierbaum R, Cowie A, Gorsevski V, Sims R. Rack M, Strapasson A, Woods J, Ravindranath N.
65. UNFCCC. *Distribution of registered projects by UNFCCC region.* Available from: <https://cdm.unfccc.int/Statistics/Public/files/201810/proj_reg_byRegion.pdf>; 2018.

66. Schlamadinger B, Jürgens I. *Bioenergy and the CDM, In:* Conference proceedings of the 2nd world conference and technological exhibition on biomass for energy. Rome: Industry and Climate Protection; 2004.
67. UNEP. *GEO-6 regional assessment for Africa.* Nairobi, Kenya: United Nations Environment Programme; 2016.
68. UNEP. *Africa environment outlook (AEO-3): our environment, our health.* Africa Environment Outlook United Nations Environment Programme (UNEP). <http://www.unep.org/pdf/aeo3.pdf>; 2013.
69. Becker K, Francis G. *Bio-diesel from Jatropha plantations on degraded land.* Department of Aquaculture Systems and Animal Nutrition, University of Hohenheim; 2003. Available at: http://www.unihohenheim.de/ ~ www480/docs/publish/jatropha_on_-degraded_land.pdf.
70. von Maltitz G, Gasparatos A, Fabricius C. The rise, fall and potential resilience benefits of Jatropha in Southern Africa. *Sustainability* 2014;**6**(6):3615–43.
71. Da Silva JAN, Da Silva CJ, De Souza CMA, Luiz AS, Serra AP. *Jatropha curcas* performance in intercropping with forages grass and grain crops species. *Afr J Agric Res* 2016;**11** (46):4718–25.
72. VENRO/GNFED/ICEED. *Prospects for Africa—rethinking biomass energy in sub-Sahara Africa energy and the millennium development goals.* Association of German Development NGOs (VENRO); 2009.
73. IRENA/DBFZ. *Biomass potential in Africa.* Abu Dhabi; 2013.
74. Kimble M, Pasdeloup MV, Spencer C. *Sustainable bioenergy development in UEMOA member countries.* Washington, DC: The UN Foundation; 2008. Available at Available from: http://www.globalproblems-globalsolutions-files.org/gpgs_files/pdf/UNF_Bioenergy/UNF_Bioenergy_full_report.pdf.
75. OECD-FAO. *Agricultural outlook 2007–2016*; 2007.
76. FLIWG (Feedstock Logistics Interagency Working Group). *Biofuel feedstock logistics: recommendation for research and commercialization, DOE, USA*; 2010.
77. Global Agriculture Information Network. *Argentine biofuels report: agricultural situation Argentina*; 2007. GAIN report, report number: AR7016.
78. Oláh J, Lengyel P, Balogh P, Harangi-Rákos M, Popp J. The role of biofuels in food commodity prices volatility and land use. *J Compet* 2017;**9**(4):81–93.
79. Von Maltitz G, Staffford W. Assessing opportunities and constraints for biofuel development in sub-Saharan Africa. In: *Working paper 58.* Bogor, Indonesia: CIFOR; 2011.
80. Nigatu G, Hjort K, Somwaru A, Hansen J. Projecting the effect of oil price regimes on biofuel markets, In: *Selected paper prepared for presentation at the 2015 AAEA & WAEA Joint Annual Meeting, San Francisco, California, 26–28 July 2015*; 2015.
81. Schwaiger H, Pena N, Maye A, Bird DN. Technologies to produce liquid biofuels for transportation: an overview. In: *Working paper 72.* Bogor: CIFOR; 2011.
82. Kousoulidou M, Laura L. Biofuels in aviation: fuel demand and CO_2 emissions evolution in Europe toward 2030. *Trans Res* 2016;**46**:166–81.
83. Gasparatos A, von Maltitz GP, Johnson FX, Lee L, Mathai M, Puppim de Oliveira JA, et al. Biofuels in sub-Sahara Africa: drivers, impacts and priority policy areas. *Renew Sustain Energy Rev* 2013;**45**:879–901.
84. Charles BL, Jumbe CBL, Mkondiwa M. Comparative analysis of biofuels policy development in sub-Saharan Africa: the place of private and public sectors. *Renew Energy* 2013;**50**:614–20.

85. Bracco S. Effectiveness of EU biofuels sustainability criteria in the context of land acquisitions in Africa. *Renew Sustain Energy Rev* 2015;**50**:130–43.
86. Zilberman D, Rajagopal D, Kaplan S. Effect of biofuel on agricultural supply and land use. In: Khanna M, Zilberman D, editors. *Handbook of bioenergy economics and policy: volume II modeling land use and greenhouse gas implications*. New York: Springer International Publishing; 2017. p. 163–82.
87. Khanna M, Zilberman D, Miao R. Innovation in agriculture: incentives for adoption and supply chain development for energy crops. In: Khanna M, Zilberman D, editors. *Handbook of bioenergy economics and policy: volume II modelling land use and greenhouse gas implications*. New York: Springer International Publishing; 2017. p. 347–72.
88. APEC (Asia Pacific Economic Cooperation). *Biofuel transportation and distribution options for APEC economies*. Singapore: APEC Energy Working Group; 2011.
89. DOE. *Biodiesel handling and use guide (fifth edition)*. U.S. Department of Energy; 2016. DOE/GO-102016-4875.
90. Lou X, Goodman LR, Singh PM, Yang D. Understanding the stress corrosion cracking of x-65 pipeline steel in fuel-grade ethanol, In: *NACE—international corrosion conference series January 2010*; 2010.
91. Ghazanfari J, Najafi B, Ardabili SF, Shamshirband S. Limiting factors for the use of palm oil biodiesel in a diesel engine in the context of the ASTM standard. *Cogent Eng* 2017;**4**:1411221.
92. IRENA. *Bioethanol in Africa: the case for technology transfer and South–South co-operation*. Abu Dhabi: International Renewable Energy Agency; 2016.
93. Popp J, Harangi-Rákos M, Gabnai Z, Balogh P, Anta G, Bai A. Biofuels and their co-products as livestock feed: global economic and environmental implications. *Molecules* 2016;**21**:285.
94. Hamelinck C. *Land grabs for biofuels driven by EU biofuel policies*; 2013.
95. Sinkala T, Johnson FX. Small-scale production of *Jatropha* in Zambia and its implications for rural development and national biofuel policies. In: Janssen R, Rutz D, editors. *Bioenergy for sustainable development in Africa*. Springer Science Business Media B.V.; 2012. p. 41–50.
96. Janssen R, Rutz D. Keynote introduction: overview on bioenergy policies in Africa. In: Janssen R, Rutz D, editors. *Bioenergy for sustainable development in Africa*. Springer Science Business Media B.V.; 2012. p. 165–82.
97. Mohamed S, Fraser G, Sawe EN. Biofuel policies in Tanzania. In: Janssen R, Rutz D, editors. *Bioenergy for sustainable development in Africa*. Springer Science Business Media B.V.; 2012. p. 183–90.
98. Rajagopal D, Zilberman D. *Review of environmental, economic and policy aspects of biofuels*. The World Bank Development Research Group Sustainable Rural and Urban Development Team; 2007.
99. Global Agriculture Information Network. *Brazil biofuels report*. GAIN report number: BR10006. Sao Paulo: ATO; 2010.
100. Global Agriculture Information Network. *Brazil biofuels report*. GAIN report number: BR16009. Sao Paulo: ATO; 2016.
101. GBEP (Global Bioenergy Partnership). *Financing options for bioenergy projects and programmes*. 2010.
102. FAO. *Walking the nexus talk: assessing the water-energy-food nexus in the context of the sustainable energy for all initiative*. Rome, Italy: Food and Agriculture Organization of the United Nations; 2014.

103. Medina L. The dynamic effects of commodity prices on fiscal performance in Latin America. In: *Working paper no. 09/56*. Washington, DC: International Monetary Fund; 2010.
104. IMF/World Bank. *Responding to global food price volatility and its impact on food security*; 2011. DC2011-0002.
105. Khanna M, Zilberman D. Bioenergy economics and policy in US and Brazil: effects on land use and greenhouse gas emissions. In: Khanna M, Zilberman D, editors. *Handbook of bioenergy economics and policy: volume II modelling land use and greenhouse gas implications*. New York: Springer International Publishing; 2014. p. 1–11.
106. Wiggins S, Henley G, Keats S. *Competitive or complementary? Industrial crops and food security in sub-Saharan Africa*. London: Overseas Development Institute (ODI); 2015. 2015.
107. Nepstad D, Stickler CM, Filho BS, Merry F. Interactions among Amazon land use, forests and climate: prospects for a near-term forest tipping point. *Philos Trans R Soc Lond B Biol Sci* 2008;**363**(1498):1737–46.
108. Walter A, Dolzan P, Quilodrán O, Garcia J, da Silva C, Piacente F, Segerstedt A. *A sustainability analysis of the Brazilian ethanol*. Campinas. Available at: <http://www.global-bioenergy.org/uploads/media/0811_Unicamp_-_A_sustainability_analysis_of_the_Brazilian_ethanol.pdf>; 2008.
109. Gao Y, Skutsch M, Masera O, Pacheco P. A global analysis of deforestation due to biofuel development. In: *Working paper 68*. Bogor: CIFOR; 2011.
110. Brown LR. *Outgrowing the Earth: the food security challenge in an age of falling water tables and rising temperatures*. New York: Norton; 2004.
111. Goldemberg J. Ethanol for a sustainable energy future. *Science* 2007;**315**(5813):808–10.
112. Goldemberg J, Guardabassi P. Are biofuels a feasible option? *Energy Policy* 2009;**2009**(37):10–14.
113. Palm Oil Today. *Palm oil biodiesel—a preferred biofuel feedstock*. Available at: <http://palmoiltoday.net/palm-oil-biodiesel-a-preferred-biofuel-feedstock/>; 2018 [accessed 10.11.18].
114. Sedano F, Silva JA, Machoco R, Meque CH, Sitoe A, Ribeiro A, et al. The impact of charcoal production on forest degradation: a case study in Tete, Mozambique. *Environ Res Lett* 2016;**11**:094020.
115. Bailis R, Ezzati M, Kammen DM. Mortality and greenhouse gas impacts of biomass and petroleum energy future inAfrica. *Science* 2005;**308**:98–103.
116. Koyuncu T, Pinar Y. The emissions from a space-heating biomass stove. *Biomass Bioenerg* 2007;**31**:73–9.
117. Mwampamba TH, Ghilardi A, Sander K, Chaix KJ. Dispelling common misconceptions to improve attitudes and policy outlook on charcoal in developing countries. *Energy Sustain Dev* 2013;**17**:158–70.
118. Maes WH, Verbist B. Increasing the sustainability of household cooking in developing countries: policy implications. *Renew Sustain Energy Rev* 2012;**16**(4):204–21.
119. Sambo AS. *Strategic developments in renewable energy in Nigeria*. International Association for Energy Economics, IAEE Energy Forum, Third Quarter; 2009.
120. Rajagopal D, Sexton SE, Roland-Holst D, Zilberman D. Challenge of biofuel: filling the tank without emptying the stomach? *Environ Res Lett* 2007;**2**:1–9.
121. IEA/World Bank. *Sustainable energy for all. Global tracking framework—progress toward sustainable energy*. Washington, DC: World Bank; 2017.
122. FAO (Food and Agriculture Organization). *Food outlook cassava, various issues*. 2008.

123. Matondi PB, Havnevik K, Beyene A. Introduction: biofuels, food security and land grabbing in Africa. In: Matondi PB, Havnevik K, Beyene A, editors. *Biofuels, land grabbing and food security in Africa*. Uppsala: Nordic Africa Institute; 2011.
124. BFAP (Bureau for Food and Agricultural Policy) BASELINE. *Agricultural outlook 2018–2027*. Available at: <http://www.bfap.co.za/wp-content/uploads/2018/08/BFAPBaseline-2018.pdf>. 2018.
125. Berndes G. Bioenergy and water the implications of large-scale bioenergy production for water use and supply. *Global Environ Change* 2002;**12**:253–71.
126. Gerbens-Leenesa W, Hoekstraa AY, van der Meer TH. The water footprint of bioenergy. *Proc Natl Acad Sci U S A* 2009;**25**:10219–23.
127. FAO. *The state of food and agriculture: biofuels: prospects, risks and opportunities*. Rome; 2008.
128. Ackom E, Ertel JK. An alternative energy approach to combating desertification and promotion of sustainable development in drought regions. *Forum der Forsch* 2005;**18**:74–8.
129. de Fraiture C, Giordano M, Yongsong L. Biofuels: implications for agricultural water use: blue impacts of green energy. *Water Policy* 2008;**10**(Suppl. I):67–81.
130. Kgathi DL, Mazonde I, Murray-Hudson M. Water implications of biofuel development in semi-arid sub-Saharan Africa: case studies of four countries. In: Janssen R, Rutz D, editors. *Bioenergy for sustainable development in Africa*. Springer Science Business Media B.V.; 2012. p. 41–50.
131. Meijerink G, Langeveld H, Hellegers P. *Biofuels and water: an exploration*. Wageningen, UR. <https://core.ac.uk/download/pdf/29256837.pdf> [accessed 13.11.18].
132. Malaviya S, Ravindranath NH. Implications of climate change on sustainable biofuel production in Africa. In: Janssen R, Rutz D, editors. *Bioenergy for sustainable development in Africa*. Springer Science Business Media B.V.; 2012. p. 281–97.
133. SCOPE. *What are the final land limits?* Available at: <http://cip.cornell.edu/DPubS/Repository/1.0/Disseminate?view = body&id = pdf_1&handle = scope/1245782016>; 2009.
134. de Fraiture C, Giordano M, Yongsong L. *Biofuels: implications for agricultural water use*. Colombo, Sri Lanka: International Water Management Institute; 2007.
135. Biofuelwatch. *Agrofuels threaten to accelerate global warming*. <www.biofuelwatch.org.uk>; 2007.
136. GBEP. *Financing options for bioenergy projects and programmes*. Available at: <http://www.globalbioenergy.org/fileadmin/user_upload/gbep/docs/BIOENERGY_INFO/1004_GBEP_-_Financing_options_for_bioenergy_projects_23april_web.pdf>; 2010.
137. Ouedraogo N. *Bioenergy for Africa: an illusion or a sustainable option to reduce the vulnerability to energy and poverty?* Place du Maréchal de Lattre-de-Tassigny, 75775 Paris Cedex 16: Université Paris-Dauphine, Centre Géopolitique de l'Energie Et des Matières Premières (GEMP); 2009. p. 1–26.
138. GBEP. *Financing options for bioenergy projects and programmes*. Available at: <http://www.globalbioenergy.org/fileadmin/user_upload/gbep/docs/BIOENERGY_INFO/1004_GBEP_-_Financing_options_for_bioenergy_projects_23april_web.pdf>; 2010.
139. Jull C, Redondo P. *The legal framework for bioenergy*. Rome: Food and Agriculture Organization (FAO); 2007. Available at Available from: http://www.globalbioenergy.org/uploads/media/0703_Legal_Framework_for_Bioenergy_Draft.pdf.

140. Jull C, Redondo PC, Mosoti V, Vapnek J. *Recent trends in the law and policy of bioenergy production, promotion and use*. Available at: <http://www.fao.org/3/a-bb101e.pdf>; 2007.

141. GIZ. *Legal frameworks for renewable energy; policy analysis for 15 developing and emerging economies*. <http://www.giz.de/fachexpertise/downloads/giz2012-enlegal-frameworks-renewable-energy.pdf>; 2012.

142. Janssen R, Rutz D, Helm P, Woods J, Chavez RD. *Bioenergy for sustainable development in Africa—environmental and social aspects*. Available from: <http://www.compete-bioafrica.net/sustainability/OD9-paper_COMPETE_Janssen_090730.pdf>; 2018.

143. AUC-ECA. *Africa bioenergy policy framework and guidelines: towards harmonizing sustainable bioenergy development in Africa, Addis Ababa*; 2013.

144. UNEP. *Atlas of Africa energy resources*. Nairobi: United Nations Environment Programme; 2017.

Chapter 4

Biohydrogen production and bagasse gasification process in the sugarcane industry

Regina Franciélle Silva Paulino and José Luz Silveira
Department of Energy, Laboratory of Optimization of Energy Systems (LOSE), School of Engineering, Guaratinguetá and Institute of Bioenergy Research (IPBEN-UNESP), Sao Paulo State University (UNESP), São Paulo, Brazil

4.1 Introduction

The development of the Brazilian sugarcane industries was based on traditional processes that use condensing steam turbines with extractions operating in cogeneration systems. Many scientists have proposed new technologies and small modifications to increase the efficiency and energy production in this important industrial sector.[1,2] Research on alternative routes to generate energy reveal that sugarcane industries have great potential for bioenergy production due to the fact that not only does it use sugarcane bagasse for cogeneration, but it is also able to produce biohydrogen from ethanol for fuel cells and can generate more bioelectricity through biomass gasification.[3] The most widely used process to obtain hydrogen worldwide is the steam reforming of natural gas, which accounts for about 50% of world production.[4–6] Natural gas has a similar origin to petroleum (nonrenewable fuel), so it is not the best source for producing hydrogen, but an alternative that is proving promising in the production of biohydrogen is the steam reforming of ethanol.[7–12] Biohydrogen has great potential for replacing fossil fuels; in its combustion process 2.5 times more energy is released than when other fuels composed of hydrocarbons are used, thus being an important route for the supply of sustainable energy and this biofuel can bring environmental, economic, and social benefits. In the near future, it may reduce Brazil's dependence on the use of foreign natural gas and reduce carbon dioxide emissions in the transport sector.[3] In this biohydrogen era,

Sustainable Bioenergy. DOI: https://doi.org/10.1016/B978-0-12-817654-2.00004-6

sugar and ethanol mills may be modified to produce hydrogen for use as fuel, in addition to the conventional products of the traditional ethanol production chain.[13] The use of more efficient energy conversion systems with biomass gasification technology can allow for cogeneration plants to increase bioelectricity production, increase their efficiency in using process wastes, and contribute to diversifying the energy grid with bioelectricity generation.[14–16] For the implantation of the gasification system into the sugarcane industry, it is indicated that the gasification of a fluidized bed—since this model presents high rates of conversion of carbon and tar and high rates of heat transfer between the fuel (sugarcane bagasse) and the material of the bed—is suitable for large capacity systems. The gas produced in this process is used as fuel in the generation of electric energy, in the production of steam in boilers for thermoelectric processes, or in the direct feeding of generating systems based on internal combustion engines.[3] The incorporation of these new processes will allow for improvements in the profitability and operational flexibility of sugarcane plants, thus corroborating the improvement of Brazil's position in the bioenergy ranking.

The aim of this chapter is to present the incorporation of the ethanol steam reforming process for biohydrogen production and the gasification system based on the biomass integrated gasification gas turbine combined cycle (BIG/GTCC) to increase the electricity supply in conventional sugar and ethanol industries. Initially, the configuration and thermodynamic parameters of the plant under study are presented. Then, a thermodynamic analysis is performed to determine the biohydrogen production capacity, the electricity generation efficiency, and the overall cogeneration efficiency. An economic analysis is also carried out to determine the cost of the production of biohydrogen and electric energy in US$/kW h. Finally, an environmental analysis is presented in order to determine the production of equivalent carbon dioxide, indicators of pollution, and the ecological efficiency of each process.

4.2 Hydrogen production from steam reforming of ethanol

Brazil is a major producer and distributor of ethanol fuel and is one of the largest producers in the sugarcane industry. Thus the steam reforming of ethanol becomes a good option for biohydrogen production countrywide. Fig. 4.1 shows a schematic representation of a conventional sugarcane industry with the incorporation of biohydrogen production through the steam reforming of ethanol.

Silva[17] proposed that biohydrogen could be produced from ethanol through a couple of stages. The first one occurs at high temperatures, between 800K and 1000K, where ethanol is transformed into a mixture of hydrogen, carbon dioxide, carbon monoxide, water, and methane. The second one occurs at low temperatures, between 473K and 673K, in which the carbon monoxide generated during the first stage reacts with water to form carbon dioxide and biohydrogen.

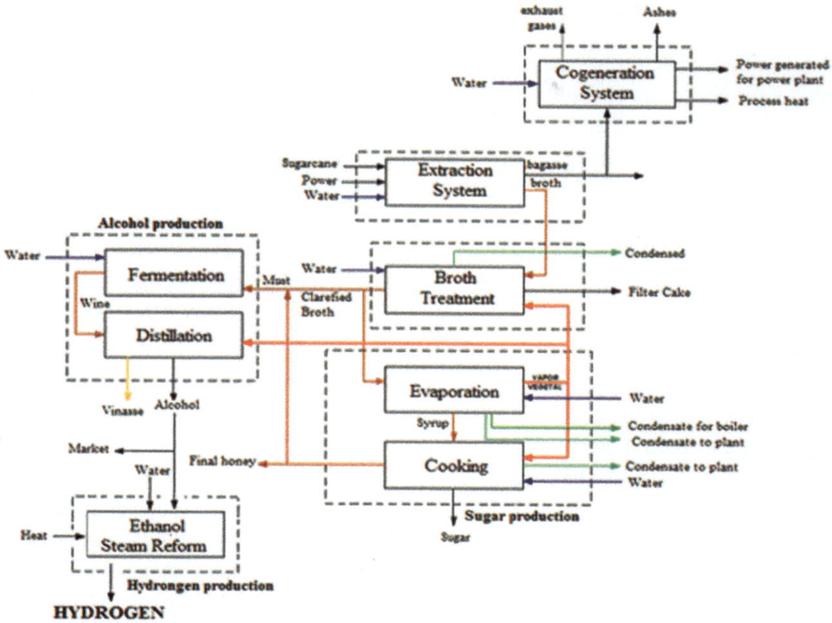

FIGURE 4.1 Schematic representation of a conventional sugarcane industry with a steam reforming system incorporated. *Adapted from Silva ME.* Experimental analysis ethanol steam reforming: technical, economic and ecological aspects *[Ph.D. dissertation]. Guaratingueta, SP, Brazil: Submitted to Sao Paulo State University; 2010 (in Portuguese) with copyright permission from Faculty of Engineering, Campus Guaratinguetá, UNESP, Brazil.*

The overall reaction of the steam reforming of ethanol is presented in Eq. (4.i)[13,18]:

$$C_2H_5OH_{(g)} + 3H_2O_{(g)} \rightarrow 2CO_2 + 6H_{2_{(g)}} \tag{4.i}$$

A thermodynamic analysis of the hydrogen production process through the steam reforming of ethanol can be carried out through these stages[13,18]:

- Steam reforming reaction:

$$C_2H_5OH_{(g)} + H_2O_{(g)} \rightarrow 2CO + 4H_{2_{(g)}} \tag{4.ii}$$

- Methane formation reaction:

$$CO_{(g)} + 3H_{2_{(g)}} \rightarrow CH_{4_{(g)}} + H_2O_{(g)} \tag{4.iii}$$

- Carbon formation reaction by carbon monoxide decomposition, which should be avoided:

$$2CO_{(g)} \rightarrow CO_{2_{(g)}} + C_{(s)} \tag{4.iv}$$

- Water–gas shift reaction:

$$CO_{(g)} + H_2O_{(g)} \rightarrow CO_{2_{(g)}} + H_{2_{(g)}} \tag{4.v}$$

4.3 Sugarcane bagasse gasification

Gasification is a thermochemical process that takes place by reacting biomass composed of carbon (sugarcane bagasse, wood, coal) with air, steam, or oxygen.[19,20] As a result of this reaction, a fuel gas known as syngas is obtained, which mainly consists of carbon monoxide, carbon dioxide, hydrogen, nitrogen, methane, and tar.[21,22] A gasifier is the reactor used for the gasification process. According to Warnecke,[23] a circulating fluidized bed gasifier is the most appropriate for sugarcane bagasse, since it has high carbon and tar conversion rates and high heat transfer rates between the fuel and bed material; moreover, it is suitable for large capacity systems (above 40 MW) operating at reaction temperatures ranging between 973K and 1273K, outlet syngas temperatures ranging between 873K and 1073K, and low tar content production.[24]

The gasification process can be incorporated into the sugarcane sector through a technology like BIG/GTCC. The system consists of a biomass dryer, a gasifier to convert the biomass into syngas, a gas cleaning system, a gas turbine set fueled by syngas combustion, a heat recovery steam generator (HRSG) to recover energy from exhaust gases by the gas turbine and generate bioelectricity, and live steam to run the steam turbine to produce more electricity and also useful heat.[25,26]

The reactions that occur in the thermochemical gasification process are described here[10,27]:

- Pyrolysis:

$$\text{biomass} + \text{heat} \rightarrow \text{coke} + \text{gases} + \text{tar} + \text{condensates} \tag{4.vi}$$

- Carbon oxidation:

$$C + \frac{1}{2}O_2 \leftrightarrow CO \tag{4.vii}$$

$$C + O_2 \leftrightarrow CO_2 \tag{4.viii}$$

- Gasification:
  - Heterogeneous reaction:

$$C + CO_2 \leftrightarrow 2CO \tag{4.ix}$$

$$C + H_2O \leftrightarrow CO \tag{4.x}$$

$$C + 2H_2 \leftrightarrow CH_4 \tag{4.xi}$$

  - Homogeneous reaction:

$$CO + H_2O \leftrightarrow CO_2 + H_2 \tag{4.xii}$$

$$CH_4 + H_2O \leftrightarrow CO + 3H_2 \tag{4.xiii}$$

- Tar cracking:

$$tar + steam + heat \rightarrow CO + CO_2 + CH_2 \quad (4.xiv)$$

- Partial oxidation of pyrolysis products:

$$CO + H_2 + CH_4 + O_2 \rightarrow CO_2 + H_2 \quad (4.xv)$$

4.4 Case study

4.4.1 Description of a conventional sugarcane industry

The industry is composed of a turbine generator capable of producing 40 MV A (turbine A). Turbine A has two points of extraction and also feeds steam into a condenser. It is also composed of other low-pressure machines powered by low-pressure steam generated by the turbine (generator A). The high-pressure steam generator (MC generator) can operate nominally and produce 150 t of steam per hour, at 6468 kPa and 803K of pressure and temperature, respectively.[28]

Data about the industry studies are shown in Table 4.1, its flow chart in Fig. 4.2, and its thermodynamic parameters in Table 4.2.

TABLE 4.1 Plant data for the 2003/2004 crop.

Parameter	Amount	Unit
Total ground sugarcane	1,300,000	t
Harvest days	215	days
Agricultural efficiency	97	%
Industrial efficiency	94	%
Actual grinding hours	4695.6	h
Grinding rate	276.9	t/h
Steam/bagasse ratio	0.47	kg/kg
Sugarcane fiber content	13.5	%
Sugarcane bagasse fiber content	47.4	%
MC generator bagasse mass flow	70.5	t/h
Total bagasse production rate	78.9	t/h
Residual bagasse flow (available)	8.4	t/h
Total amount of residual bagasse after harvest	39,213	t

Source: Adapted from Fiomari MC. *Energetic and exergetic analyses of sugarcane industry of Paulista West with expansion of the cogeneration system* [Ph.D. dissertation]. Ilha Solteira, Brazil: Submitted to Sao Paulo State University; 2004 (in Portuguese) with copyright permission from Faculty of Engineering, Campus Ilha Solteira, UNESP, Brazil.

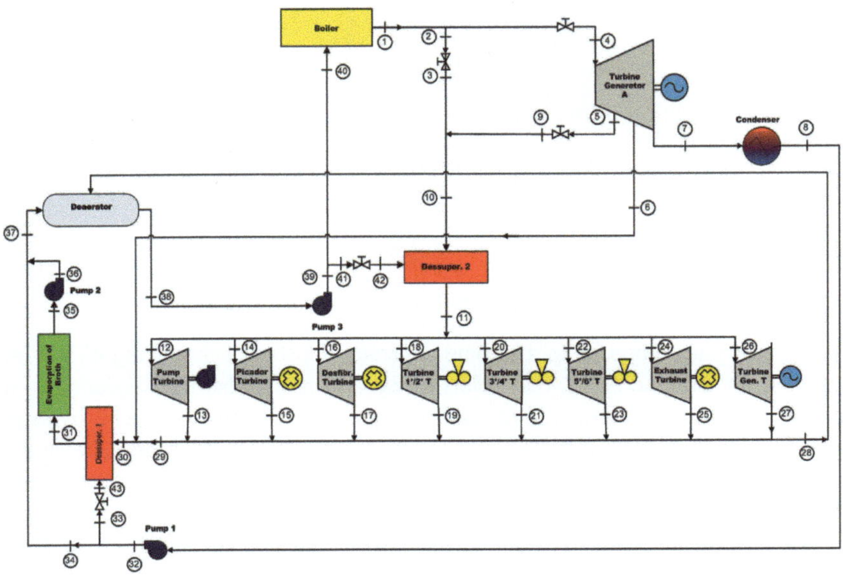

FIGURE 4.2 Schematic representation of conventional plant. *Adapted from Fiomari MC.* Energetic and exergetic analyses of sugarcane industry of Paulista West with expansion of the cogeneration system *[Ph.D. dissertation]. Ilha Solteira, Brazil: Submitted to Sao Paulo State University; 2004 (in Portuguese) with copyright permission from Faculty of Engineering, Campus Ilha Solteira, UNESP, Brazil.*

4.4.2 Thermodynamic analysis of the incorporation of the biohydrogen production process

By incorporating the steam reforming process of ethanol, the industry can simultaneously produce sugar, alcohol, and biohydrogen, as shown in Fig. 4.3.

In order to incorporate such a process, it is necessary to reduce surplus electricity production so as to supply bagasse for biohydrogen production. In such a case, there must be energy balance in turbine A, as shown in Fig. 4.4, in order to investigate the availability of sugarcane bagasse for electricity production. Turbine A, model VE 32, has some limitations in both inlet and outlet flow and steam properties. These limitations are presented in Table 4.3.

It is observed that the steam flow toward extractions and condensation can be reduced. In extraction 1 and 2, it could not be reduced because it would not ensure the thermal matching of the cogeneration system operation in order to guarantee the thermal energy of processes. Therefore it is possible to reduce the steam flow toward the condenser from 9.44 to 2.23 kg/s (point 7 in Fig. 4.4) (thus making sugarcane bagasse available for hydrogen production). Initially, a thermodynamic analysis was conducted so as to investigate

TABLE 4.2 Plant operation parameters.

Points	$\dot{m}$ (kg/s)	P (kPa)	T (K)	h (kJ/kg)	s (kJ/kg K)
1	41.67	6468	803.0	3488.5	6.931
2	8.14	6468	803.0	3488.5	6.931
3	8.14	2156	283.2	3488.5	7.424
4	33.53	6468	803.0	3488.5	6.931
5	10.00	2650	699.0	3295.1	7.070
6	17.78	245	420.2	2759.3	7.165
7	5.75	8,5	315.7	2578.2	8.205
8	5.75	8,5	315.6	178.4	0.607
9	10.00	2156	695.7	3295.1	7.163
10	180.56	2156	735.0	3381.9	7.284
11	20.33	2156	593.0	3064.8	6.804
12	–	–	–	–	–
13	–	–	–	–	–
14	4.08	2156	583.0	3041.6	6.765
15	4.08	245	458.0	2837.3	7.343
16	4.08	2156	583.0	3041.6	6.765
17	4.08	245	458.0	2837.3	7.343
18	4.17	2156	583.0	3041.6	6.765
19	4.17	245	467.5	2856.6	7.385
20	4.00	2156	583.0	3041.6	6.765
21	4.00	245	467.5	2856.6	7.385
22	4.00	2156	583.0	3041.6	6.765
23	4.00	245	467.5	2856.6	7.385
24	–	–	–	–	–
25	–	–	–	–	–
26	–	–	–	–	–
27	–	–	–	–	–
28	0.83	245	463.7	2848.8	7.368
29	19.50	245	463.7	2848.8	7.368
30	37.28	245	442.8	2806.1	7.274

(*Continued*)

TABLE 4.2 (Continued)

Points	$\dot{m}$ (kg/s)	P (kPa)	T (K)	h (kJ/kg)	s (kJ/kg K)
31	38.33	245	408.0	2733.5	7.103
32	5.75	490	315.7	179.3	0.608
33	1.06	490	315.7	179.3	0.608
34	4.69	490	315.7	179.3	1.307
35	38.33	245	373.0	419.2	1.307
36	38.33	490	373.1	419.6	1.307
37	43.03	490	366.8	393.4	1.237
38	43.86	245	378.0	440.3	1.363
39	43.86	8820	379.8	454.1	1.376
40	41.67	8820	379.8	454.1	1.376
41	2.19	8820	379.8	454.1	1.376
42	2.19	2156	380.8	454.1	1.394
43	1.06	245	315.8	179.3	0.609

Source: Adapted from Fiomari MC. *Energetic and exergetic analyses of sugarcane industry of Paulista West with expansion of the cogeneration system* [Ph.D. dissertation]. Ilha Solteira, Brazil: Submitted to Sao Paulo State University; 2004 (in Portuguese) with copyright permission from Faculty of Engineering, Campus Ilha Solteira, UNESP, Brazil.

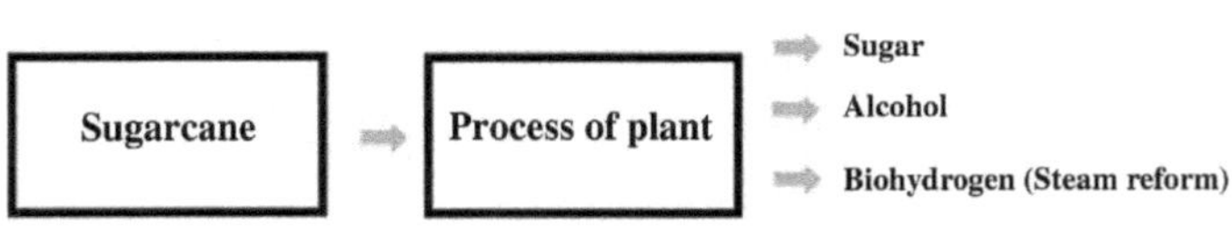

FIGURE 4.3 Schematic representation proposed for the sugarcane industry. *Adapted from Silva ME.* Experimental analysis ethanol steam reforming: technical, economic and ecological aspects *[Ph.D. dissertation]. Guaratingueta, SP, Brazil: Submitted to Sao Paulo State University; 2010 (in Portuguese) with copyright permission from Faculty of Engineering, Campus Guaratinguetá, UNESP, Brazil.*

the total electricity surplus produced by the industry. In this case, initially, the condenser operated at the minimum steam flow (2.23 kg/s) and then at the actual steam flow toward the condenser, which was 5.75 kg/s, as shown in Fig. 4.2.

In order to obtain the amount of surplus electricity reduction to make bagasse available for the hydrogen production process, shaft work and electrical power of turbine A in actual operation conditions are calculated by considering the minimum flow toward the condenser. For actual operation

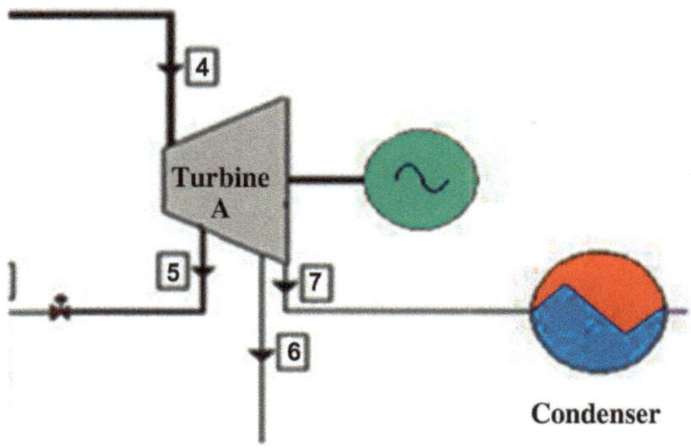

FIGURE 4.4 Condensation turbine A with two extraction points. *Adapted from Fiomari MC.* Energetic and exergetic analyses of sugarcane industry of Paulista West with expansion of the cogeneration system *[Ph.D. dissertation]. Ilha Solteira, Brazil: Submitted to Sao Paulo State University; 2004 (in Portuguese) with copyright permission from from Faculty of Engineering, Campus Ilha Solteira, UNESP, Brazil.*

TABLE 4.3 Operating data on turbine A.

Mode	Steam flow (kg/s)		Pressure (bar abs.)		Temperature (K)	
	Minimum	Maximum	Minimum	Maximum	Minimum	Maximum
Inlet	–	33.53	–	66.0	–	803.0
Extraction 1	0	10	26.5	30.2	698.0	711.0
Extraction 2	17.77	33.34	2.37	2.37	407.5	420.2
Condenser	2.23	9.44	0.055	0.115	307.6	321.6

Source: Adapted from Fiomari MC. *Energetic and exergetic analyses of sugarcane industry of Paulista West with expansion of the cogeneration system* [Ph.D. dissertation]. Ilha Solteira, Brazil: Submitted to Sao Paulo State University; 2004 (in Portuguese) with copyright permission from Faculty of Engineering, Campus Ilha Solteira, UNESP, Brazil.

conditions (steam flow rate toward the condenser at 5.75 kg/s), Eqs. (4.1) and (4.2) can be used[3]:

$$\dot{W}_{ST_{C.R.}} = \dot{m}_{4C.R.} \times h_4 - \dot{m}_5 \times h_5 - \dot{m}_6 \times h_6 - \dot{m}_{7C.R.} \times h_7 \qquad (4.1)$$

$$E_{PST_{AC.R.}} = \dot{W}_{ST_{AC.R.}} \times 0.95 \qquad (4.2)$$

where $E_{PST_{AC.R.}}$ is the electrical power of steam turbine in actual operation conditions (turbine A) (kW), h_4 is the enthalpy at point 4 (3488.5 kJ/kg), h_5 is the enthalpy at point 5 (3295.1 kJ/kg), h_6 is the enthalpy at point 6 (2759.3 kJ/kg), h_7 is the enthalpy at point 7 (2578.2 kJ/kg), $\dot{m}_{4C.R.}$ is the steam flow at point 4 (33.53 kg/s), $\dot{m}_5$ is the steam flow at point 5 (10 kg/s), $\dot{m}_6$ is the steam flow at point 6 (17.78 kg/s), $\dot{m}_{7C.R.}$ is the steam flow at point 7 (5.75 kg/s), $\dot{W}_{ST_{AC.R.}}$ is the steam turbine mechanical power (turbine A) (kW).

For minimum steam flow through the condenser (steam flow toward the condenser at 2.23 kg/s), Eqs. (4.3) and (4.4) can be used[3]:

$$\dot{W}_{ST_{A_{min.}}} = \dot{m}_{4min.} \times h_4 - \dot{m}_5 \times h_5 - \dot{m}_6 \times h_6 - \dot{m}_{7min.} \times h_7 \quad (4.3)$$

$$E_{pST_{A_{min.}}} = \dot{W}_{ST_{A_{total}}} \times 0.95 \quad (4.4)$$

where $E_{pST_{A_{min.}}}$ is the minimum electrical power allowed in the steam turbine (turbine A) (kW), h_4 is the enthalpy at point 4 (3488.5 kJ/kg), h_5 is the enthalpy at point 5 (3295.1 kJ/kg), h_6 is the enthalpy at point 6 (2759.3 kJ/kg), h_7 is the enthalpy at point 7 (2578.2 kJ/kg), $m_{4min.}$ is the steam flow at point 4 (31.30 kg/s), $\dot{m}_5$ is the steam flow at point 5 (10 kg/s), $\dot{m}_6$ is the steam flow at point 6 (17.78 kg/s), $\dot{m}_{7min.}$ is the steam flow at point 7 (2.23 kg/s), $W_{ST_{A_{min.}}}$ is the steam turbine minimum work (generator A) (kW).

The amount of electricity production reduction by the turbine (turbine A) $\left(E_{pST_{A_{red.}}}\right)$ can be determined by Eq. (4.5):

$$E_{pST_{A_{red.}}} = E_{pST_{A_{C.R.}}} - E_{pST_{A_{min.}}} \quad (4.5)$$

By means of Eq. (4.6), it is possible to determine the bagasse flow available for hydrogen production by the steam reforming of ethanol[17]:

$$\dot{m}_b = \frac{\dot{m}_s \times \Delta h_c}{\eta_{c_I} \times LHV_{b.u}} \quad (4.6)$$

where $LHV_{b.u}$ is the lower heating value of wet bagasse (kJ/kg), $\dot{m}_b$ is the bagasse flow (kg/s), $\dot{m}_s$ is the generator outlet steam flow (kg/s), η_{c_I} is the generator energy efficiency (%), Δh_c is the generator water enthalpy difference (kJ/kg).

According to Silva,[17] the energy supplied by 1.23 kg/h of sugarcane bagasse ($LHV_{b.u}$ of 7320 kJ/kg) provides enough heat to produce 1 N m^3/h of hydrogen through the steam reforming process of ethanol. Thus 0.8597 L/h of hydrous ethanol or 0.7961 L/h of anhydrous ethanol is required to obtain 1 N m^3/h of hydrogen.

The electricity generation efficiency of the cogeneration plant can be determined by Eq. (4.7):

$$\eta_e = \frac{E_{pST} - W_p}{E_{fuel}} \quad (4.7)$$

In order to determine the overall efficiency of the cogeneration plant (η_g), Eq. (4.8) can be used:

$$\eta_g = \frac{E_{pST} + E_m + H_p - W_p}{E_{fuel}} \quad (4.8)$$

TABLE 4.4 Conventional cogeneration system data.

Electrical power of steam turbine in actual operation conditions (turbine A) ($E_{pST_{AC.R}}$)	19,125.19 kW
Mechanical power (E_m)	3919.28 kW
Process thermal energy flow (H_p)	88,714.83 kW
Minimum electrical power allowed in steam ($E_{pST_{A_{min.}}}$)	16,081.14 kW
Electricity production reduction by the turbine (turbine A) ($E_{pST_{A_{red.}}}$)	3044.05 kW
Electricity generation efficiency (η_e)	15.24%
Cogeneration plant overall efficiency (η_g)	77.54%

where E_M is the mechanical power (kW), E_{pST} is the electric power—steam turbine (kW), H_p is the useful heat in the process (kW), $\dot{W}_P$ is the electric power supplied to pump (kW), η_e is the conventional plant electrical efficiency (%), η_g is the overall efficiency of conventional cogeneration plant (%).

Table 4.4 presents the thermodynamic analysis data on the conventional cogeneration system.

4.4.2.1 Results of the thermodynamic analysis of biohydrogen production process incorporation

The largest biohydrogen production for the plant under study is 5125 N m^3/h of hydrogen, which also ensures thermal matching during the cogeneration system operation. Table 4.5 shows a scenario with no production (0 N m^3/h) and one with the largest production (5125 N m^3/h) of biohydrogen.

It is observed that a reduction in steam flow toward the condenser leads to lower electrical power generation by turbine A and the amount of heat rejected goes to the condenser. Table 4.6 shows the cogeneration system performance by considering the minimum and maximum hydrogen production capacity.

4.4.3 Thermodynamic analysis of the biomass integrated gasifier/gas turbine combined cycle incorporation

For incorporating the gasification process into the plant; a bagasse dryer, a gasifier, a cleaning system consisting of a cyclone and a filter, an air compressor for gasification, a syngas compressor, and a gas turbine cycle were integrated, as shown in Fig. 4.5. The plant has been kept in its original conditions, except for the replacement of the MC generator for a HRSG with the same superheated steam production capacity. Therefore points 1–43 (mentioned in Fig. 4.2) are the same as those of the conventional plant, whose

TABLE 4.5 Biohydrogen production and its respective calculations.

H_2 production (N m^3/h)	Biohydrogen production bagasse flow ($\dot{m}_b$) (kg/s)	Steam reduction at condenser ($\dot{m}_v$) (kg/s)	Volume of anhydrous ethanol (L/h)	Volume of hydrous ethanol (L/h)
0	–	–	–	–
300	0.10	0.22	238.83	257.91
600	0.20	0.44	477.66	515.82
900	0.30	0.65	716.49	773.73
1200	0.41	0.87	955.32	1031.64
1500	0.51	1.10	1194.15	1289.55
2000	0.68	1.45	1592.20	1719.40
3000	1.00	2.18	2688.30	2579.10
4000	1.36	2.90	3184.40	3438.80
5125	1.72	3.52	4080.76	4406.77

Source: Adapted from Paulino RFS. *Incorporation of new technologies in the sugarcane industry: hydrogen production and increase of electricity supply.* Guaratinguetá, SP, Brazil: GraduateWork, Sao Paulo State University; 2014 [in Portuguese]. p. 1–92 with copyright permission from Faculty of Engineering, Campus Guaratinguetá, UNESP, Brazil.

thermodynamic properties are shown in Fig. 4.2 and Table 4.2. Other thermodynamic properties are shown in Fig. 4.5, such as:

- *Dryer*

  Bagasse is inserted into the dryer at 50% moisture content with an ambient temperature and pressure of 298K and 101.3 kPa, respectively. As proposed by Olivares,[29] bagasse for gasification should have a moisture content ranging between 15% and 30%. For this work, a 30% moisture content was adopted for the gasifier inlet.

  Moreover, Graebin[30] suggested that the operating temperature should range between 333K and 373K. For the present work, it was concluded that the dryer should operate at 373K and 101.3 kPa of pressure. It was also found that the condensate (point 45) is released at ambient pressure and temperature.
- *Gasifier*

  Part of the bagasse is converted into ash and Sousa recommends an ash content of less than 20% for biomass gasification processes (circulating fluidized bed gasifier), however, 5% was adopted for the present work. For gasification reactions to take place, a gasifying agent is required, which may be air, steam, or oxygen. In the present study, air was used as a

TABLE 4.6 Thermodynamic parameters of a conventional cogeneration plant through hydrogen production incorporation.

H_2 production ($N\ m^3/h$)	Electrical power (kW)	Heat released by the condenser (kW)	E_{comb} (kW)	Electricity generation efficiency (%)	Cogeneration plant overall efficiency (%)
0	19,125.19	13,798.85	143,350.00	13.34	77.54
300	18,936.59	13,275.49	142,599.70	13.27	77.81
600	18,747.99	12,752.13	141,849.40	13.21	78.09
900	18,559.40	12,228.76	141,099.10	13.15	78.37
1200	18,370.80	11,705.40	140,348.80	13.08	78.66
1500	18,182.20	11,182.04	139,598.50	13.02	78.94
2000	17,867.88	10,309.77	138,348.00	12.91	79.43
3000	17,239.22	8565.24	135,847.00	12.69	80.43
4000	16,610.56	6820.70	133,346.00	12.45	81.47
5125	15,902.73	4856.45	130,530.01	12.18	82.68

Source: Adapted from Paulino RFS. *Incorporation of new technologies in the sugarcane industry: hydrogen production and increase of electricity supply.* Guaratinguetá, SP, Brazil: GraduateWork, Sao Paulo State University; 2014 [in Portuguese]. p. 1–92 with copyright permission from Faculty of Engineering, Campus Guaratinguetá, UNESP, Brazil.

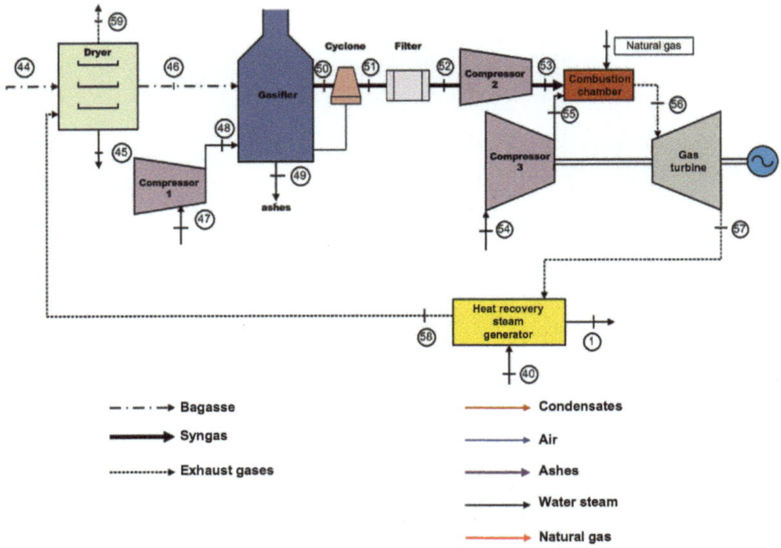

FIGURE 4.5 Schematic representation proposed for combined cycle. *Adapted from Paulino RFS.* Incorporation of new technologies in the sugarcane industry: hydrogen production and increase of electricity supply. *Guaratinguetá, SP, Brazil: GraduateWork, Sao Paulo State University; 2014 [in Portuguese]. p. 1–92 with copyright permission from Faculty of Engineering, Campus Guaratinguetá, UNESP, Brazil.*

gasifying agent; however, the air/biomass ratio in circulating fluidized bed gasifiers should be in the range of 0.25–0.4.[31,32] In order to achieve the highest efficiency, an air mass flow of 0.25 was used.

In accordance with the literature, the gasifier mass balance was calculated through Eq. (4.9):

$$\dot{m}_{d.b} + \dot{m}_{air} = \dot{m}_{ashes} + \dot{m}_{syngas} \tag{4.9}$$

where $\dot{m}_{air}$ is the air mass flow (point 48) (kg/s), $\dot{m}_{ashes}$ is the ash mass flow (point 49) (kg/s), $\dot{m}_{d.b.}$ is the dry bagasse mass flow (point 46) (kg/s), $\dot{m}_{syngas}$ is the syngas mass flow (point 50) (kg/s).

The air mass flow in a gasification process depends on the composition of the bagasse, air volume, air factor (air/biomass ratio), air density under normal conditions, and bagasse flow through the gasifier. The composition of sugarcane bagasse varies according to where it is grown. For this case study, the average elemental composition presented in Table 4.7 was considered.

Air volume (V_A) can be calculated using Eq. (4.10).[31]

$$V_A = (0.0889 \times (C_C + (0.375 \times C_S))) + (0.265 \times C_H) - (0.0333 \times C_O) \tag{4.10}$$

TABLE 4.7 Typical average elemental composition of bagasse.

Element	C	H	N	O	S
Composition (%)	45.65	5.98	0.26	45.61	0.08

Therefore air mass flow can be calculated by Eq. (4.11).

$$\dot{m}_{air} = V_A \times FA \times \delta_{air} \times \dot{m}_{d.b} \tag{4.11}$$

where *FA* is the air/biomass ratio (−), $\dot{m}_{d.b}$ is the dry bagasse mass flow (point 46) (kg/s), V_A is the air volume (N m^3/kg), δ_{air} is the air density (kg/m^3).

- *Cyclone*

  A cyclone is able to remove and bring larger particles of syngas flow back to the gasifier. Removed particles are mainly composed of ash and tar.[24]
- *Filter*

  A filter is able to remove smaller and lighter particles that could not be removed by the cyclone. These are tar, nitrogen oxides (NO_x), sulfur oxides (SO_x), and soot, which can cause damage to the equipment if inserted into the combustion chamber. Syngas is inserted into the filter at high temperatures, thus it is necessary to use a ceramic filter resistant to abrasion at high temperatures.[33]
- *Syngas compressor*

  The air and syngas must be inserted into the combustion chamber at the same pressure so that they do not interfere with the inlet flow. Thus it is necessary to use compressor 2 (syngas compressor). The syngas output temperature in compressor 2 can be determined using Eq. (4.12), with its outlet pressure being the same as that of compressor 3.

$$T_{53} = T_{52} \times \left(\frac{P_{53}}{P_{52}}\right)^{\lambda_{syngas}-1/\lambda_{syngas}} \tag{4.12}$$

where λ_{syngas} is the ratio between specific heat at constant pressure and specific heat at constant volume of syngas (kJ/kg K).

- *Gas turbine set*

  The most suitable gas turbine set to meet the plant needs is shown in Table 4.8.
- *Air compressor*

  Compressors 1 and 3 (air compressor) are responsible for maintaining inlet temperature and ambient pressure, i.e., 298K and 101.3 kPa,

TABLE 4.8 Properties used in the gas turbine set.

Turbine	Power (MW)	Heat hate (Btu/kW h)	Efficiency (%)	Pressure ratio (−)	Exhaust gas flow (kg/s)	Exhaust gas temperature (K)
Hitachi PG 7241(E)	171,700	9360	36.46	15.5	431.81	873.89

respectively. A pressure ratio of 15.5 was used and the compressor outlet air temperature can be calculated using Eq. (4.13):

$$T_s = T_e \times \left(\frac{P_s}{P_e}\right)^{(\lambda_{air}-1)/\lambda_{air}} \tag{4.13}$$

where T_e is the compressor inlet air temperature (K), T_s is the compressor outlet air temperature (K), P_e is the compressor inlet air pressure (kPa), P_s is the compressor outlet air pressure (kPa), λ_{air} is the ratio between specific heat at constant pressure and specific heat at constant air volume (kJ/kg K).

- *Combustion chamber*

  For calculating the air mass flow (m_{air}) required for the combustion chamber by considering that it is carried out with syngas and an additional fuel (natural gas), Eq. (4.14) can be used.

$$\dot{m}_{air} = \dot{m}_{EG} - \dot{m}_{syngas} - \dot{m}_{NG} \tag{4.14}$$

where $\dot{m}_{air}$ is the air mass flow (kg/s), $\dot{m}_{EG}$ is the exhaust gas mass flow (kg/s), $\dot{m}_{NG}$ is the additional natural gas mass flow required for additional burning (kg/s), $\dot{m}_{syngas}$ is the syngas mass flow (kg/s).

Natural gas mass flow can be calculated using Eq. (4.15). According to Villela,[34] the lower heating value of natural gas (PCI_{NG}) is 47.966 kJ/kg.

$$E_{NG} = \dot{m}_{NG} \times PCI_{NG} \tag{4.15}$$

where E_{NG} is the energy in natural gas (kW), LHV_{NG} is the lower heating value of natural gas (kJ/kg), $\dot{m}_{NG}$ is the natural gas mass flow required for additional burning (kg/s).

The total power supplied by the syngas together with that of the additional fuel (natural gas) can be determined by Eq. (4.16):

$$E_{fuel_{total}} = \frac{\dot{W}_{GT}}{\eta_{TGT}} \tag{4.16}$$

where $E_{fuel_{total}}$ is the total power supplied by fuel to gas turbine (kW), $\dot{W}_{GT}$ is the gas turbine power (kW), $\eta_{T_{GT}}$ is the gas turbine efficiency (−).

In order to estimate the combustion chamber outlet temperature of exhaust gases (point 56), Eq. (4.17) can be used:

$$T_{56} = \frac{E_{fuel_{total}} + \dot{m}_{air} \times c_{pair} \times T_{55}}{\dot{m}_{EG} \times c_{p_{EG}}} \tag{4.17}$$

where $E_{fuel_{total}}$ is the total power supplied by fuel to gas turbine (kW), $\dot{m}_{air}$ is the air mass flow (kg/s), $\dot{m}_{NG}$ is the natural gas mass flow required for additional burning (kg/s), T_{55} is the combustion chamber inlet temperature (point 55) (K), T_{56} is the combustion chamber outlet air temperature (point 56) (K).

- *HRSG*

  The thermodynamic properties of superheated steam (point 1) and HRSG feedwater (point 40) are the same as those shown in Table 4.1. The HRSG inlet exhaust gas pressure (point 57) can be calculated using Eq. (4.18):

$$P_{57} = \frac{P_{58}}{\Delta P_{HRSG}} \tag{4.18}$$

  where P_{57} is the HRSG inlet exhaust gas pressure (kPa), P_{58} is the HRSG outlet exhaust gas pressure (kPa), ΔP_{HRSG} is the HRSG pressure loss, i.e., 5%[35].

- *Thermodynamic efficiency*

  The efficiency of electricity generation (η_e) and the overall efficiency of the cogeneration plant (η_g) by incorporating the BIG/GTCC can be determined by Eqs. (4.19) and (4.20), respectively.

$$\eta_e = \frac{E_{p_{ST}} + E_{p_{GT}} - \dot{W}_P}{E_{fuel_{total}}} \tag{4.19}$$

$$\eta_g = \frac{E_{p_{ST}} + E_{p_{GT}} + E_m + Q_p - \dot{W}_P}{E_{fuel_{total}}} \tag{4.20}$$

  where E_M is the mechanical power (kW), $E_{p_{GT}}$ is the electric power–gas turbine (kW), $E_{p_{ST}}$ is the electric power–steam turbine (kW), H_p is the useful heat in the process (kW), $\dot{W}_P$ is the electric power supplied to pump (kW), η_e is the conventional plant electrical efficiency (–), η_g is the overall efficiency of conventional cogeneration plant (–).

4.4.3.1 Results of the thermodynamic analysis of biomass integrated gasification gas turbine combined cycle incorporation

For the Hitachi PG7241 turbine to generate as much power as possible, additional combustion is required. From this choice, the mass flow, temperature, and pressure calculations presented in Table 4.9 were performed.

TABLE 4.9 Thermodynamic results of the plant by considering gasification with additional combustion of natural gas.

Point	$\dot{m}$ (kg/s)	T (K)	P (kPa)
44	21.92	298.00	101.30
45	6.58	298.00	101.30
46	15.34	373.00	101.30
47	18.54	298.00	101.30
48	18.54	614.62	1276.38
49	1.69	298.00	101.30
50	32.19	973.00	700.00
51	32.19	973.00	700.00
52	32.19	973.00	700.00
53	32.19	1130.90	1570.15
Natural gas	6.13	–	–
54	393.26	298.00	101.30
55	393.26	691.43	1570.15
56	431.81	1340.65	1739.77
57	431.81	876.89	112.24
58	431.81	502.23	106.63
59	431.81	390.32	101.3

In order to obtain optimal electricity production through gasification incorporation, it is necessary to burn an additional 6.13 kg/s of natural gas in the combustion chamber or to gasify 42.14 kg/s of sugarcane bagasse. In this work, the additional burning of natural gas was considered. Table 4.10 shows the results of the parameters calculated for the plant by incorporating the gasification process.

After the incorporation of the gasification process, an 8.56-fold electricity supply increase was observed in comparison to the electricity generated by the conventional plant by considering the additional burning of natural gas. When only syngas produced from available bagasse is considered, it generates 55,824.27 kW of electricity, which corresponds to 2.91 times the electricity produced by the conventional plant. This indicates that a larger surplus can be sold to the agency.

TABLE 4.10 Calculated parameters for the plant with the gasification process incorporation.

Parameters	Results
$E_{comb_{total}}$ (kW)	447,335.69
PCI_{syngas} (kJ/kg)	4755.47
$PCI_{b.s.}$ (kJ/kg)	14,974,00
$C_{p_{syngas}}(973\ K)$ (kJ/kg K)	3.44
$C_{v_{syngas}}(973\ K)$ (kJ/kg K)	2.48
λ_{syngas} (−)	1.38
E_{pTG} (kW)	163,115
η_e (%)	40.73
η_g (%)	61.31

4.5 Economic analysis of incorporating the steam reforming of ethanol for hydrogen production in the plant

Economic calculations are aimed at determining the cost of producing hydrogen (US$/kW h) through the steam reforming of ethanol. The cost of hydrogen production can be determined using Eq. (4.21).[17,36–38]

$$C_{H_2} = \frac{I_{REF.} \times f}{H \times E_{H_2}} + C_{O_{REF.}} + C_{M_{REF.}} \tag{4.21}$$

where C_{H_2} is the cost of hydrogen production (US$/kW h), $C_{M_{REF.}}$ is the maintenance cost of steam reformer (US$/kW h), $C_{O_{REF.}}$ is the operational cost of steam reformer (US$/kW h), E_{H_2} is the power supplied by hydrogen $\left(\dot{m}_{H_2} \times LHV_{H_2}\right)$ (kW), f is the annuity factor (1/year), H is the equivalent period of utilization (h/year), $I_{REF.}$ is the steam reformer investment cost (US$).

Through Eq. (4.22), it is possible to determine the investment cost of steam reformers $\left(I_{nv_{REF}}\right)$ according to their hydrogen production capacity.[13]

$$I_{REF.} = 9900 \times \left(\frac{m_{H_2}}{5800}\right)^{0.84} \times 10^3 \tag{4.22}$$

where m_{H_2} is the volume of hydrogen produced (N m^3/h).

In this case, sugarcane bagasse is used as a source of heat, thus the operational cost of the steam reformer for ethanol can be obtained using Eq. (4.23).

$$C_{O_{REF.}} = \frac{E_{b.u} \times C_b}{E_{H_2}} + \frac{E_{EtOH} \times C_{EtOH}}{E_{H_2}} \tag{4.23}$$

where C_b is the cost of fuel (sugarcane bagasse) (US\$/kW h), C_{EtOH} is the cost of ethanol (US\$/kW h), $C_{O_{REF.}}$ is the operational cost of steam reformer (US\$/kW h), $E_{b.u}$ is the power supplied by burning wet sugarcane bagasse (kW), E_{EtOH} is the power supplied by ethanol ($\dot{m}_{EtOH} \times LHV_{EtOH}$) (kW), E_{H_2} is the power supplied by hydrogen $\left(\dot{m}_{H_2} \times LHV_{H_2}\right)$ (kW).

According to Kothari,[39] the maintenance cost of a steam reformer is calculated by Eq. (4.24).

$$C_{M_{REF.}} = 0.03 \times \frac{I_{REF.} \times f}{H \times E_{H_2}} \tag{4.24}$$

where

$$f = \frac{q^k \times (q-1)}{q^k - 1} \tag{4.25}$$

$$q = 1 + \frac{r}{100} \tag{4.26}$$

4.5.1 Results of the economic analysis of the steam reforming of ethanol incorporation for hydrogen production

In order to obtain the economic analysis results, these parameters need to be considered:

- Ethanol consumption during the steam reforming system is 0.7961 L/h of anhydrous ethanol or 0.8597 L/h of hydrous ethanol to produce 1 N m^3/h of hydrogen.
- Amortization period ranging from 4 to 12 years.
- According to UNICA (2014) the cost of anhydrous ethanol production is US\$0.40/L and hydrous ethanol is US\$0.36/L.
- The cost of sugarcane bagasse was considered zero in this work.

Fig. 4.6 shows the cost of producing hydrogen for optimal hydrogen and anhydrous and hydrous ethanol production with an annual interest rate of 4% and 12%, respectively. It is observed that the cost of hydrogen decreases as the amortization period (k) increases, but it increases as the annual interest rate does (Fig. 4.6).

4.6 Economic analysis of incorporating the gasification process into the plant

In the economic analysis of the cost of producing electricity through the gasification process; equipment, operational and maintenance costs, interest

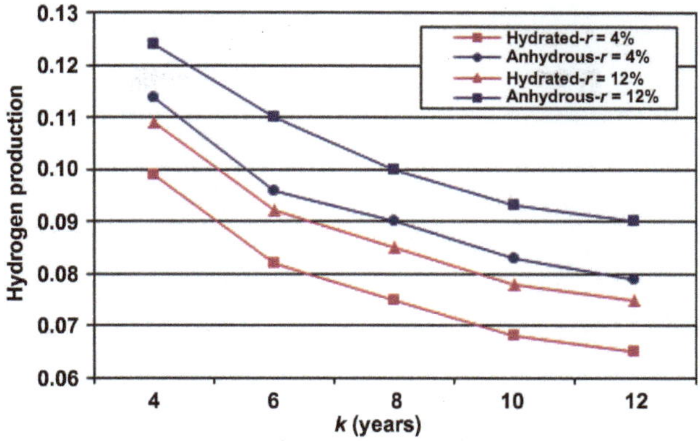

FIGURE 4.6 Comparison of the cost of hydrogen production using different types of ethanol as a function of amortization period in an optimal production scenario.

rate, return on investment time, fuel cost, and equipment lifespan, were taken into account.

Through Eq. (4.27), it is possible to determine the cost of syngas production.

$$C_{syngas} = \frac{I_{GASEIF.} \times f}{H \times E_{syngas}} + \frac{C_b \times LHV_{d.b}}{E_{syngas}} + C_{M_{GASEIF.}} + C_{O_{GASEIF..}} \tag{4.27}$$

where C_b is the price of sugarcane bagasse (US$/kW h), $C_{M_{GASEIF.}}$ is the maintenance cost of gasifier (US$/kW h), $C_{O_{GASEIF..}}$ is the operational cost of gasifier (US$/kW h), C_{syngas} is the cost of syngas production (US$/kW h), E_{syngas} is the power supplied by syngas combustion (kW), f is the annuity factor (1/year), H is the equivalent period of utilization (h/year), $I_{GASEIF.}$ is the gasifier investment cost (US$), $LHV_{d.b}$ is the lower heating value of dry bagasse (kJ/kg).

The gasifier investment ($I_{GASEIF.}$) was estimated as being 30% of the cost of the gas turbine set. The gasifier maintenance and operational cost are 3% and 6% of the total investment costs of the gasification system, respectively,[40] which can be obtained through Eqs. (4.28) and (4.29).

$$C_{M_{GASEIF.}} = 0.03 \times \frac{I_{GASEIF.} \times f}{H \times E_{syngas}} \tag{4.28}$$

$$C_{O_{GASEIF.}} = 0.06 \times \frac{I_{GASEIF.} \times f}{H \times E_{syngas}} \tag{4.29}$$

For calculating the cost of the electricity generation of the steam turbine (C_{elGT}) and the gas turbine set, mechanical energy $\left(C_{E_M}\right)$, and process heat $\left(C_{H_p}\right)$ in US$/kW h; Eqs. (4.30)–(4.33), respectively, were used.

$$C_{elST} = \frac{I_{ST} \times f}{H \times E_{pST}} + C_{M_{fuel}} \times FP_{E_{pST}} + CM_{ST} + (C_O \times FP_{E_{pST}}) \tag{4.30}$$

$$C_{elGT} = \frac{I_{GT} \times f}{H \times E_{pGT}} + C_{M_{fuel}} \times FP_{E_{pGT}} + CM_{GT} + (C_O \times FP_{E_{pGT}}) \tag{4.31}$$

$$C_{E_M} = \frac{I_{MT} \times f}{H \times E_M} + C_{M_{fuel}} \times FP_{E_M} + CM_{MT} + (C_O \times FP_{E_M}) \tag{4.32}$$

$$C_{H_p} = \frac{I_{HRSG} \times f}{H \times H_p} + C_{M_{fuel}} \times FP_{H_p} + CM_{HRSG} + (C_O \times FP_{H_p}) \tag{4.33}$$

where $C_{M_{fuel}}$ is the average fuel cost in the combustion chamber (US \$/kW h), CM_{HRSG} is the maintenance cost of HRSG (US\$/kW h), CM_{GT} is the maintenance cost of gas turbine (US\$/kW h), CM_{MT} is the maintenance cost of machine turbine (US\$/kW h), CM_{ST} is the maintenance cost of steam turbine (US\$/kW h), C_O is the personnel cost (US\$/kW h), E_M is the mechanical power (kW), $E_{p_{GT}}$ is the electric power—gas turbine (kW), E_{pST} is the electric power—steam turbine (kW), FP_{E_M} is the weighting factor of mechanical power production (−), $FP_{E_{pGT}}$ is the weighting factor of electric power production—gas turbine (−), $FP_{E_{pST}}$ is the weighting factor of electric power production—steam turbine (−), FP_{H_p} is the weighting factor of useful heat in the process (−), H_p is the useful heat in the process (kW), $I_{GASEIF.}$ is the investment in gasifier (US\$), I_{GT} is the investment in gas turbine (US\$), I_{HRSG} is the investment in HRSG (US\$), I_{MT} is the investment in turbomachinery (US\$), I_{ST} is the investment in steam turbine (US\$).

The investment cost of the steam turbine and turbomachinery were considered zero, since those already in operation at the plant would be used. The FOB (free on board) investment cost of a Hitachi PG 7241 turbine is US \$234/kW.[41] In order to convert it to the CIF (Cost, Insurance, and Freight) price, 50% of the FOB price was added and the investment in the selected gas turbine set can be determined using Eq. (4.34). However, the HRSG investment can be calculated using Eq. (4.35).[40]

$$I_{GT} = \left(234 \times E_{pGT}\right) \times 1.5 \tag{4.34}$$

$$I_{HRSG} = 4745 \times \left(\frac{h_1}{\log(T_1 - T_{57})}\right)^{0.8} + (11{,}820 \times m_1) + (658 \times m_{57}) \tag{4.35}$$

The steam turbine maintenance costs US\$0.017, US\$0.015/kW h for the gas turbine set, and US\$0.008/kW h for the HRSG. US\$0.017/kW h was considered as the maintenance cost for turbomachinery.[36]

Through Eq. (4.36), developed by Sales,[40] it is possible to calculate the cost of personnel in US\$/kW h.

$$C_O = \frac{156,000}{72,000 \times (E_{pST} + E_{pGT} + E_M + H_p)} \tag{4.36}$$

For the plant under study, it was necessary to add fuel (natural gas), thus it is necessary to calculate the average cost of fuel [Eq. (4.37)] used in the combustion chamber:

$$C_{M_{fuel}} = \left(C_{syngas} \times P_{syngas}\right) + (C_{NG} \times P_{NG}) \tag{4.37}$$

In this case, $P_{syngas} = 0.8023$ and $P_{NG} = 0.1977$ refers the proportion of syngas and natural gas, respectively, burned in the combustion chamber and to obtain the maximum output of electricity in gas turbine.[3]

The weighting factors of the steam turbine ($FP_{E_{pST}}$), gas turbine (FP_{pGT}), turbomachinery (FP_M), and process heat (FP_{H_p}) can be calculated using Eqs. (4.38)–(4.41), respectively. This calculation is necessary so that each item of equipment compensates for the percentage of fuel needed to produce energy.

$$FP_{E_{pST}} = \frac{E_{pST}}{E_{pST} + E_{pGT} + E_M + H_p} \tag{4.38}$$

$$FP_{pGT} = \frac{E_{pGT}}{E_{pST} + E_{pGT} + E_M + H_p} \tag{4.39}$$

$$FP_M = \frac{E_{pMT}}{E_{pST} + E_{pGT} + E_M + H_p} \tag{4.40}$$

$$FP_{H_p} = \frac{H_p}{E_{pST} + E_{pGT} + E_M + H_p} \tag{4.41}$$

The cost of electricity (C_{el}) produced by the plant is calculated by Eq. (4.42).

$$C_{el_{annual}} = \frac{\left(C_{elGT} \times E_{pGT}\right) + (C_{elST} \times E_{pST})}{E_{pGT} + E_{pST}} \tag{4.42}$$

Through Eq. (4.43), the annual revenue (R_{annual}) can be determined by considering the sale of surplus electricity.

$$R_{annual} = \left(E_{ptotal} - E_r\right) \times H \times \left(P_{s.surplus.} - C_{el_{annual}}\right) \tag{4.43}$$

where E_{ptotal} is the total electric energy produced by the gasification incorporation (kW), E_r is the electric energy required at the plant (kW), $P_{s.surplus}$ is the sale price of surplus electricity (US$/kW h).

where

$$E_{ptotal} = E_{pST} + E_{pGT} \tag{4.44}$$

The plant under study requires 16,081.14 kW of electricity for its operation.

4.6.1 Results of the economic analysis by incorporating the gasification process into the plant

Some parameters that were used to perform the calculations include:

- Annual interest rates: 4%, 8%, and 12%.
- Operating hours: 2500 and 4000 h/year.
- Amortization period ranging from 2 to 12 years.
- According to COMGAS[42], the cost of industrial natural gas is US\$0.44/m^3.

Figs. 4.7 and 4.8 present the cost of electricity produced by the gas turbine and process heat generated by the HRSG as a function of amortization period, respectively. It is observed that there are higher production costs in comparison with other energy generation systems, as shown in Figs. 4.9 and 4.10, since there is a need to purchase the gas turbine set and HRSG, furthermore the investment cost of these items of equipment were considered in accordance with the price set by the CIF.

Figs. 4.9 and 4.10 present the cost of electricity produced by the steam turbine and the mechanical energy generated by turbomachinery as a function of amortization period, respectively. It can be observed that the energy production costs of these are lower and the curves are more stable in comparison to other energy generation systems shown in Figs. 4.6 and 4.7, since there is no need to purchase the gas turbine set and turbomachinery, since those belonging to the plant itself are used. Therefore the equipment investment cost is zero, thus reducing energy production costs.

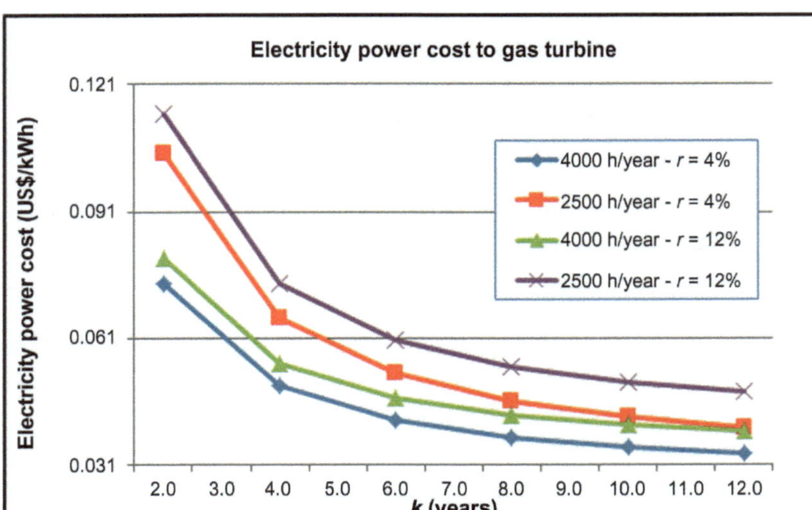

FIGURE 4.7 Cost of electricity as a function of amortization period (k) for the gas turbine set.

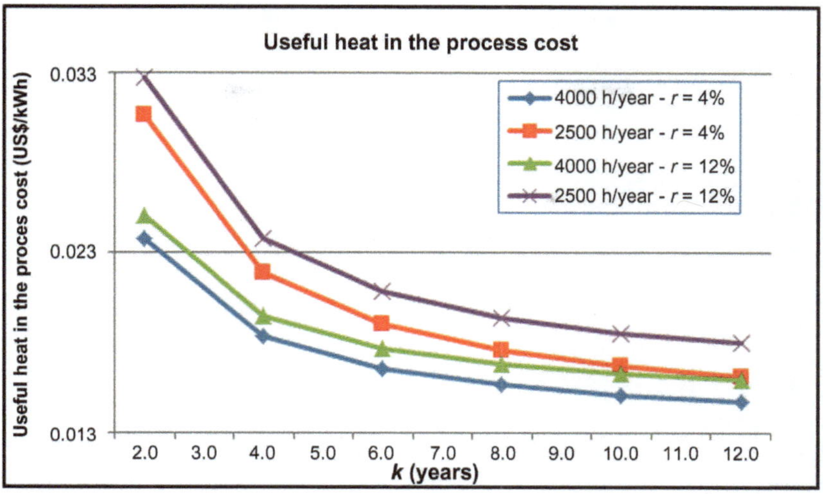

FIGURE 4.8 Cost of electricity as a function of the amortization period (*k*) for useful heat.

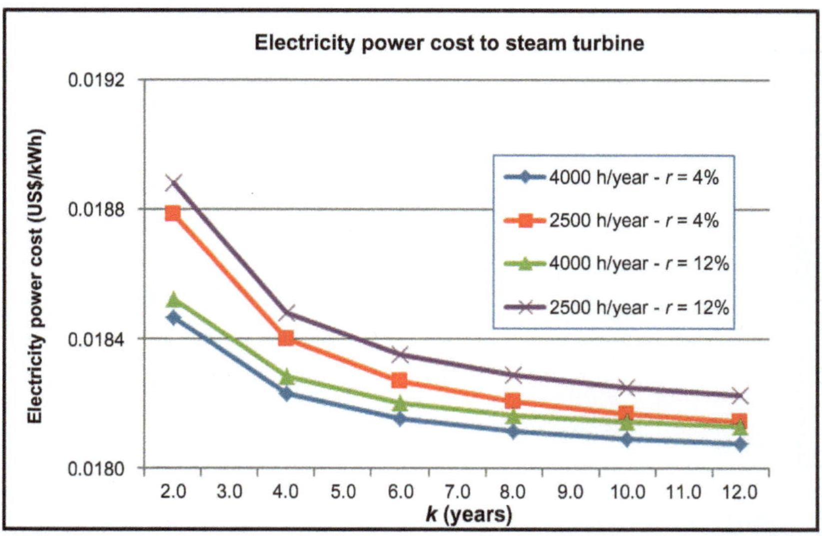

FIGURE 4.9 Cost of electricity as a function of amortization period (*k*) for the steam turbine.

Fig. 4.11 shows the annual revenue as a function of amortization period for an operating period of 4000 hours and an annual interest rate of 12% according to the possible sale prices of surplus electricity. Through this figure, it is possible to observe that as annual revenue increases, the amortization period becomes longer, thus the best option for the sale price of electricity is US$0.070/kW h for a return on investment within 2 years.

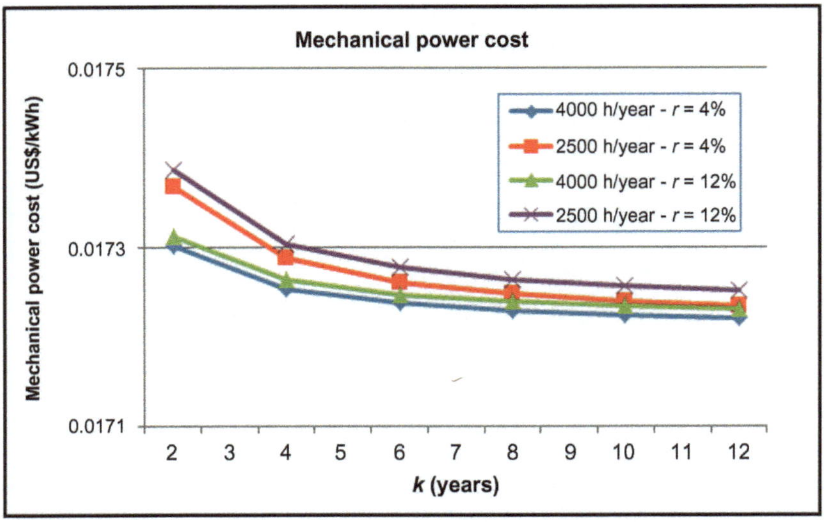

FIGURE 4.10 Cost of electricity as a function of amortization period (k) for turbomachinery.

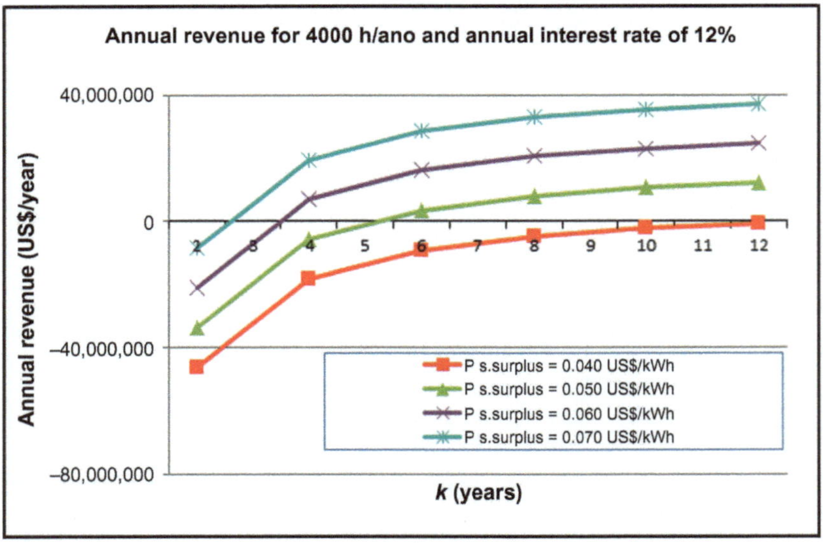

FIGURE 4.11 Annual revenue as a function of amortization period (k).

4.7 Environmental analysis

Given the growing concern for the environment and for the reduced emission of pollutants as imposed by international agreements, an environmental analysis of projects is of utmost importance. The study of ecological efficiency is important in order to determine how pollutive the energy production process is. The most harmful gases to the environment are carbon monoxide (CO), nitrogen oxide (NO_x), sulfur dioxide (SO_2), particulate matter (PM), and carbon dioxide (CO_2).[43] An environmental analysis of each process includes the calculation of equivalent carbon dioxide, pollutant indicator, and energy or overall efficiency in order to determine their ecological efficiency.

The ecological efficiency quantifies the level of pollution of a system, which ranges from 0 to 1 or 0 to 100%. A system that achieves an efficiency of 1 means that it is not pollutive and a system that achieves a 0 is very pollutive.[34]

- *Equivalent carbon dioxide*

  Eq. (4.45) was used to calculate equivalent carbon dioxide.[34,44–46] This equation considers the standards established by of the National Environment Council, Resolution 3 on June 28, 1990, which determines that the PM concentration must not exceed 150 μg/m^3 and the CO_2 concentration cannot be higher than 10,000 mg/m^3.[47,48] It also considered the air quality standards set by the World Health Organization, which recommends 200 μg/m^3 of NO_x and 125 μg/m^3 of SO_2 for 1 hour of exposure.

$$(CO_2)_e = M_{CO_2} + 80 \times M_{SO_2} + 50 \times M_{NO_x} + 67 \times M_{MP} \quad (4.45)$$

where $(CO_2)_e$ is the equivalent carbon dioxide (kg/kg$_{comb}$), M_{CO_2} is the carbon dioxide emission (kg_{CO_2}/kg_{comb}), M_{MP} is the PM emission (kg_{MP}/kg_{comb}), M_{NO_x} is the nitrogen oxide emission (kg_{NO_x}/kg_{comb}), $MM_{biogás}$ is the biogas molar mass (kmol/kg).

According to Villela,[34] CO_2 emissions can be calculated using Eq. (4.46):

$$M_{CO_2} = \frac{(w \times 44)_{CO_2}}{MM_{fuel}} \quad (4.46)$$

According to Carvalho and Lacava,[48] NO_x and PM emissions can be calculated using Eqs. (4.47) and (4.48):

$$M_{NO_x} = \frac{(2120.0.0224)/10^6}{MM_{fuel} \times 10^{-3}} \quad (4.47)$$

$$M_{MP} = \frac{(224.0.0224)/10^6}{MM_{fuel} \times 10^{-3}} \quad (4.48)$$

For calculating CO_2, NO_x, and PM emissions, the fuel molar mass was used.

- *Pollutant indicator*

Pollutant indicator is the ratio between equivalent carbon dioxide and the fuel heating value, which calculates environmental impact through Eq. (4.49).[13,34,49,50]

$$\Pi_g = \frac{(CO_2)_e}{PCI_{b.g.u}} \tag{4.49}$$

$(CO_2)_e$ is the equivalent carbon dioxide (kg/kg$_{comb}$), $LHV_{b.g.u}$ is the lower heating value of wet bagasse (kJ/kg), Π_g is the pollutant indicator (kg/MJ).

- *Ecological efficiency*

  The ecological efficiency allows for the assessment of the environmental impact of gaseous emissions from thermochemical processes by considering the system pollutant indicator and energy efficiency, which can be calculated using Eq. (4.50).[17]

$$\varepsilon = \left[\frac{0.204 \times \eta_{sistema} \times \ln(135 - \Pi_g)}{\eta_{sistema} + \Pi_g}\right]^{0.5} \tag{4.50}$$

where ε is the efficiency of electrolysis or steam reforming process (−), $\eta_{sistema}$ is the energy efficiency of hydrogen production process (−).

4.7.1 Analysis of the steam reforming of ethanol incorporation into the plant for hydrogen production

According to Silva,[17] the pollutants emissions caused by the incorporation of the steam reforming process of ethanol are presented Table 4.11.

TABLE 4.11 Emissions of CO_2, NO_x, SO_2, and PM from bagasse combustion.

Compounds	Bagasse combustion
$NO_x(kg_{NO_x}/kg_{comb})$	0.0012
$SO_2(kg_{SO_2}/kg_{comb})$	–
$MP(kg_{MP}/kg_{comb})$	0.0071
$CO_2(kg_{CO_2}/kg_{comb})$	1.8239

PM, Particulate matter.
Source: Adapted from Silva ME. *Experimental analysis ethanol steam reforming: technical, economic and ecological aspects* [Ph.D. dissertation]. Guaratingueta, SP, Brazil: Submitted to Sao Paulo State University; 2010 (in Portuguese) with copyright permission from Faculty of Engineering, Campus Guaratinguetá, UNESP, Brazil.

4.7.1.1 Results of the environmental analysis of incorporating the steam reforming of ethanol for hydrogen production into the plant

The results for the equivalent carbon dioxide and cogeneration pollutant indicator of incorporating the process of the steam reforming of ethanol are shown in Table 4.12.

The data obtained for the ecological efficiency analysis of the plant for cogeneration by incorporating the steam reforming of ethanol are presented in Table 4.13.

The greater the hydrogen production is the better the ecological efficiency of the cogeneration system becomes. As ecological efficiency is calculated through the overall efficiency of the cogeneration system, this increase is normal due to the fact that once hydrogen production is increased, there is a smaller quantity of bagasse combustion in the cogeneration system since the plant process is the same.

TABLE 4.12 Result of equivalent carbon dioxide and pollutant indicator.

	Result
Equivalent carbon dioxide $[(CO_2)_e]$ (kg_{CO_2}/kg_{comb})	2.8953
Pollutant indicator (Π_g) (kg/MJ)	0.3955

TABLE 4.13 Ecological efficiency of the conventional cogeneration plant with hydrogen production incorporation.

H_2 production (N m^3/h)	Ecological efficiency (%)
0	84.1
300	84.1
600	84.1
900	84.2
1200	84.2
1500	84.3
2000	84.4
3000	84.5
4000	84.7
5125	84.8

TABLE 4.14 Emissions of CO_2, NO_x, SO_2 and PM of bagasse combustion.

Compounds	Bagasse combustion
$NO_x(kg_{NO_x}/kg_{comb})$	0.000608
$SO_2(kg_{SO_2}/kg_{comb})$	–
MP (kg_{MP}/kg_{comb})	0.000064
CO_2 (kg_{CO_2}/kg_{comb})	0.563853

PM, Particulate matter.
Source: Adapted from Silva ME. Experimental analysis ethanol steam reforming: technical, economic and ecological aspects [Ph.D. dissertation]. Guaratingueta, SP, Brazil: Submitted to Sao Paulo State University; 2010 (in Portuguese) with copyright permission from Faculty of Engineering, Campus Guaratinguetá, UNESP, Brazil.

4.7.2 Environmental analysis of incorporating the gasification process into the plant

Eq. (4.xvi) presents the chemical composition of the exhaust gas from the combustion chamber.[24]

$$0.638H_2 + 0.6408CO + 0.34CO_2 + 0.374H_2O + 0.019CH_4 + 1.315N_2 + 1.4(2.5491N_2 + 0.6774O_2) \rightarrow w\,CO_2 + 4.8795N_2 + 1.049H_2O + 0.266O_2 \tag{4.xvi}$$

Eq. (4.xvi) produces one mole of CO_2 (w) and 40% excessive normalized air is considered.[24]

4.7.2.1 Results of the environmental analysis of incorporating the gasification process into the plant

According to Carvalho and Lacava,[48] the percentage of sulfur is low, which leads to a reduced emission factor. Therefore sulfur emission was considered null in the present work. Table 4.14 shows CO_2, PM, SO_2, and NO_x emissions.

Table 4.15 shows the results of the equivalent carbon dioxide and pollutant indicator of the cogeneration plant with the incorporation of the gasification process.

4.8 Discussion

The incorporation of the steam reforming process for biohydrogen production and of gasification for electricity production alter electricity generation, overall cogeneration, and ecological efficiencies, as well as energy

TABLE 4.15 Results of equivalent carbon dioxide, pollutant indicator, and ecological efficiency.

	Result
Equivalent carbon dioxide $[(CO_2)_e]$ (kg_{CO_2}/kg_{comb})	0.5986
Pollutant indicator (Π_g) (kg/MJ)	0.0818
Ecological efficiency (ε) (%)	93.2

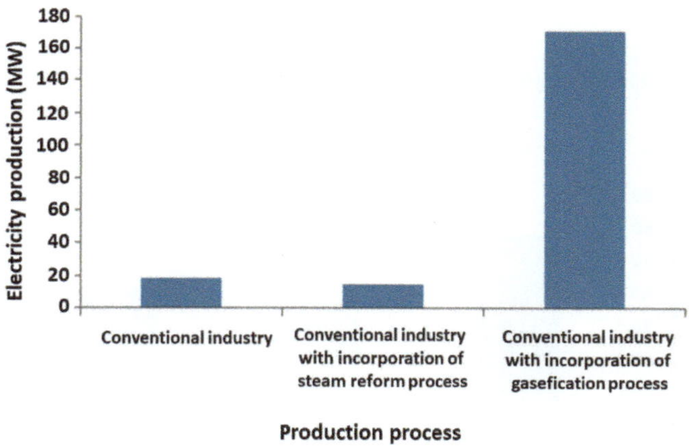

FIGURE 4.12 Electricity production using different processes.

generation. Figs. 4.12–4.15 show comparisons between the studied aspects of a conventional plant when incorporating the steam reforming and gasification processes. Optimal hydrogen production (5125 N m^3/h) was considered for plotting the graphs.

Fig. 4.13 shows that the incorporation of the gasification process achieves the aim of increasing electricity production, from which the surplus can be sold to agencies, thus increasing the industry's revenues.

While electricity production is greater through the incorporation of the gasification process, Fig. 4.14 demonstrates that it also has the highest electricity generation efficiency. On the other hand, the incorporation of the steam reforming process lowers electricity generation on account of the need to reduce energy production so that there is enough bagasse to be used for hydrogen production.

The overall efficiency of cogeneration is increased by incorporating the steam reforming process for electricity and hydrogen production using the same amount of fuel that the conventional plant uses to produce just

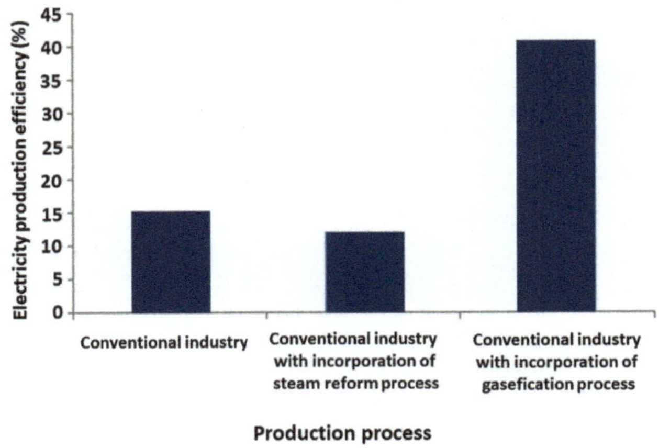

FIGURE 4.13 Electricity generation efficiency using different processes.

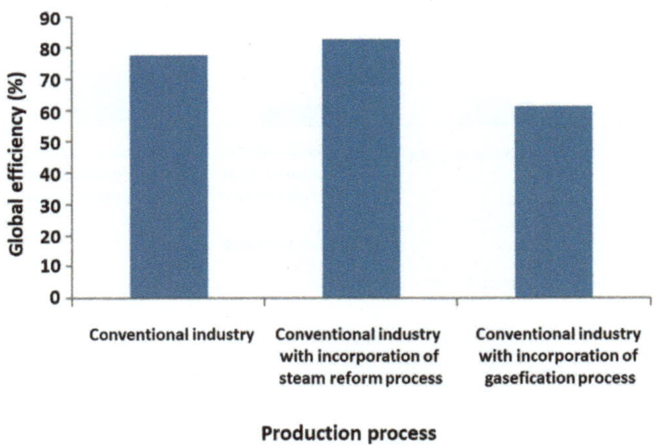

FIGURE 4.14 Cogeneration plant overall efficiency using different processes.

electricity. Furthermore, the overall efficiency of cogeneration is lower by incorporating the gasification process due to it requiring an additional amount of fuel to meet the needs of the gas turbine set (Fig. 4.15).

The sugarcane industry has achieved the highest ecological efficiency by incorporating the process of the steam reforming of ethanol, due to it being as pollutive as conventional methods, but generating two products. Moreover, ecological efficiency is lower with the incorporation of the gasification process on account of the increased emissions of equivalent carbon dioxide from additional natural gas combustion.

In economic terms, both the steam reforming of ethanol and the gasification process are feasible, since the cost of producing hydrogen is relatively

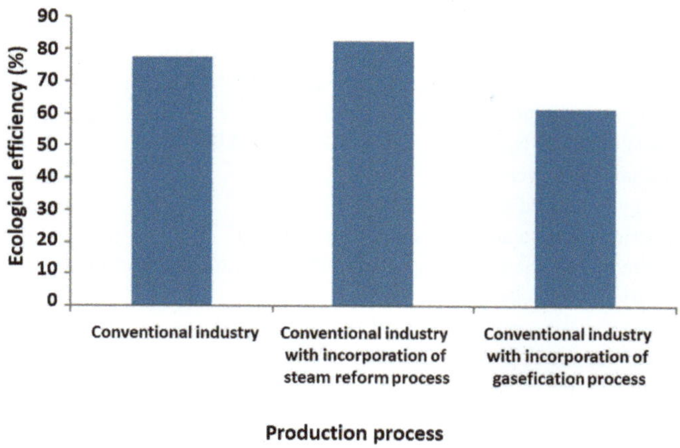

FIGURE 4.15 Ecological efficiency using different processes.

low. Furthermore, it is possible to obtain a return on investment within 2 years by selling electricity at US$0.070/kW h.

4.9 Conclusion

The incorporation of technologies for the steam reforming of ethanol and bagasse gasification have added technical and economic possibilities and new products to the plant under study, that is, the production of hydrogen and the increase of electric energy supply without competition with the production of the original products, ethanol and sugar. From an ecological point of view, the gasification process is more feasible. Therefore the incorporation of these new processes allows for improvements in the profitability and operational flexibility of sugar and ethanol plants, corroborating the improvement of Brazil's position in the bioenergy ranking.

Nomenclature

C_{EtOH}	price of ethanol (US$/kW h]
C_{H_2}	cost of biohydrogen production (US$/kW h)
$C_{M_{fuel}}$	fuel price in the combustion used in combustion chamber (US$/kW h)
$C_{M_{GASEIF.}}$	maintenance cost of gasifier (US$/kW h)
CM_{GT}	maintenance cost of gas turbine (US$/kW h)
CM_{HRSG}	cost of HRSG maintenance (US$/kW h)
CM_{MT}	maintenance cost of turbomachinery (US$/kW h)
$C_{M_{REF.}}$	maintenance cost of reformer (US$/kW h)
CM_{ST}	maintenance cost of steam turbine (US$/kW h)
C_O	personnel cost (US$/kW h)
$C_{O_{GASEIF..}}$	operational cost of gasifier (US$/kW h)
C_{syngas}	syngas production cost (US$/kW h)

$C_{O_{REF.}}$	operational cost of steam reformer (US$/kW h)
E_{EtOH}	power supplied by ethanol ($\dot{m}_{EtOH} \times LHV_{EtOH}$) (kW)
E_{fuel}	power supplied by fuel ($\dot{m}_{fuel} \times LHV_{fuel}$) (kW)
$E_{fuel_{total}}$	total power supplied by the gas turbine (kW)
E_{H_2}	power supplied by hydrogen ($\dot{m}_{H_2} \times LHV_{H_2}$) (kW)
E_M	mechanical power (kW)
E_{NG}	energy contained in natural gas (kW)
$E_{pST_{A_{C.R.}}}$	electric power steam turbine (turbine A) (kW)
$E_{pST_{A_{min.}}}$	minimum electric power capacity of the steam turbine (turbine A) (kW)
E_{pGT}	electric power—gas turbine (kW)
E_{pST}	electric power—steam turbine (kW)
E_{syngas}	power supplied by syngas burning (kW)
f	annuity factor (1/year)
FA	air/biomass ratio (−)
FP_{E_M}	poderation factor of mechanical power production (−)
$FP_{E_{pGT}}$	ponderation factor of electric power production—gas turbine (−)
$FP_{E_{pST}}$	ponderation factor of electric power production—steam turbine (−)
FP_{H_p}	ponderation factor of useful heat production (−)
h_4	enthalpy at point 4 (kJ/kg)
h_5	enthalpy at point 5 (kJ/kg)
h_6	enthalpy at point 6 (kJ/kg)
h_7	enthalpy at point 7 (kJ/kg)
H	equivalent period of utilization (h/year)
H_p	useful heat in the process (kW)
$I_{GASEIF.}$	gasifier investment cost (US$)
I_{GT}	gas turbine investment cost (US$)
I_{HRSG}	heat recovery steam generator (HRSG) investment cost (US$)
I_{MT}	mechanical turbines investment cost (US$)
$I_{REF.}$	ethanol steam reformer investment cost (US$)
I_{ST}	steam turbine investment cost (US$)
$LHV_{b.u}$	lower heating value of wet bagasse (kJ/kg)
LHV_{fuel}	lower heating value of fuel (blend of natural gas and syngas) (kJ/kg)
LHV_{NG}	lower heating value of natural gas (kJ/kg)
m_{H_2}	hydrogen production produced (N m^3/h)
$\dot{m}_b$	available bagasse mass flow (kg/s)
$\dot{m}_{4C.R.}$	steam mass flow at point 4 (kg/s)
$m_{4min.}$	steam mass flow at point 4 (kg/s)
$\dot{m}_5$	steam mass flow at point 5 (kg/s)
$\dot{m}_6$	steam mass flow at point 6 (kg/s)
$\dot{m}_{7C.R.}$	steam mass flow at point 7 (kg/s)
$\dot{m}_{7mín.}$	steam mass flow at point 7 (kg/s)
$\dot{m}_{air}$	air mass flow (point 48) (kg/s)
$\dot{m}_{ashes}$	mass flow rate of ashes (point 49) (kg/s)
$\dot{m}_{d.b.}$	dry bagasse mass flow (point 46) (kg/s)
$\dot{m}_{EG}$	exhaust gases mass flow (kg/s)
$\dot{m}_{fuel}$	fuel mass flow (kg/s)
$\dot{m}_{NG}$	natural gas mass flow required for additional burning (kg/s)

$\dot{m}_s$	steam generator outlet mass flow (kg/s)
$\dot{m}_{syngas}$	syngas mass flow rate (point 50) (kg/s)
M_{CO_2}	carbon dioxide emission (kg_{CO_2}/kg_{comb})
M_{NO_x}	nitrogen oxide emission (kg_{NO_x}/kg_{comb})
M_{MP}	particulate material emission (kg_{MP}/kg_{comb})
MM_{fuel}	fuel masss molar (kmol/kg)
P_{57}	exhaust gas pressure at HRSG inlet (kPa)
P_{58}	exhaust gas pressure at HRSG outlet (kPa)
P_e	compressor inlet air pressure (kPa)
P_s	compressor outlet air pressure (kPa)
T_{55}	combustion chamber inlet air (gases) temperature (point 55) (K)
T_{56}	combustion chamber outlet air temperature (point 56) (K)
T_e	compressor inlet air temperature (K)
T_s	compressor outlet air temperature (K)
V_A	volume of air used in gasifier (N m^3/kg)
$\dot{W}_{GT}$	electric power of gas turbine (kW)
$\dot{W}_P$	electric power supplied to pump (kW)
$\dot{W}_{ST_{A_{C.R.}}}$	steam turbine mechanical power generation (turbine A) (kW)
$W_{ST_{A_{min.}}}$	steam turbine power generation (generator A) (kW)
δ_{air}	air density (kg/m^3)
ε	ecological efficiency of ethanol steam reforming or gasification process (−)
λ_{air}	ratio specific heat at constant pressure and specific heat at constant of air (kJ/kg K).
λ_{syngas}	ratio specific heat at constant pressure/specific heat at constant of syngas (kJ/kg K)
Π_g	pollutant indicator (kg_{CO_2}/MJ)
η_{c_I}	steam generator efficiency (%)
η_e	electrical efficiency of conventional cogeneration system (%)
η_g	overall efficiency of conventional cogeneration system (%)
$\eta_{sistema}$	energetic efficiency of biohydrogen production process (−)
$\eta_{T_{GT}}$	gas turbine turbine efficiency (−)
Δh_c	enthalpy difference of water in the steam generator (kJ/kg)
ΔP_{HRSG}	HRSG pressure loss (−)
$(CO_2)_e$	equivalent carbon dioxide (kg_{CO_2}/kg_{comb})

References

1. Silveira JL, Tuna CE, Pedroso DT, da Silva ME, Machin EB, Braga LB, et al. Technological advancements in biohydrogen production and bagasse gasification process in the sugarcane industry with regard to Brazilian conditions. In: Da Silva SS, Chandel AK, editors. *Biofuels in Brazil*. Cham: Springer; 2014. p. 393–411.
2. Paulino RFS, Silveira JL, Vane LF. Technical and economic analysis of the bagasse gasification process applied in sugarcane industry, In: *23rd ABCM international congress of mechanical engineering*; 2015.
3. Paulino RFS. *Incorporation of new technologies in the sugarcane industry: hydrogen production and increase of electricity supply*. Guaratinguetá, SP, Brazil: GraduateWork, Sao Paulo State University; 2014. p. 1–92 (in Portuguese).

4. Anzelmo B, Wilcox J, Liguori S. Natural gas steam reforming reaction at low temperature and pressure conditions for hydrogen production via Pd/PSS membrane reactor. *J Membrane Sci* 2017;**522**:343–50.
5. Alhamdani YA, Hassim MH, Ng RTL, Hurme M. The estimation of fugitive gas emissions from hydrogen production by natural gas steam reforming. *Int J Hydrogen Energy* 2017;**42**(14):9342–51.
6. Braga LB, Silveira JL, Silva ME, Tuna CE, Machin EB, Pedroso DT. Hydrogen production by biogas steam reforming: a technical, economic and ecological analysis. *Renew Sustain Energy Rev* 2013;**28**:166–73.
7. Mariño F, Boveri M, Baronetti G, Laborde M. Hydrogen production from steam reforming of bioethanol using Cu/Ni/K/γ-Al_2O_3 catalysts. Effect of Ni. *Int J Hydrogen Energy* 2001;**26**(7):665–8.
8. Sharma YC, Kumara A, Prasad R, Upadhyay SN. Ethanol steam reforming for hydrogen production: latest and effective catalyst modification strategies to minimize carbonaceous deactivation. *Renew Sustain Energy Rev* 2017;**74**:89–103.
9. Zanchet D, Santos JBO, Damyanova S, Gallo JMR, Bueno JMC. Toward understanding metal-catalyzed ethanol reforming. *ACS Catal* 2015;**5**:3841–63.
10. Contreras JL, Salmones J, Colín-Luna JA, Nuño L, Quintana B, Córdova I, et al. Calatysts for H_2 production using the ethanol steam reforming (a review). *Int J Hydrogen Energy* 2014;**39**(33):18835–53.
11. Sun J, Wang Y. Recent advances in catalytic conversion of ethanol to chemicals. *ACS Catal* 2014;**4**:1078–90.
12. Hwang JJ. Sustainability study of hydrogen pathways for fuel cell vehicle applications. *Renew Sustain Energy Rev* 2013;**19**:220–9.
13. Silveira JL, Martinelli VJ, Vane LF, Freire Junior JC, Zanzi Vigouroux RA, Tuna CE, et al. Incorporation of hydrogen production process in a sugar cane industry: steam reforming of ethanol. *Appl Therm Eng* 2014;**71**(1):94–103.
14. Sikarwar SV, Ming Z, Fennell PS, Shah N, Anthony EJ. Progress in biofuel production from gasification. *Prog Energy Combust* 2017;**61**:189–248.
15. Passolongo R. *Thermodynamic, thermoeconomic and economic product line of the integration of biomass gasification systems in a sugarcane industry* [Ph.D. dissertation]. Ilha Solteira: Sao Paulo State University; 2011 (in Portuguese).
16. Sansaniwal SK, Pal K, Rosen MA, Tyagi SK. Recent advances in the development of biomass gasification technology: a comprehensive review. *Renew Sustain Energy Rev* 2017;**72**:363–84.
17. Silva ME. *Experimental analysis ethanol steam reforming: technical, economic and ecological aspects* [Ph.D. dissertation]. Guaratingueta, SP, Brazil: Submitted to Sao Paulo State University; 2010 (in Portuguese).
18. Iulianelli A, Palma V, Bagnato G, Ruocco C, Huang Y, Nejat T, et al. From bioethanol exploitation to high grade hydrogen generation: steam reforming promoted by a Co-Pt catalyst in a Pd-based membrane reactor. *Renew Energy* 2018;**119**:834–43.
19. Parraga J, Khalilpour KR, Vassallo A. Polygeneration with biomass-integrated gasification combined cycle process: review and prospective. *Renew Sustain Energy Rev* 2018;**92**:219–34.
20. Perkins G. Underground coal gasification—Part I: Field demonstrations and process performance. *Prog Energy Combust* 2018;**67**:158–87.
21. Coronado CR, Yoshioka JT, Silveira JL. Electricity, hot-water and cold-water production from biomass. Energetic and economical analysis of the compact system of cogeneration run with wood gas from a small downdraft gasifier. *Renew Energy* 2011;**36**(6):1861–8.

22. Kaushal P, Tyagi R. Advanced simulation of biomass gasification in a fluidized bed reactor using ASPEN PLUS. *Renew Energy* 2017;**101**:629–36.
23. Warnecke R. Gasification of biomass: comparison of fixed bed and fluidized bed gasifier. *Biomass Bioenergy* 2000;**18**:489–97.
24. Ku X, Jin H, Lin J. Comparison of gasification performances between raw and torrefied biomasses in an air-blown fluidized-bed gasifier. *Chem Eng Sci* 2017;**168**:235–49.
25. Filho PTD, Silveira JL, Tuna CE, Lamas WQ. Energetic, ecologic and fluid-dynamic analysis of a fluidized bed gasifier operating with sugar cane bagasse. *Appl Therm Eng* 2013;**57**(1-2):116–24.
26. Pedroso DT, Machin EB, Pérez NP, Braga LB, Silveira JL. Technical assessment of the Biomass Integrated Gasification/Gas Turbine Combined Cycle (BIG/GTCC) incorporation in the sugarcane industry. *Renew Energy* 2017;**114**:464–79.
27. Nicodème T, Berchem T, Jacquet N, Richel A. Thermochemical conversion of sugar industry by-products to biofuels. *Renew Sustain Energy Rev* 2018;**88**:151–9.
28. Fiomari MC. *Energetic and exergetic analyses of sugarcane industry of Paulista West with expansion of the cogeneration system* [Ph.D. dissertation]. Ilha Solteira, Brazil: Submitted to Sao Paulo State University; 2004 (in Portuguese).
29. Olivares E. *Projeto, Construção e Avaliação Preliminar de um Reator de Leito Fluidizado para Gaseificação de Bagaço de Cana-de-açúcar* [Ph.D. dissertation]. Campinas, Brazil: University of Campinas; 1996 (in Portuguese).
30. Graebin GO. *Dryng bagasse from sugar cane and study of its application in removing heavy metal* [Ph.D. dissertation]. Brazil: Universidade Estadual do Oeste do Paraná; 2014 (in Portuguese).
31. Pérez NP, Machin EB, Pedroso DT, Antunes JS, Silveira JL. Fluid-dynamic assessment of sugarcane bagasse to use as feedstock in bubbling fluidized bed gasifiers. *Appl Therm Eng* 2014;**73**(1):238–44.
32. Manatura K, Lu JH, Wu KT, Hsu HT. Exergy analysis on torrefied rice husk pellet in fluidized bed gasification. *Appl Therm Eng* 2017;**111**:1016–24.
33. Filho PTD. *Increase in the supply of electricity in the sugar and alcohol sector: incorporation of combined cycle associated to fluidized bed gasifier* [Ph.D. dissertation]. Guaratinguetá, Brazil: Sao Paulo State University; 2014 (in Portuguese).
34. Villela IAC, Silveira JL. *Thermoeconomic model considering the environment impacts on thermoelectric power plants: natural gas versus diesel. J Energy Resour ASME*, 2. 2008. p. 945–51.
35. Silveira JL, Tuna CE. Thermoeconomic analysis method for optimization of combined heat and power system. Part II. *Prog Energy Combust* 2004;**30**(6):673–8.
36. Silveira JL. *Sustainable hydrogen production processes*. Cham: Springer; 2017. p. 185.
37. Silveira JL. *A contribution for thermoeconomic modelling: optimization of operating and energy systems design* ["Full Professor" dissertation]. Guaratinguetá, Brazil: Sao Paulo State University; 1998 (in Portuguese).
38. Souza ACC. *Technical and economic analysis of an ethanol reforming for hydrogen production* [Ph.D. dissertation]. Brazil: Sao Paulo State University; 2005 (in Portuguese).
39. Kothari R, Buddhi D, Sawhney RL. Comparison of environmental and economic aspects of various hydrogen production methods. *Renew Sustain Energy Rev* 2008;**12**:553–63.
40. Sales CAVB. *Techno economic assessment of conventional and advanced technological routes for electricity production through biomass gasification* [Ph.D. dissertation]. Brazil: Universidade Federal de Itajubá, UNIFEI; 2006 (in Portuguese).
41. Farmer R., Gas Turbine World. *Pequot. Gas turbine world handbook*, vol. 29. 2012. 212.

42. COMGAS. Piped Natural Gas Rates. <http://www.comgas.com.br/pt/nossosServicos/Tarifas/Paginas/industrial.aspx> [accessed 20.12.16].
43. Costa AAS. *Characterization of an aeronautical internal combustion engine using gasoline ethanol blends as fuel: fluid dynamics, thermodynamic and ecological aspects* [Ph.D. thesis]. Brazil: Sao Paulo State University; 2011 (in Portuguese).
44. Silveira JL, Carvalho-Junior JA, Villela IAC. Combined cycle versus one thousand diesel power plants: pollutant emissions, ecological efficiency and economic analysis. *Renew Sustain Energy Rev* 2007;**11**:524–35.
45. Villela IAC, Silveira JL. Ecological efficiency in thermoelectric power plants. *Appl Therm Eng* 2007;**27**(5-6):840–7.
46. Coronado-Rodriguez CJ, Leal EM, Villela IAC, Silveira JL. Ecological efficiency in CHP: biodiesel case. *Appl Therm Eng* 2009;**30**(5):458–63.
47. Silveira JL, Lamas WQ, Tuna CE, Villela IAC, Miro LS. Ecological efficiency and thermoeconomic analysis of a cogeneration system at a hospital. *Renew Sustain Energy Rev* 2012;**16**(1):2894–906.
48. Carvalho JR, Lacava PT. *Emissões em Processos de Combustão*. São Paulo, Brasil: Editora UNESP; 2003. p. 135 p.
49. Cardu M, Baica M. Regarding a new variant methodology to estimate globally the ecologic impact of thermopower plants. *Energy Convers Manage* 1999;**40**(14):1569–75.
50. Cardu M, Baica M. Regarding a global methodology to estimative the energy ecologic efficiency of thermopower plants. *Energy Convers Manage* 1999;**40**(1):71–87.

Chapter 5

Production of biojet fuels from biomass

Xuebing Zhao[1], Xiaoying Sun[2], Xingkai Cui[1] and Dehua Liu[1]
[1]*Key Laboratory for Industrial Biocatalysis, Department of Chemical Engineering, Ministry of Education of China, Institute of Applied Chemistry, Tsinghua University, Beijing, P.R. China,* [2]*Institute of New Energy Technology, Academy of Military Science of Chinese, Beijing, P.R. China*

5.1 Introduction

Conventional aviation fuel, which is a specialized type of petroleum-based fuel used to power aircrafts primarily includes two categories: aviation gasoline that is suitable for piston engines and aviation kerosene that is a jet fuel for turbine engines, and there are international specifications for each. According to data provided by the US Energy Information Administration (EIA) (https://www.eia.gov/energyexplained/index.php?page = oil_refining), 4 gallons out of every 42-gallon barrel of crude oil is converted into jet fuel. The worldwide aviation industry consumes approximately 1.5–1.7 billion barrels of conventional jet fuel per year, and global airline operations consumed approximately 1.5 billion barrels of Jet A-1 fuel in 2013, producing 705 million metric tons of CO_2.[1] With the increasing number of passengers carried by airplanes, the demand for aviation fuels is also increasing rapidly. It is estimated that the global total fuel consumption of commercial airlines will be over 100 billion gallons in 2020 (https://www.statista.com). However, as there are great concerns about environmental pollution and global warming caused by the large-scale combustion of fossil fuels, biofuels produced from renewable biomass have attracted great interest in recent decades. Biojet fuel, as one of the most important biofuel candidates that can be produced from various biomass feedstocks, has also definitely been considered with a potential for large-scale application.

The aviation transport sector requires fuels with high energy density and so it depends mainly on liquid hydrocarbon fuels. Therefore alternative aviation fuels must possess some specific qualities, such as good cold flow

Sustainable Bioenergy. DOI: https://doi.org/10.1016/B978-0-12-817654-2.00005-8

properties, thermal stability, and a low freezing point.[2] Some people get confused between biodiesel and biojet fuel by thinking that they are same thing. The preparation process of traditional biodiesel is mainly achieved by the transesterification of oils and fats with short-chain alcohols in catalytic or supercritical systems to obtain fatty acid alkyl (methyl, ethyl, or propyl) esters, namely biodiesel.[3] Actually, various kinds of biodiesel existing in aviation biofuels will increase the freezing point, which will result in fuel solidification during navigation. In addition, biodiesel will affect the stability of aviation biofuels. Generally the shelf life of aviation fuel is only 6 months if biodiesel is available, and the viscosity of stored jet fuel will increase causing turbidity and precipitation when biodiesel exists.[4] Therefore aircraft engines require higher quality biofuels, especially in terms of low-temperature performance, for example, the ice point should not be higher than −47°C. Strict standards have been made for aviation fuels. The international standard, American Society for Testing Materials (ASTM) D7566, established in 2009, specifies the technical requirements for synthetic jet fuels, including biojet fuels and blends with petroleum kerosene. Table 5.1 lists the ASTM standards of conventional jet fuel Jet A and JP-8, and a comparison of fuel properties with some representative alternative jet fuels.[5] Only those fuels meeting the standard are considered to be equivalent to conventional jet fuels and then can be blended into aircraft and supply infrastructure without the need for separate tracking or approval.[6] Conventional petroleum-derived jet fuel is a mixture of hundreds of hydrocarbons, including *n*-paraffins, *iso*-paraffins, *cyclo*-paraffins, and aromatics. The oxygen content of the fuel is required to be very low for the safety and stability of the jet fuel.

Many airline companies have shown great interest in using biojet fuels, due to the frequent cost fluctuations of fossil-based jet fuels and the possible levying of aviation carbon tax in the future. Moreover, the abundance of biomass feedstock for the production of biojet fuels makes it promising for obtaining a sustainable supply of fuels. Since 2008, many airlines have been trying to use aviation biofuels for trial flights as shown in Table 5.2.[7] In 2008, Virgin Atlantic carried out the world's first flight of a commercial aircraft powered with biofuels in an effort to show that it can produce less CO_2 than normal jet fuels. The flight duration was 1.5 hours and the aero engines used blended fuels consisting of common fuels and aviation biofuels (ABC NEWAs, Feb 25, 2008). In 2011, American Boeing and Honeywell UOP carried out a pilot test flight with a 50%/50% blend of aviation biofuels and conventional jet fuel in Beijing, China. All of these demonstration flights (Table 5.2) have indicated that aviation biofuels fully meet the performance requirements of large aircraft in terms of flight height, acceleration, and engine restart. In some cases, the fuel was even praised for having a low flash point and sufficiently low freezing point.

TABLE 5.1 American Society for Testing Materials (ASTM) standard and properties of conventional and alternative jet fuels.[5]

Property	ASTM Standard	Jet A	JP-8	S-8	Shell GTL	Sasol IPK	R-8	CHRJ	THRJ
POSF number	–	4658	3773	4734	5172	5642	5469	6152	6308
Composition									
n-Paraffins (wt.%)	–	28	19	17.7	53.3	2.1	13.1	10.2	8.8
Iso-Paraffins (wt.%)	–	29	38.2	82	45.7	88	–	78.9	89.4
Cyclo-paraffins (wt.%)	–	20	24.1	<0.4	<1.0	9	–	11.1	2.1
Aromatics (wt.%)	Report	20	13.5	<0.1	<0.2	<0.5	–	<0.2	<0.4
Total sulfur (wt.%)	Max 0.3	–	0.0064	<0.001	<0.001	<0.001	<0.001	<0.0018	<0.003
Distillation									
Initial boiling point (°C)	Report	158	152	182	146	149	158	151	165
10% recovered (°C)	Max 205	184	173	195	162	166	175	161	179
20% recovered (°C)	Report	192	179	–	162	170	185	166	185
50% recovered (°C)	Report	213	198	228	169	180	215	182	210
90% recovered (°C)	Report	248	239	–	184	208	260	237	243
Final boiling point (°C)	Max. 300	269	260	280	198	228	274	259	255
Flash point (°C)	Min. 38	47	48	49	44	44	48	43	55
Freezing point (°C)	Max. −47	−49	−49	−59	−54	<−78	−49	<−77	−62
Density at 15°C (kg/m^3)	775–840	806	790	757	737	762	762	751	758
Viscosity at −20°C (cSt)	Max 8.0	5.2	4.1	4.6	2.6	3.6	5.5	3.3	5.3
Neat heat of combustion (MJ/kg)	Min 42.8	42.8	43	44.1	44.2	44	44.1	44.3	44.1
Smoke point (mm)	Min 19.0	21	25	>43	40	>40	>40	50	>40
H/C molar ratio	–	1.957	1.937	2.152	2.202	2.119	2.152	2.169	2.152
Molecular weight (g/mol)	–	142	153	168	146	156	177	160	174

Note: *CHRJ*, camelina-based hydroprocessed renewable jet fuel; *S-8*, Syntroleum S-8 synthetic jet fuel from gas by FT synthesis; *Sasol IPK*, *iso*-paraffinic kerosene; *Shell GTL*, shell gas-to-liquid jet fuel; *THRJ*, tallow-based hydroprocessed renewable Jet fuel.

TABLE 5.2 Some reported demonstration flights powered with aviation biofuels.[7]

Date	Airline Company	Aircraft	Biofuels produced from	Notes
Oct 2007	GreenFlight International	Aero L-29 Delfin	Waste vegetable oil	This was the very first flight of an aircraft powered entirely by 100% biofuel
Feb 2008	Virgin Atlantic	Boeing 747	Coconut and babassu	Virgin flew a biofuel test flight between London and Amsterdam using a 20% blend of biofuels in one of its engines
Jan 2009	Continental Airlines	Boeing 737	Algae and jatropha	Continental Airlines ran the first flight of an algae-fueled jet. The pilots on board executed a series of tests at 38,000 ft. (12,000 m), including a mid-flight engine shutdown
Apr 2010	US Navy	F/A-18	Camelina	The navy tested this biofuel blend on the F/A-18 Super Hornet aka "Green Hornet." Results from those tests indicated the aircraft performed as expected through its full flight envelope with no degradation of capability
Jun 2011	Boeing	Boeing 747-8F	Camelina	Boeing flew its new model 747-8F to the Paris Air Show with all four engines burning a 15% mix of biofuel from camelina
Aug 2011	US Navy	T-45	Camelina	Successfully flew a T-45 training aircraft using a biofuel mixture of 50/50 petroleum-based JP-5 jet fuel and plant-based camelina
Sep 2011	US Navy	AV-8B	Camelina	The first biofuel flight test in an AV-8B Harrier from Air Test and Evaluation Squadron 31
Oct 2011	Air China	Boeing 747-400	Jatropha	Air China flew China's first flight using aviation biofuels. The flight was conducted using Chinese grown jatropha oil from PetroChina. The flight was 2 h in duration above Beijing, and used 50% biofuel in one engine

(Continued)

TABLE 5.2 (Continued)

Date	Airline Company	Aircraft	Biofuels produced from	Notes
Apr 2012	Jeastar Airways	Airbus A320	Refined cooking oil	JQ flight 705 departed Melbourne at 09:50 and arrived in Hobart at 11:05 supplied by SkyNRG
Mar 2013	KLM	Boeing 777-206ER	Waste vegetable oil	KLM begins weekly flights by a Boeing 777-200 between John F. Kennedy Airport in New York City, USA and Amsterdam's Schiphol Airport, Netherlands using biofuel supplied by SkyNRG
Aug 2014	Gol Transportes Aéreos	Boeing 737-700	Inedible corn oil and waste vegetable oil	Gol Flight 2152 took off from Rio Santos Dumont Airport (SDU) toward Brasilia (BSB) with a 4% mix of biojet fuel
Mar 2015	Hainan Airlines	Boeing 737-800	Waste vegetable oil	Hainan Airlines conducted China's first commercial biofuel flight carrying 156 passengers from Shanghai to Beijing. The fuel, supplied by Sinopec, was a fuel blend of approximately 50% aviation biofuel mixed with conventional petroleum jet fuel
Sep 2016	KLM	Boeing 737-400	Waste vegetable oil	SkyNRG supplies the sustainable biofuel, which is made from used cooking oil by AltAir Fuels in Los Angeles
May 2017	Singapore Airlines	Airbus A350-900	HEFA (hydroprocessed esters and fatty acids)	Singapore Airlines begun a series of biofuel flights using an A350-900 aircraft on nonstop *trans*-Pacific flights between Singapore and San Francisco

5.2 Conversion technologies of biojet fuels from biomass feedstocks

There are several advantages to producing biojet fuels from biomass feedstocks. First, biomass is renewable and abundant, which can substantially reduce the net emission of CO_2. Current researches on biojet fuel production have involved many biomass feedstocks, including oil or lipid-containing biomass, such as vegetable oil, waste cooking oil, and microbial oils; lignocellulosic biomass, such as forestry and agricultural residuals; and sugar or starch. Second, the synthetic routes for obtaining biojet fuels are varied depending on the feedstock used, and thus the technical routes are flexible. Third, since biojet fuels are primarily produced from biomass feedstocks, less dependence on petroleum supplying countries can be achieved.[2] Generally, the conversion technologies for biojet fuel production can be classed into thermochemical and biochemical approaches. However, due to the complexity of biomass, especially lignocellulosic biomass, a combination of these approaches is usually used. There are generally six families of conversion technologies as summarized by Mawhood et al.[6]: hydroprocessed esters and fatty acids (HEFA); Fischer–Tropsch synthesis (FTS); direct sugars to hydrocarbons (DSHC); hydrotreated depolymerized cellulosic jet; alcohol to jet (ATJ); and aqueous phase reforming (APR), but only four pathways are at present certified by the ASTM for biojet fuel production, namely FTS, HEFA, DSHC, and ATJ.[8] Fig. 5.1 summarizes the different technical routes to produce biojet fuels from different types of biomass. Oils or lipids are the most frequently used feedstock for biojet fuel production with a relatively simple and mature technical route. Most of the biojet fuels used for powering demonstration flights were produced from oil or lipids, as shown in Table 5.2, such as camelina, jatropha, algae, and waste oils.[2] When using lignocellulosic biomass as the feedstock, the conversion processes are usually complicated with multistep processing. After pretreatment, the lignocellulose can be pyrolyzed or liquefied using the thermochemical process to obtain bio-oil, which is further upgraded to obtain liquid hydrocarbon fuels. Another pathway is to gasify the lignocellulosic biomass to obtain syngas, which is further used as a feedstock for FTS to obtain synthetic liquid

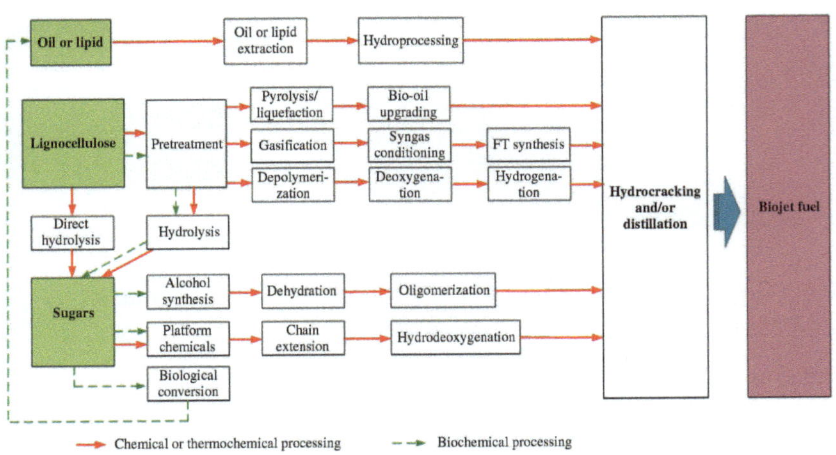

FIGURE 5.1 Possible routes for biojet fuel synthesis from different biomass feedstocks.

hydrocarbon fuels. These two routes are relatively mature with the most promising potential for commercialization. Lignocellulosic biomass can also first be converted into sugars for further conversion into biojet fuels, either by chemical or biochemical processing or with a combination of the two. For sugar conversion into biojet fuels, one of the most mature routes is to convert the sugar into alcohols, such as ethanol or butanol, followed by dehydration to form corresponding olefins, which can be oligomerized for carbon chain extension to meet the carbon chain length range for jet fuels. Sugars also can be converted chemically or biologically into platform chemicals, such as furfural or 5-hydroxymethylfurfural (HMF), followed by carbon–carbon coupling, hydrogenation, and hydrodeoxygenation to form hydrocarbon fuels.[9] Sugars can also be used as carbon sources for microbial fermentation to accumulate microbial lipids, which can be further converted into jet fuel by hydroprocessing. As can be seen from this discussion, the routes for converting biomass into biojet fuels are diverse. More details are described in here.

5.2.1 Hydrotreatment of oil-containing feedstocks to jet fuels: hydroprocessed esters and fatty acids

Oil-to-jet technology is relatively mature for commercialization, which is known as hydroprocessed esters and fatty acids (HEFA) and has been certified by the ASTM. A representative flow scheme of the process is shown in Fig. 5.2.[7]

In terms of chemistry, catalytic hydrogenation, decarboxylation (DCOX), hydrodeoxygenation (HDO), hydrodecarbonylation (HDC), and/or catalytic isomerization are involved in the reaction mechanisms with the presence of

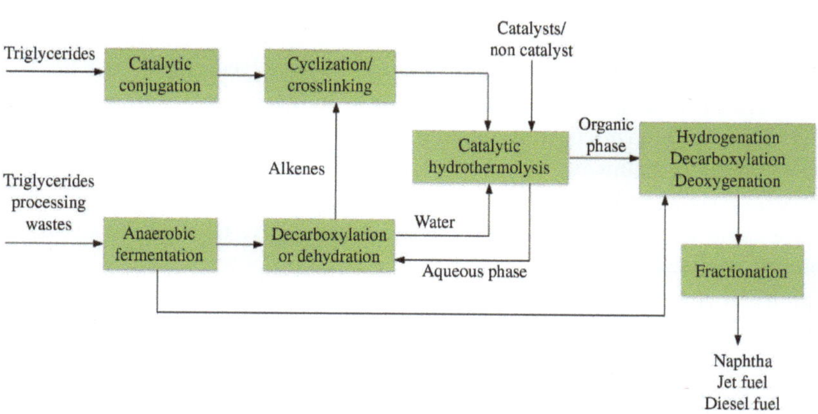

FIGURE 5.2 Hydrotreatment of oil-containing feedstocks to jet fuels. *Adapted from Wang WC, Tao L. Bio-jet fuel conversion technologies.* Renew Sust Energy *Rev 2016;53:801–22 with copyright permission from Elsevier.*

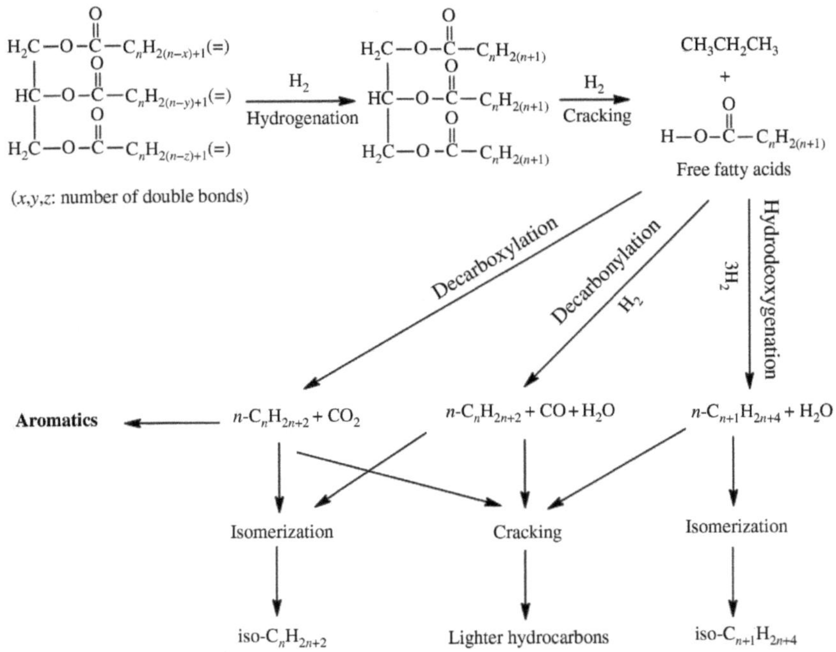

FIGURE 5.3 Reactions involved in the conversion of triglyceride feedstocks to hydrocarbon fuels under catalytic hydrotreatment. *Adapted from Veriansyah B, Han JY, Kim SK, Hong SA, Kim YJ, Lim JS, et al. Production of renewable diesel by hydroprocessing of soybean oil: effect of catalysts.* Fuel *2012;**94**(1):578–85 with copyright permission from Elsevier.*

hydrogen (Fig. 5.3).[10] The double bonds in the oil molecule (triglyceride) get saturated through the catalytic addition of hydrogen (hydrogenation) at certain temperatures and pressures. Further hydrotreatment of the saturated triglyceride under the action of catalysts results in the addition of hydrogen to the carbonyl group, causing the cleavage of the glycerol unit and the formation of propane and free fatty acid chains. In the following steps, the fatty acids can be converted into liquid hydrocarbon compounds by generally three chemical pathways, namely DCOX, HDO, and HDC. In the DCOX process, a hydrocarbon with one carbon atom less than the fatty acid chain and a mole of CO_2 is produced.[7] Stoichiometrically, no external hydrogen sources are needed for this reaction. In the HDO process, the fatty acid reacts with hydrogen to produce a hydrocarbon chain with the same number of carbon atoms and two moles of water. Stoichiometrically, three moles of H_2 are needed for one mole of fatty acid. The HDC route, in which one mole of fatty acid reacts with one mole of H_2, produces a hydrocarbon with one carbon atom less, as well as a mole of CO and water.[8] Therefore these routes remove lighter hydrocarbon species, like propane in addition to CO, CO_2, and water from oil feedstocks. They proceed with the formation of even and odd-numbered chains of paraffins, respectively. As reported by Zhang et al.[11] when studying the catalytic HDO reaction of palm oil, all the hydrogenation, DCOX, HDO, and HDC reactions are exothermic. Direct HDO released the maximum heat and HDC released the least. C15–C18 paraffins were the main products from the HDO of palm oil. However, the obtained straight chains (*n*-alkanes) need to be subjected to isomerization to convert the linear paraffins into corresponding isomers (*iso*-alkanes) in order to enhance the properties of the obtained biojet fuels.

The reaction yields of the target products vary with the oil feedstocks and catalysts used as well as the reaction conditions. As summarized by Vasquez et al.,[8] various oil feedstocks have been studied for hydrocarbon fuel production, including carinata, macauba, palm, jatropha, and coconut, at a temperature range of 300°C–600°C and a pressure range of 10–50 bar for between 2 and 5 hours with various catalysts, and the hydrocarbon fuel yields are in a wide range of 30%–100%. The catalyst used is the most important factor to achieve good degrees of hydrogenation, DCOX, HDO, and HDC. Two types of catalysts are usually employed for the catalytic hydrotreatment of oil feedstocks, including monometallic catalysts, such as Ni, Pd, Pt, Rh and bimetallic sulfide catalysts, such as $NiMoS_2$, $CoMoS_2$, and $NiWS_2$ supported on Al_2O_3.[12] Among the metal catalysts, noble metals, including Pd, Pt and Rh, are the main industrial materials for HDO and HDC reactions. Wang[13] compared different catalysts for the hydrotreatment of soybean oil to produce hydrocarbon biofuels. When commercialized ZSM-5 was used as a catalyst, though a high yield (21%) of jet fuel was obtained, significant amounts of tar and coke were formed during the reaction, while the bifunctional Ru/ZSM-5 catalyst resulted in the formation of less tar and coke. A lower yield of jet fuel (16%) was obtained compared to commercialized sulfided NiMo catalysts at a much lower pressure of 650 psi. A much higher soybean oil conversion could be obtained by bifunctional catalysts; NiMo carbide or nitride supported on ZSM-5, zeolite β, USY, γ-alumina oxide, and Al-SBA-15 under 650 psi and 360°C–450°C. γ-alumina and Al-SBA-15 supported catalysts mainly produced green diesel fuels. In particular, Al-SBA-15 supported NiMo carbide catalyst gave the highest yield of organic liquid product (96%) with the highest selectivity (97%) to hydrocarbons in the boiling range of the diesel fraction. Veriansyah et al.[10] studied the effects of various supported catalysts on the hydroprocessing of soybean oil. The results indicated that the hydroprocessing conversion followed the order, from highest to lowest: sulfided NiMo/γ-Al_2O_3 (92.9%); 4.29 wt.% Pd/γ-Al_2O_3(91.9%); sulfided CoMo/γ-Al_2O_3 (78.9%); 57.6 wt.% Ni/SiO_2–Al_2O_3 (60.8%); 4.95 wt.% Pt/γ-Al_2O_3 (50.8%); and 3.06 wt.% Ru/Al_2O_3 (39.7%) at a catalyst/oil weight ratio of 0.044 with the most abundant composition in the liquid product being straight-chain n-C_{17} and n-C_{15} alkanes. However, the CoMo catalyst showed enhanced isomerization and cracking reaction activity, thus producing lighter and more isomerized hydrocarbons. Generally, the catalytic activity of the noble metals followed the order: Pd > Pt > Ru; however, Ni and Pd based catalysts are reported as the most efficient, and the former has been proven to have good selectivity for jet fuel production depending on the acidic levels of the support. Nevertheless, carbonaceous depositions due to cracking reactions are usually the main cause for the deactivation of catalysts, especially for Ru.[14] Researchers are also trying to develop cheaper catalysts, for example, using transition metals instead of noble metals. Ni, Mo, Co, W, and their composites are the most commonly used transition metals for the hydrotreatment of oil feedstocks or bio-oils. More details on the effects of catalyst design and support materials on the efficiency of catalysts can be found in a review paper by Galadima and Muraza.[15]

The HEFAs obtained by the hydrotreatment of oil feedstocks are high energy biofuels that can be used even without blending with conventional jet-fuels. Most demonstration flights are powered by fuel mixtures of fossil jet fuel and HEFAs as shown in Table 5.2. These fuels are free of aromatics and sulfur with high cetane numbers, high thermal stability, and low tailpipe emissions, and are stable for storage and resistant to microbial growth.[16] The most widely and commercially deployed HEFA technologies are those developed by Honeywell UOP (UOP Green Fuel Process) and Neste (NexBTL).[6] UOP green fuel process technology (Ecofining Process) produces green diesel or jet fuel based on market demand and price from a range of sustainable feedstocks. It primarily consists of two sections, namely the deoxygenation section and the isomerization/cracking (or hydrocracking) section based on the hydroprocessing technology commonly used in today's refineries.[17] Neste's NexBTL technology was initially developed for producing renewable diesel based on the hydrogen treatment of vegetable oils and waste animal fat. However, by adjusting the technology, renewable aviation fuel can be produced, and the produced jet fuel is compatible with all aircraft engines currently in use and fully complies with the requirements of the ASTM D7566-11a standard.

The technologies for the hydrotreatment of oil feedstocks into jet fuels are now at a relatively mature stage with high commercialization potential. However, this process is limited by the relatively high cost and unstable supply of oil or lipid-containing feedstocks. The production cost of biojet fuel, which is majorly dependent on feedstock cost, is still relatively high, and thus in the current market environment, it may still be difficult for aviation to secure large-scale production capacity due to the existence of incentives for road transport biodiesel and the comparatively low price of kerosene.[6]

5.2.2 From lignocellulose to biojet fuel

Compared to vegetable oil feedstock, lignocellulose is cheaper, more abundant, and a nonedible biomass feedstock. Lignocellulosic biomass, thus, should be the most promising feedstock for future biofuel production without the food–fuel debate. However, because of the complex structure of lignocellulose, the conversion process is relatively complicated. Generally, six pathways have been developed for converting lignocellulose into biojet fuel, namely: (1) gasification followed by FTS; (2) pyrolysis or liquefaction followed by upgrading; (3) direct catalytic depolymerization and conversion with hydrotreatment; (4) hydrolysis to sugars and conversion into alcohol followed by ATJ; (5) hydrolysis to sugars and conversion into platform chemicals followed by carbon chain extension and HDO; and (6) hydrolysis to obtain sugars followed by further conversion into microbial lipids by biological conversion followed by the HEFA process.

5.2.2.1 Gasification of lignocellulose with Fischer–Tropsch synthesis to biojet fuel

The gasification of biomass to obtain syngas or synthesis gas is one of the most important pathways for the production of gas fuels from lignocellulose. Prior to gasification, lignocellulosic biomass must usually undergo pretreatment to make it suitable for gasification, especially for nonwoody biomass. Pretreatment processes include size reduction, drying, and densification, such as torrefaction and pelleting.[18] The gasification process is usually conducted with partial oxidation at a high temperature (generally in the range of 600°C–900°C or even higher) in the presence of a gasifying agent (air, oxygen, steam, CO_2, or mixtures) which acts as an oxidant. During gasification, due to the incomplete combustion of the biomass solid, combustible gases, like CO, H_2, and methane are produced by the reactions of water vapor and CO_2 through a glowing layer of charcoal[18,19]:

$$\text{Char oxidation: } C + 1/2O_2 = CO\left(-110.6\ \text{kJ/mol}\right)$$
$$C + O_2 = CO_2\left(-393.6\ \text{kJ/mol}\right)$$
$$\text{Boudouard reaction: } C + CO_2 = CO\left(+172\ \text{kJ/mol}\right)$$
$$\text{Steam reforming of charcoal: } C + H_2O = CO + H_2\left(+122.9\ \text{kJ/mol}\right)$$
$$\text{Methanation reaction: } C + 2H_2 = CH_4\left(+74.9\ \text{kJ/mol}\right)$$

Devolatilization gas and cracking gas species can also react with water vapor:

$$\text{Water gas shift (WGS) reaction: } CO + H_2O = CO_2 + H_2\left(-41.1\ \text{kJ/mol}\right)$$
$$\text{Steam reforming of methane: } CH_4 + H_2O = CO + 3H_2\left(+206\ \text{kJ/mol}\right)$$

Thus the reactions of the gasification of biomass ($CH_\alpha O_\beta$) can be written as[20]:

$$CH_\alpha O_\beta(\text{biomass}) + O_2(21\%\ \text{of air}) + H_2O(\text{steam}) \rightarrow$$
$$CH_4 + CO + CO_2 + H_2 + H_2O(\text{unreacted steam}) + C(\text{char}) + \text{tar}$$

The biomass characteristics, such as biomass type, moisture content, particle size, and ash content, as well as gasification parameters, including bed materials, operational parameters, gasifying agents, equivalence ratio, and steam to biomass ratios all show significant influence on the yield and quality of syngas. Generally, the biomass syngas from a bubbling fluidized bed (BFB) contains 5%–26% H_2, 13%–27% CO, 12%–40% CO_2, <18% H_2O, 2%–11% CH_4, <3% C_{2+}, <0.11% tars, ppm magnitude of H_2S, <0.2% O_2, and 13%–56% N_2.[21] Cellulose, hemicellulose, and lignin are the main carbon sources for the reactions taking place during gasification.[19] The obtained syngas mainly consists of H_2 and CO, which can be further converted by FTS into aliphatic straight-chain hydrocarbons (C_xH_y), such as the light hydrocarbons methane (CH_4), ethene (C_2H_4), and ethane (C_2H_5), LPG (C_3–C_4, propane and butane), gasoline (C_5–C_{12}), diesel fuel (C_{13}–C_{22}), and waxes (C_{23}–C_{33}).[22] These products show wide applications in transportation and industries.

FTS already has about 100 years of history since it was developed in Germany by Franz Fischer and Hans Tropsch during the 1920s.[23] FTS can be divided into high-temperature (300°C–350°C) FTS and low-temperature (200°C–240°C) FTS according to the operating conditions. Gasoline, diesel, solvent oil, and olefins can be produced by high-temperature FTS, while stable kerosene, diesel oil, lube base oil, and naphtha distillate can be produced by low-temperature FTS.[7] There are at least three kinds of FTS technologies according to the feedstock material used, including coal to oil using coal as the raw material, gas to liquid using natural gas as the raw material, and biomass to oil using biomass as the feedstock. The preparation of aviation biofuels from coal and natural gas may be a pathway to solve the shortage of petroleum-based jet fuel, whereas it still causes a net increase in CO_2 and does not make energy sustainable from a long-term goal point of view. The preparation of aviation biofuel from renewable biomass can make full use of low-quality biomass waste and convert it into clean energy that has significant environmental benefits.

During FTS, these chemical reactions are involved[23]:

The formation of straight-chain alkanes with a heat of reaction of −231 kJ/mol:

$$(2n+1)H_2 + n\,CO \rightarrow C_nH_{2n+2} + n\,H_2O$$

The formation of straight chains:

$$2n\,H_2 + n\,CO \rightarrow C_nH_{2n} + (n-1)H_2O$$

The formation of alcohols:

$$2n\,H_2 + n\,CO \rightarrow C_nH_{2n+1}OH + (n-1)H_2O$$

and

$$(n+1)H_2 + (2n-1)CO \rightarrow C_nH_{2n+1}OH + (n-1)CO_2$$

In addition, some ketones, aldehydes, or carboxylic acids may also be produced.

Liquid fuel obtained by FTS is usually called syncrude and the properties of syncrude are determined by the specifics of each FTS technology. The composition and properties of FT hydroprocessed synthesized paraffinic kerosene (FT-SPK) must meet the requirements of the ASTM standard (ASTM D 7566-11a) in order to be used as jet fuel as shown in Table 5.3. Various factors, including the catalyst type, operating conditions, gas loop design, catalyst deactivation profile, and the reactor type, can all play important roles in determining the composition of syncrude or the final biojet fuel product.[24] Generally, a notable feature of FTS is the wide distribution of products, and the selectivity of the catalyst is the key to obtaining the target hydrocarbon. The active components of the catalyst are Fe, Co, and Ru with an activity order of $Ru > Co > Fe$, and a chain growth probability order of $Ru > Co \approx Fe$.[25] In general, Co-based FT catalysts are more hydrogenating than Fe-based FT catalysts, and when used in a similar type of reactor, Co-based FT results in a syncrude containing less olefins and oxygenates.[24] High-temperature FT processes usually employ iron-based catalysts, while low-temperature FT processes generally employ either iron or cobalt-based catalysts.

Iron catalysts used in FTS can be prepared by precipitation, sintering, or fusion of molten oxides. The Fe_3O_4 based catalyst is the most commonly used as a molten iron catalyst auxiliary in Fe-based catalysts, which are beneficial to the modulation of activity and selectivity. Among them, the most studied are alkali metals, alkaline earth metals, Cu, and other excessive metals. The role of K in the microsphere $Fe/Cu/K/SiO_2$ catalyst has been reported in the literature.[26–28] It is considered that the addition of K increases the activity of FTS and the activity of the water vapor reaction (WGS). It has been found that the selectivity of hydrocarbons for CH_4 and C_{2-4} hydrocarbons decreases with an increase in K content, while the selectivity of C_{5+} hydrocarbon increases by up to 84% with an increase in K. In addition, adding a Cu promoter to Fe-based catalysts can both promote the reduction of Fe and reduce the reduction temperature of the catalyst. It was shown that the induction period was significantly shortened with little effect on the activity when Cu was added to an $Fe-Mn-K/SiO_2$ catalyst.[29] Compared to the catalyst without Cu, more high carbon hydrocarbons tend to be formed. Co-based catalysts have good carbon chain growth ability with high selectivity of high carbon straight-chain saturated hydrocarbons and less oxygen compounds in products. It has the characteristics of not easily accumulating carbon in the reaction process and having low WGS reaction activity.[30] A zirconium promoter can significantly change the adsorption behavior of $Co-SiO_2$ FT catalysts and reactants by weakening the adsorption strength of CO, which facilitates the formation of heavy hydrocarbons. Hydrophobicity modification of the catalyst surface and improving the

TABLE 5.3 Composition and properties of FT hydroprocessed synthesized paraffinic kerosene (FT-SPK) required by American Society for Testing Materials (ASTM) standards (ASTM D 7566-11a).

Property		FT-SPK	ASTM test method
Composition			
Acidity, total mg KOH/g	Max		
Hydrocarbon composition			
Cyclo-paraffins, mass	Max	15	D 2425
Aromatics mass	Max	0.5	D 2425
Paraffins, mass		Report	D 2425
Carbon and hydrogen, mass	Min	99.5	D 5291
Nonhydrocarbon composition	Max		
Nitrogen (mg/kg)	Max	2	D 4269
Water (mg/kg)	Max	75	D 6304
Sulfur (mg/kg)	Max	15	D 5453
Volatility			
Distillation			
1. Physical distillation			D 86
Distillation temperature (°C)			
10% recovered, temperature (T10)	Max	205	
50% recovered, temperature (T50)		Report	
90% recovered, temperature (T90)		Report	
Final boiling point, temperature	Max	300	
T90−T10 (°C)	Min	22	
Distillation residue (%)	Max	1.5	
Distillation loss (%)	Max	1.5	
2. Simulated distillation			D 2887
Distillation temperature (°C)			
10% recovered, temperature (T10)		Report	
50% recovered, temperature (T50)		Report	
90% recovered, temperature (T90)		Report	
Final boiling point, temperature		Report	
Flash point (°C)	Min	38	D 56 or D 3828*E*
Density at 15°C (kg/m^3)		730−770	D 1298 or D 4052
Freezing point (°C)	Max	−40	D 5972, D 7153, D 7154, or D 2386

structure of mesoporous materials can significantly improve the selectivity of a catalyst. Type I cobalt-based catalyst that has been industrialized has the characteristics of high activity, low methane selectivity (5%–6%), and high straight-chain heavy hydrocarbon selectivity (90%).[31]

As seen from the description, FT pathways for converting biomass to biojet fuel combine biomass gasification with FTS technologies and catalytic cracking to produce SPK. The obtained SPK is of high quality and can directly substitute conventional kerosene jet fuel. FTS technology has been applied to coal and natural gas feedstocks in the industry for several decades. Thus it is fully commercial, yet even so, research and development are still ongoing to improve the catalyst performance. Nevertheless, FT-SPK technology with biomass as the starting material still requires considerable development, particularly regarding the handling of biomass feedstocks and syngas cleaning. Moreover, this capital-intensive process is only cost-effective at a large scale, thus increasing the cost of feedstock logistics and investment risk.[6]

5.2.2.2 Upgrading of lignocellulose-derived bio-oil to biojet fuel

Another thermochemical pathway to convert lignocellulose to biojet fuel refers to the pyrolysis or liquefaction of biomass to obtain bio-oil followed by an upgrading with hydrotreatment (Fig. 5.4). Being similar to gasification technology, pyrolysis is also one of the most important processes for the thermochemical conversion of biomass. It refers to the thermal decomposition of a feed biomass into smaller fragments by heating without the presence of oxygen. The main benefit of the pyrolysis process over gasification is that the liquid product can be easily stored and transported.[32] The products of pyrolysis include char (biochar), liquid (bio-oil), and gas (bio-gas also called syngas) containing methane, hydrogen, carbon monoxide, and carbon

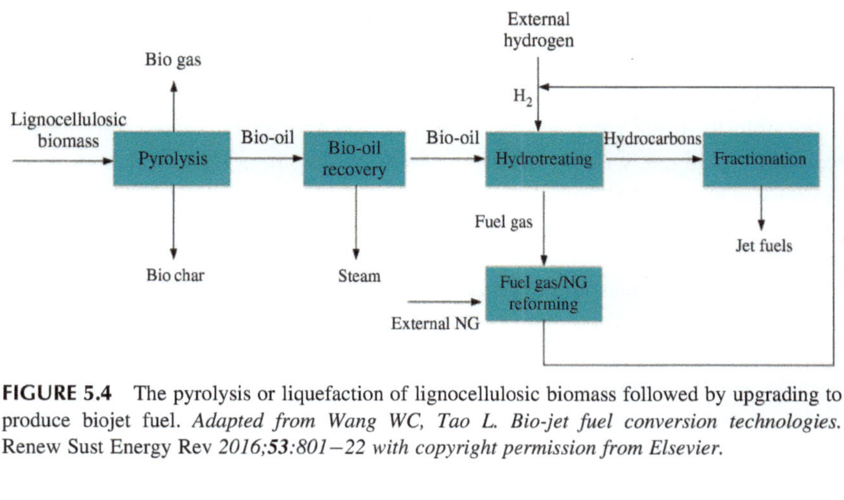

FIGURE 5.4 The pyrolysis or liquefaction of lignocellulosic biomass followed by upgrading to produce biojet fuel. *Adapted from Wang WC, Tao L. Bio-jet fuel conversion technologies.* Renew Sust Energy Rev *2016;**53**:801–22 with copyright permission from Elsevier.*

dioxide, and the distribution of the products strongly depends on the reaction conditions. The fast pyrolysis of biomass at a rapid heating rate and a short hot vapor residence time (<1 second) at an intermediate temperature usually produces liquid with a yield of up to 75 wt.%, while slow pyrolysis at a slow heating rate and long residence time is usually beneficial to produce biochar.[33] The yield of bio-oil is most significantly affected by the carrier gas flow rate, residence time, and heating rate. As found by Sensoz and Can,[34] when bark-free chips of Turkish red pine (*Pinus brutia* Ten.) were pyrolyzed in a laboratory-scale fixed bed reactor, the highest liquid yield was obtained at 500°C pyrolysis temperature with a heating rate of 40°C/min and a sweep gas flow rate of 50 cm^3/min. Apparent increases in the pyrolysis conversion at the temperature interval of 450°C–500°C was observed, which was due to the rapid devolatilization of cellulose and hemicellulose. However, for nonwoody biomass, such as agricultural residues, the optimum pyrolysis temperatures are lower. Biswas et al.[35] found that the pyrolysis of corn cob, wheat straw, rice straw, and rice husk showed optimum temperatures of 450°C, 400°C, 400°C, and 450°C with bio-oil yields of 47.3, 36.7, 28.4, and 38.1 wt.%, respectively.

Biomass direct liquefaction is another process to convert solid biomass into liquid fuels. It is a thermochemical process performed at low temperatures (250°C–400°C) and high-pressures (5–20 MPa), by which biomass is converted into three fractions, namely a bio-oil product (target product), a gas, and a solid residue, in water or another suitable solvent.[36] The biomass undergoes depolymerization to form monomers, which are further decomposed by dehydration, decarboxylation, and deamination to form light fragments of small molecules. The rearrangement of the light fragments results in the formation of new and complicated compounds through condensation, cyclization, and polymerization.[37,38] Liquefaction solvents show significant effects on bio-oil yields and product distribution. Two systems, namely water and organic solvent systems, are usually used for biomass liquefaction. Water is the most widely used liquefaction solvent, because not only is it environment-benign, cheap, and produces no environmental pollution, but it can also facilitate the recovery and recycling of inorganic matter contained in biomass in ionic form.[36] Using water as a liquefaction solvent also makes it possible to direct treat wet biomass, such as freshly harvested agricultural residues and algae, without drying.[39] However, biomass liquefaction in water systems also suffer shortcomings, such as low yield of water-insoluble bio-oil and low-heating values of the resulting bio-oil.[36] Organic solvents, such as methanol, ethanol, acetone, 1,4-dioxane, phenol, glycol, and glycerol are also used for biomass liquefaction. Some liquefaction solvents have much lower critical points than that of water so they can succeed in converting biomass into bio-oils under mild reaction conditions. However, though using organic solvents can obtain higher yields of water-insoluble bio-oil with higher heating values, the high cost of organic solvents make it unfeasible for large-scale application.

TABLE 5.4 Comparison of biocrudes from hydrothermal processing, bio-oil from fast pyrolysis and heavy petroleum fuel.[40]

Characteristics	Hydrothermal bio-oil		Fast pyrolysis bio-oil		Heavy petroleum fuel
Water content (wt.%)	3–5		15–25		0.1
Insoluble solids (%)	1		0.5–0.8		0.01
HHV (MJ/kg)	30		17		40
Density (g/mL)	1.10		1.23		0.94
Viscosity (cp)	3000–17,000		10–150		180
	Wet	**Dry**	**Wet**	**Dry**	
Carbon (%)	73.0	77.0	39.5	55.8	85.2
Hydrogen (%)	8.0	7.8	7.5	6.1	11.1
Oxygen (%)	16.0	13.0	52.6	37.9	1.0
Nitrogen (%)	<0.1	<0.1	<0.1	<0.1	0.3
Sulfur (%)	<0.05	<0.5	<0.05	<0.5	2.3
Ash (%)	0.3–0.5	0.3–0.5	0.3–0.5	0.2–0.3	<0.1

Bio-oil obtained by pyrolysis or direct liquefaction, also known as biocrude, contains a large number of unsaturated bonds and phenols, aldehydes, and ketones. As shown in Table 5.4,[40] the oxygen and water contents of bio-oil are relatively high, which reduces the calorific value and can easily lead to polymerization, thus affecting the bio-oil stability. Therefore bio-oil must be further upgraded in order to reduce its oxygen content and improve the quality so as to meet the requirements. In addition, sulfur and nitrogen also need to be removed. Therefore the hydrotreatment of bio-oil for upgrading needs to be further conducted in order to obtain a jet fuel standard range of hydrocarbon fuel. Bio-oil upgrading can be achieved by physical methods, such as emulsification, and through chemical approaches, including esterification, hydrogenation, cracking, and steam reforming.[41] For preparing jet-fuel range hydrocarbon compounds, hydrogenation and cracking are the most proper approaches. Hydrogenation can increase the saturation and remove the organic acids and aldehydes so as to improve the stability and energy density of bio-oil. Carbon–oxygen bonds are broken, which removes oxygen in the form of H_2O or CO_2. Cracking can effectively remove oxygen and water and promote the esterification of acidic components and alcohols in bio-oils, consequently reducing the viscosity and acidity.[41]

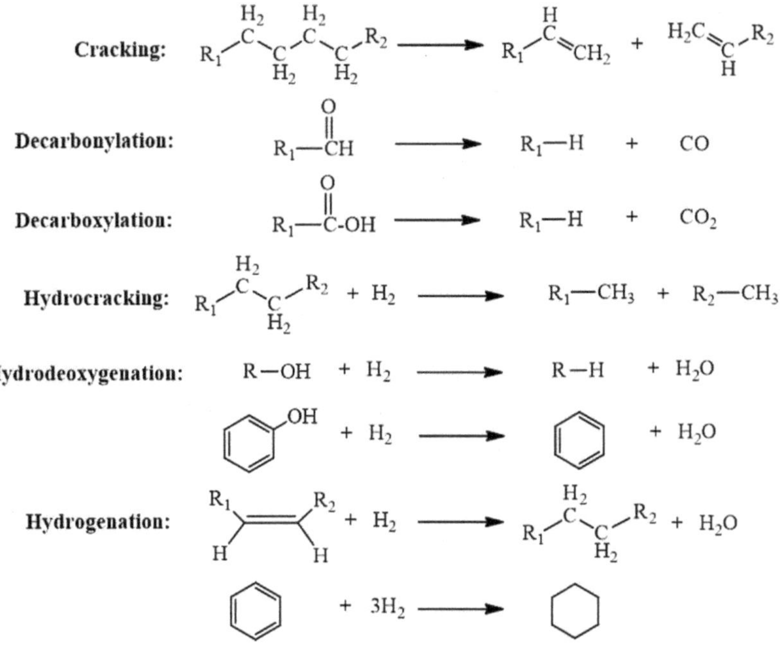

FIGURE 5.5 Representative reactions involved in the catalytic upgrading of bio-oil. *Adapted from Mortensen PM, Grunwaldt JD, Jensen PA, Knudsen KG, Jensen AD. A review of catalytic upgrading of bio-oil to engine fuels.* Appl Catal A—Gen *2011;**407**(1–2):1–19 with copyright permission from Elsevier.*

Catalytic upgrading with hydrotreatment is the most frequently used approach to produce hydrocarbon fuels for engines from bio-oil. The process involves a complex reaction network of HDO and zeolite cracking, including cracking, decarbonylation, decarboxylation, hydrocracking, hydrodeoxygenation, hydrogenation, and polymerization as shown in Fig. 5.5.[42]

The catalyst is the most important factor for the efficiency of upgrading. Noble-metal catalysts are preferred, while some transition metals also show good catalytic efficiency. Wildschut et al.[43] reported an experimental study on the upgrading of fast pyrolysis oil by catalytic hydrotreatment with a variety of heterogeneous noble-metal catalysts (Ru/C, Ru/TiO_2, Ru/Al_2O_3, Pt/C, and Pd/C), and compared these to those obtained with typical hydrotreatment catalysts (sulfided NiMo/Al_2O_3 and CoMo/Al_2O_3). The Ru/C catalyst was found to be superior to the classical hydrotreating catalysts with an oil yield of up to 60 wt.% and a deoxygenation level of up to 90 wt.%. The upgraded pyrolysis oil had lower contents of organic acids, aldehydes, ketones, and ethers than the feed, whereas the amounts of phenolics, aromatics, and alkanes were considerably higher. Nickel and Co–Mo–P based catalysts have also been reported for bio-oil upgrading.[44,45] The application of Mo–10Ni/γ-Al_2O_3 as a catalyst in the catalytic hydrogenation of bio-oil can increase the pH value of bio-oil from 2.16 to 2.84 and the hydrogen element from 6.61% to 6.93%, while the viscosity of bio-oil decreases to a certain extent. The experiment with Co–Mo–P as a catalyst showed that the oxygen content of the modified bio-oil decreased from 41.8% before modification to 3%, and the calorific value was also increased. The most typical catalytic hydrogenation technologies used include the UOP process developed by the Honeywell UOP Company and the Bio-Synfining process developed by the Syntroleum Company. In the UOP process, the oxygen in the biomass oil is first removed by hydrogenation and then the selective pyrolysis and isomerization reactions take place to obtain the components of aviation biofuels. The Bio-Synfining process removes impurities and water from the raw oil first, then the biomass oil is converted to long carbon chain saturated alkanes by catalytic hydrogenation. Finally, the biofuels are prepared by hydrocracking/isomerization.

Cracking is a process in which large hydrocarbon molecules are broken up and "cracked" into smaller, more valuable hydrocarbon molecules. As the name implies, the catalytic cracking process takes place in the presence of a catalyst, where C–C and C–O bonds are cracked so that the oxygen is removed in the form of H_2O, CO, or CO_2 and the viscosity and oxygen content of the biomass oil are reduced. The molecular sieve catalyst HZSM-5 having the characteristics of strong acidity, high activity, and high shape selectivity, is the most commonly used catalyst for cracking. Not only can it promote the dehydration or deoxidization of biomass oil into hydrocarbons, but it can also promote the reaction of acidic components and alcohols in bio-oil, thus reducing the acidity and viscosity of the oil products.[46] The coke yields of the HZSM-5 and H-Y catalyst in the process of catalytic cracking have been studied and it was found that the HZSM-5 catalyst produced less coking and deactivation than the H–Y catalyst.[47,48] The studies showed that the stronger the acidity of the catalyst, the better the pyrolysis and conversion of the bio-oil. Catalytic cracking has great prospects for development because the equipment required for the process are not strict as compared to the catalytic hydrogenation process. Low yield, high coking rate, and easy deactivation of the catalyst are the main issues that need to be solved. In addition, high temperatures may cause the polymerization of biomass oil. Therefore it is crucial to develop new catalysts that can catalyze the reaction at low temperatures.

Biojet fuel production through the upgrading of biomass bio-oil is also near the commercialization stage with many pilot demonstrations being reported. However, the fuel has not yet been approved by the ASTM. In addition, the process is still facing issues, such as high coking, poor quality of fuels, and relatively high cost.[49]

5.2.2.3 Catalytic depolymerization and hydrodeoxygenation

Lignocellulosic biomass can also be directly depolymerized by thermal or chemical processes in the presence of hydrogen under the catalysis of various catalysts. However, this process is usually only applied to the isolated fraction of lignocellulose, typically isolated lignin. In this process, the depolymerization of lignin to obtain monomeric compounds is a key step. Various thermal and chemical processes have been used for lignin depolymerization. Pyrolysis and direct liquefaction can similarly be applied to produce bio-oil from lignin. Chemical depolymerization refers to the use of chemical catalysts in either water or organic solvents to break down lignin macromolecules into smaller fragments. Base, acid, and metallic catalysts have been employed in aqueous, organic, ionic liquid, or supercritical fluids for lignin depolymerization.[50] Lignin molecules can also be broken down by oxidative or enzymatic degradation.[51] In general, base-catalyzed lignin depolymerizations are performed at a temperature above 300°C and a high pressure with the formation of catechol, syringol, and derivatives. These mechanisms involve the cleavage of the aryl–alkyl bond, which usually occurs above 270°C. The acid-catalyzed process employs a mineral acid, such as HCl, H_2SO_4, or organic acid, such as formic acid, to achieve the depolymerization of lignin. A relatively low temperature in the range of 78°C–200°C is usually used in this process, but it is not severe enough to break the complex lignin structure into monomeric compounds.[50] The cleavage of alkyl–aryl ether linkages, such as α-O-4, β-O-4 bonds, is primarily responsible for the depolymerization of lignin.[52]

Further hydrodeoxygenation of the lignin depolymerization products may produce hydrocarbon fuels, like alkanes or cycloalkanes. Catalysts used for the catalytic upgrading of bio-oil can also be applied in the HDO of lignin depolymerization products. The type of catalyst, temperature, pressure, and the type of reactor all show significant effects. Two basic parallel reactions have been determined by studying the HDO of lignin model compounds, namely direct HDO and hydrogenation. In direct HDO, the aromatic character of molecules is retained by the hydrogenolysis of C–O bonds in side chains, leaving aromatic hydrocarbons as the primary products. In hydrogenation, the benzene ring is saturated first, followed by the hydrogenolysis of C–O bonds in side chains primarily producing cycloalkanes, such as cyclohexane and methyl-cyclohexane.[53]

The process of hydrocarbon fuels production by the catalytic depolymerization of biomass (lignin) followed by HDO is still at the laboratory stage. The obtained mixture still needs further upgrading to meet the requirements of jet fuel. However, using lignin as the starting materials for jet fuel production still shows some merits. Lignin is usually disposed of as a waste in the pulp and biorefinery industry and, thus, is available at a low price. Also, aromatic hydrocarbons can easily be obtained because the monomeric unit of lignin is phenylpropanoid. Aromatic hydrocarbons play several roles in the properties of conventional jet fuels, such as high energy ratings, high knock ratings, and better performance on elastomer (O-ring) swelling than alkanes. However, aromatic hydrocarbons are precursors for secondary organic aerosol formation.[53] Similarly, lignin can also be used as a feedstock, just like agricultural residual biomass, to produce bio-oil via thermal pyrolysis or direct liquefaction, followed by catalytic upgrading through hydrotreatment to produce biojet fuel. This pathway appears to be more promising than the catalytic depolymerization route.

5.2.2.4 Hydrolysis to sugars followed by alcohol to jet

The main chemical components of lignocellulosic biomass are cellulose, hemicellulose, and lignin. Cellulose and hemicelluloses are polysaccharides and can be converted into monosaccharides, which can be further converted into alcohols, such as ethanol and butanol, by microbial fermentation. After catalytic dehydration, alcohols can be converted to the corresponding alkenes. The oligomerization of the alkenes results in the extension of the carbon chain to jet fuel range. Further hydrogenation leads to the formation of hydrocarbon fuels suitable for jet engines. Using butanol as an example, the primary chemical reactions involved in the process are shown in Fig. 5.6.[54] However, this pathway has faced several bottlenecks, primarily including the efficient release of fermentable sugars from lignocelluloses, cheap conversion of the sugars into alcohols, and the precise control of oligomerization to obtain a carbon length in the jet-fuel range.

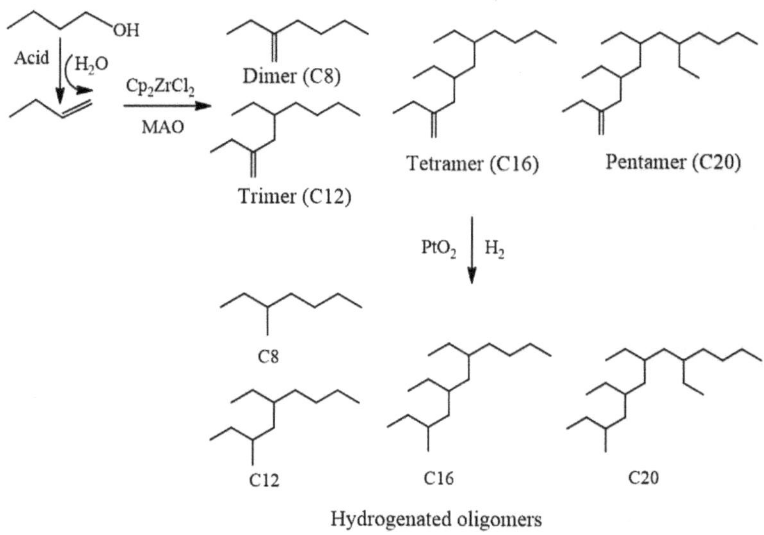

FIGURE 5.6 Primary chemical reactions involved in the conversion of lignocellulose into biojet fuel via ATJ with *n*-butanol as an intermediate. *Adapted from Wright ME, Harvey BG, Quintana RL. Highly efficient zirconium-catalyzed batch conversion of 1-butene: a new route to jet fuels.* Energy Fuel *2008;**22**(5):3299–302 with copyright permission from American Chemical Society.*

The conversion of lignocellulose into fermentable sugars is an old chemistry process, which can be done by a chemical or enzymatic hydrolysis or a hybrid process. The biomass can be directly hydrolyzed under the catalysis of acids, typically with mineral acids, such as sulfuric, hydrochloric, and phosphoric acids. The advantages of acid hydrolysis over enzymatic hydrolysis are that the acid can penetrate the lignin without pretreatment for delignification, and the rate of acid hydrolysis is much faster than that of enzymatic hydrolysis.[55] The hemicelluloses fraction of lignocellulose is easy to hydrolyze, while cellulose cannot be well hydrolyzed unless more severe conditions (higher temperature or acid concentration) are used. Depending on the catalyst employed, the chemical hydrolysis of lignocellulosic biomass can be classified into liquid acid hydrolysis and solid acid hydrolysis, and both of them can be conducted in either an aqueous system or a nonaqueous system. In the concentrated acid hydrolysis (CAH) process, the acids commonly used are sulfuric acid (50%–80%), hydrochloric acid (37%–42%), and phosphoric acids (82%–84%), or a mixture of sulfuric acid and phosphoric acid (40:60–70:30 H_2SO_4:H_3PO_4 ratio) is used.[56] The process is usually done in two steps at normal atmospheric pressure and at temperatures below 100°C.[57] In the first step, the raw material is pretreated with concentrated acids to disrupt the cellulose crystalline structure and dissolve the fibers, and in the second step, the acid is diluted to hydrolyze the glycosidic bonds for the formation of monosaccharides. CAH can be performed at low temperatures with a high sugar yield (usually 90% theoretical) and little sugar degradation. However, the high corrosion of concentrated acids requires either expensive alloys or specialized nonmetallic constructions, such as ceramic or carbon-brick lining. The acid recovery is also an energy-intensive process. Moreover, a large amount of gypsum is produced when lime is used to neutralize H_2SO_4. The CAH process is also found to have a significantly negative impact on the environment, particularly for concentrated hydrochloric acid hydrolysis.[58] Compared to CAH, dilute acid hydrolysis (DAH) seems to be more promising since the process consumes much less acid. Generally, the hydrolysate can be directly used after neutralization with lime, which avoids using various processes to separate the acid and sugar. It is usually conducted with a 0.1%–5% acid concentration and the reaction temperature can be as high as 260°C. Mineral acids, such as H_2SO_4, HCl, and others, can be used, while H_2SO_4 is the most commonly employed. The sugar yield is greatly dependent on temperature, acid concentration, and retention time. Since the acid concentration is low, a high temperature, usually >180°C, should be employed to hydrolyze cellulose. However, such severe conditions may lead to the formation of inhibitors, such as furfural and 5-HMF.[59] The enzymatic hydrolysis of lignocellulosic biomass, primarily the cellulose fraction, through a fungal cellulase system is achieved by a synergism of cellulase components: including endoglucanases that internally cleave glucan chains in less-ordered cellulose regions; cellobiohydrolases that hydrolyze cellulose chains processively from the reducing and nonreducing ends; and β-glucansidases that hydrolyze cellobiose to glucose, thus eliminating the product inhibition.[60] Enzymatic hydrolysis can be performed under much milder conditions (50°C) than the DAH process with no formation of inhibitors. However, due to the recalcitrance of the biomass constructed by the lignocellulose cell wall architecture, the substrates must undergo pretreatment to increase cellulose accessibility.[61,62] The sugar yield from pretreated substrates is generally in the range of 40%–90% depending on the pretreatment approaches.[63] Therefore the efficient release of sugars from the plant cell wall has become a bottleneck in the production of second-generation bioethanol.

Ethanol and butanol are the most important liquid biofuels that can be used to substitute gasoline. Fermentable sugars can be biologically converted into ethanol with a theoretical yield of 0.51 g of ethanol per gram of glucose. Typical microorganisms used in ethanol industries are *Saccharomyces cerevisiae* and *Zymomonas mobilis*.[64] Significant progress has been made in the development of second-generation bioethanol in terms of pretreatment, cellulase production, and pentose fermentation, which brings the process closer to realizing commercialization.[65,66] Butanol can also be produced by fermentation. *n*-Butanol can be produced by *Clostridia* strains, such as *Clostridium beijerinckii* and *Clostridium acetobutylicum* via acetone–butanol–ethanol (ABE) fermentation, which produces acetone, *n*-butanol, and ethanol from biomass in a typical molar ratio of 3:6:1.[67] Isobutanol can be produced by engineered *Escherichia coli*, or by *Bacillus subtilis* and *Cornebacterium glutamicum* in aerobic conditions with a theoretical yield of 100%.[68]

Alcohols can be easily dehydrated to form corresponding alkenes under the catalysis of acids.[69,70] Homogenous acid, such as H_2SO_4 and *p*-toluenesulfonic acid (PTSA), and solid acid, such as alumina-based, ZSM-5 zeolites, γ-type zeolites, and Amberlyst acidic resins, can be used as catalysts for dehydration, among which alumina-based catalysts are used in most current industrial processes. For ethanol dehydration, the ethanol concentration in the feed is usually 95% (w/w) and the process is performed at a temperature of 300°C–500°C, pressure of 0.1–0.2 MPa, and space velocity of 0.1^{-1} $hour^{-1}$ with an ethylene yield of 94%–99%.[69] For the dehydration of *iso*-butanol, as studied by Taylor et al.,[71] a yield of 99.1% and a selectivity of 95.1% for isobutene, 1.6% for 1-butene, 0.5% for *trans*-2-butene, and 1.9% for *cis*-2-butene were obtained at 325°C, 60 psig, with γ-alumina as a catalyst at a space velocity of 2 $hour^{-1}$. As disclosed in the patent of Wright,[72] *n*-butanol can be efficiently dehydrated to 1-butene with a silane-modified γ-alumina catalyst at 380°C, 30 psig pressure, and a space velocity of 0.01 $hour^{-1}$, with the highest butane yield of 98%, and 95% selectivity for 1-butene, 2% for *cis*-2-butene, and 3% for *trans*-2-butene.

The oligomerization of the obtained alkenes results in the formation of the precursors of jet-fuel. Frequently used catalysts are Ziegler–Natta catalysts[72] and zeolite-based catalysts.[73,74] Wright et al.[72] achieved a quantitative conversion of 1-butene to a Schultz–Flory distribution of oligomers using Group 4 transition-metal catalysts in the presence of methylaluminoxane (MAO) (Fig. 5.6). The resulting olefins mixture contained 27% C8, 26% C12, 18% C16, 12% C20, 8% C24, 5% C28, and 4% C32. After removing the dimer, followed by catalytic hydrogenation of PtO_2, 100% saturated hydrocarbon fuel with a density of 0.78 g/mL, a viscosity of 12.5 cSt at −20°C (ASTM 445), and a calculated heat of combustion of 44 + MJ/kg was obtained. When *n*-butanol from ABE fermentation was dehydrogenated under a Pd/C-K_3PO_4 catalyst, a 38% carbon yield of C7–C15 ketones and a 20% carbon yield of C4–C6 products were obtained.[75] Kim et al.[73] compared three types of zeolites, HZSM-5, H-beta, and H-Y, for evaluating the performance during the synthesis of jet fuel via the oligomerization of a butene mixture composed of 1-butene, *cis*-2-butene, and *trans*-2-butene. The results indicated that HZSM-5 showed a good butene conversion rate with high selectivity to jet-fuel range hydrocarbons, which was primarily attributed to its high resistance to coking.[73]

ATJ routes may produce hydrocarbon fuels of jet fuel range. The reported biojet fuels can meet the ASTM standards. However, this process is not economically feasible yet, especially when the sugars are from lignocellulosic biomass. Reducing the cost of bioalcohol production can be beneficial to improving the market competitiveness of ARJ. Nevertheless, the ARJ route is still promising, because the infrastructure and technology for bioalcohol production, especially for bioethanol, have already been well established, but the quantifying yields, capital and operating requirements, and external energy requirements, as well as the process integration, need further investigation for the ATJ route.

5.2.2.5 Platform chemical routes

The platform chemical route for biojet fuel production refers to the conversion of biomass into platform chemicals followed by the chemical or biological conversion of the chemicals into hydrocarbon compounds in the range of jet-fuel. The process for biojet fuel production via the platform chemicals route should include or consider these aspects: (1) Pretreatment to degrade the crosslinked structure of lignocellulose to achieve a fractionation of the biomass components; (2) Cellulose, hemicellulose, and lignin conversion into monomers, such as hexoses, pentoses, and phenols by enzymatic, biotransformation, or chemical conversion pathways; (3) The conversion of the resulting monomers into platform chemicals, such as furfural,[76] HMF,[76] levulinic acid (LA),[77,78] 2-methylfuran,[77,78] γ-valerolactone,[79] small molecule alcohols, ketones,[67] and other platform compounds with different functional groups by chemical conversion, such as dehydration, oxidation, or microbial fermentation; (4) The carbon atoms in lignocellulose-based platform chemicals are generally insufficient to meet the requirements of jet fuel, and therefore, carbon chain extension, for example, hydroxyalkylation/alkylation (HAA), aldol condensation, ketonization,[80] and oligomerization,[81] should be employed to obtain oxygenate precursors; (5) The oxygen-containing precursors must then undergo HDO to produce hydrocarbon liquid products; and (6) the obtained hydrocarbon products may be not fully satisfy the physical and chemical properties of current fossil-based jet fuel and further upgrading may be needed. In this situation, a catalytic reconstruction of the obtained HDO products should be employed. Actually, the ATJ route is a specific example of a platform chemical route because alcohols are usually important platform chemicals derived from lignocellulose. In this section, platform chemicals that can be used for jet fuel production refer to those nonalcohol compounds derived from lignocellulosic biomass, such as furfural, HMF, and γ-valerolactone.

Glucose can be obtained from the chemical or enzymatic hydrolysis of cellulose and can be further converted into various platform chemicals (see Fig. 5.7). Glucose can be converted into 5-HMF by dehydration under the catalysis of synthetic multifunctional catalysts, metal chlorides, ionic liquids, or other catalysts.[82] A one-pot process to directly convert cellulose into HMF has also been reported.[83] HMF has both hydroxy and aldehyde groups, and it can be used for aldol condensation to extend the carbon chain due to the high reactivity of the aldehyde group. Furthermore, HMF can be catalytically dehydrated to form LA that contains both a keto group and a carboxyl group. The selective hydrogenation of LA, generally under the catalysis of

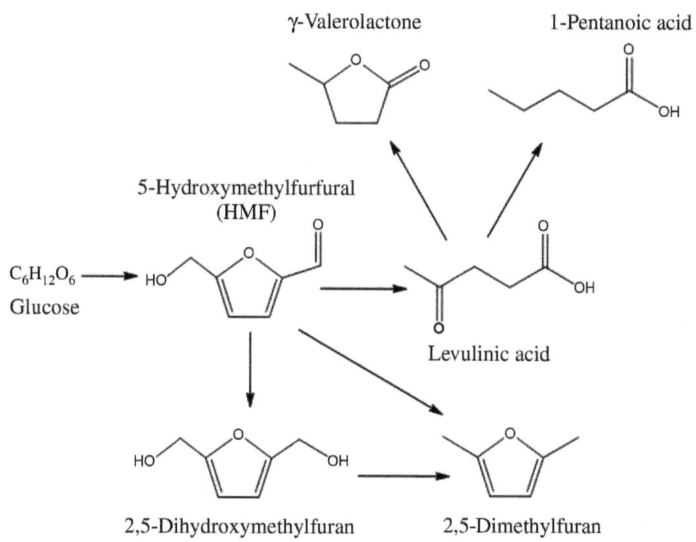

FIGURE 5.7 Typical glucose or fructose derived platform chemicals.

an Ni-based catalyst followed by its subsequent esterification and dehydration can produce γ-valerolactone[84]; 1-pentanoic acid can also be obtained from levulinic acid through selective hydrodeoxygenation.[85] HMF can also be partially hydrogenated to 2,5-dimethylolfuran, and after further dehydrogenation, another platform compound, 2,5-dimethylfuran (2,5-DMF), with good carbon–carbon coupling reactivity can be obtained. In addition, 2,5-DMF can also be obtained by the direct hydrodeoxygenation of HMF by bimetallic catalysts, such as Ni–Co/C,[86] Ni–W carbide catalysts,[87] and so on. As reported by Zu et al.,[88] with Ru/Co_3O_4 as a catalyst at 130°C and 0.7 MPa, the conversion of 5-HMF was higher than 99.0% and the yield of 2,5-DMF was 93.4%.

The preparation of hemicellulose-based platform compounds, for example, furfural (Fig. 5.8), is more mature. Hemicellulose is currently the main feedstock for furfural production in industry. The dehydration process is generally catalyzed by dilute acid under high-temperature and high-pressure conditions. Commonly used acid catalysts include sulfuric acid, hydrochloric acid, phosphoric acid, acetic acid, and formic acid. Another widely used hemicellulose-based platform compound, 2-methylfuran, can be further obtained through the catalytic hydrodeoxygenation of furfural. The catalysts used are generally bifunctional catalysts with both acidic sites and hydrogenation sites, such as Ru/RuOx/C,[89] Cu/SiO_2,[90] Ni–Cu/Al_2O_3,[91] Ru/$NiFe_2O_4$,[92] and so on. When Ni–Cu/Al_2O_3 was used as a catalyst, the 2-methylfuran yield reached 92%.[91] Cyclopentanone is also a widely used platform compound and can be obtained through the selective hydrogenation of furfural under Cu, Co, or other metal catalysts.[93,94] Furfural can be reduced by a Cu-based catalyst under mild conditions (about 120°C and atmospheric conditions) to give furfuralcohol. On one hand, furfuralcohol can be used as a raw material for synthetic resins, but on the other hand, it can also be used as a feedstock for the production of another important platform compound,

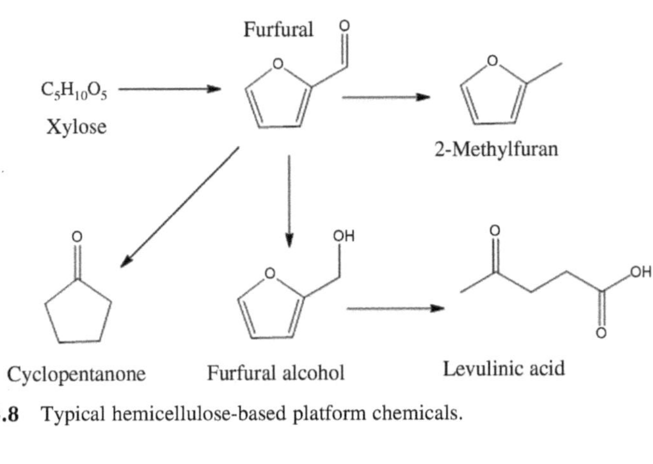

FIGURE 5.8 Typical hemicellulose-based platform chemicals.

levulinic acid. For example, under mild conditions, furfuralcohol could be hydrolyzed by dilute acid with a molar yield of levulinic acid of up to 80%.[95]

Since the carbon chain length of the platform compound molecule is generally short, an appropriate carbon chain extension is necessary for obtaining an intermediate product having the target carbon chain distribution to meet the requirements of commercial jet fuel. Currently, commonly used carbon chain extension reactions for lignocellulose-based platform compounds include alkali or acid-catalyzed aldol condensation,[96,97] acid-catalyzed hydroxyalkylation/alkylation (HAA),[98–100] ketonization[101,102], and oligomerization.[103] Platform compounds with an aldehyde or ketone group, for example, acetone and acetoin,[104] are reactive to both aldol condensation and HAA. They can react with furfural or HMF through aldol condensation or with 2-methylfuran or 2,5-DMF via HAA to achieve carbon–carbon coupling. In addition, there are also other methods that can be used to extend the carbon chain of platform compounds. For example, furfural or HMF can be converted into bimolecular condensation products with a 10–12 carbon atom length through benzoic condensation under the catalysis of organic catalysts.[105] The chemical reactions involved in the aldol condensation, HAA, ketonization, and oligomerization processes are summarized in Table 5.5.[97,106–108]

Of the processes listed in Table 5.5, aldol condensation and HAA are the most commonly used. The aldol condensation of lignocellulose platform compounds for extending the carbon chain is mainly catalyzed by basic catalysts. Until now, the reaction systems used generally included two-phase systems with organic solvents and direct condensation reactions in solvent-free systems, both of which have advantages and disadvantages. In general, the catalysts used in the solvent system are strong bases, such as NaOH and KOH, and the reaction is typically performed under mild conditions, such as room temperature and atmospheric pressure, but the subsequent separation and recycling of the organic solvents is required. The catalysts used in solvent-free systems are generally weak bases, mostly solid bases like CaO and MgO with relatively simple subsequent separations of the product, but the reaction is usually performed at high temperature. Huber et al. comprehensively studied the aldol condensation of furfural with acetone in tetrahydrofuran (THF) system with sodium hydroxide solution as a catalyst.[97,108] Furfural and acetone are concentrated in the THF phase, and NaOH in the aqueous phase, thereby preventing the side reactions that generate heavy components. The key to the solvent-free system lies in the choice of catalyst. Too strong alkalinity will lead to the production of heavy by-products, but too weak alkalinity will obtain a less efficient conversion. There have been many studies on aldol condensation reactions in solvent-free systems with catalysts like metal oxides (such as ZnO, MgO, ZrO_2, Li_2O, TiO_2), metal hydroxides (e.g., $Ca(OH)_2$), and some synthetic base catalysts. A typical

TABLE 5.5 Typical carbon chain extension reactions to obtain precursors for biojet fuel production from platform chemicals.

Carbon chain extending techniques	Typical application examples
Aldol condensation	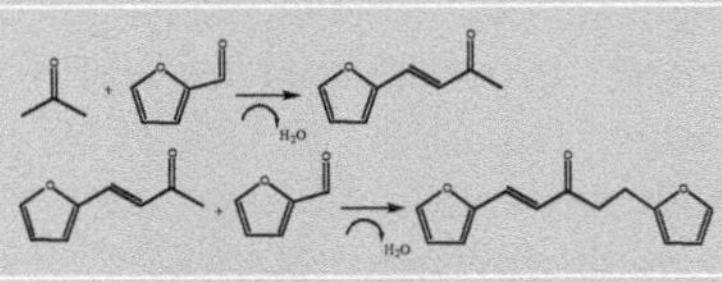 Aldol condensation of furfural and acetone[97]
HAA	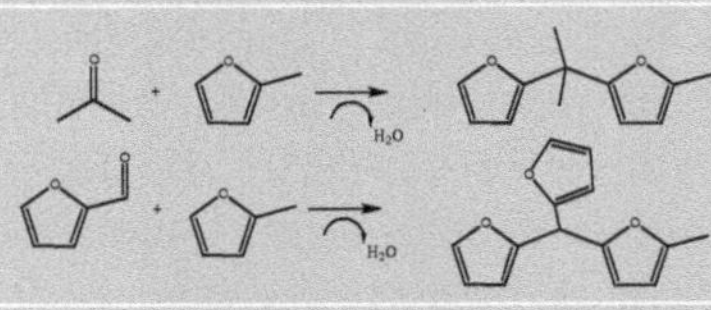 HAA of 2-methylfuran with acetone or furfural[106]
Ketonization	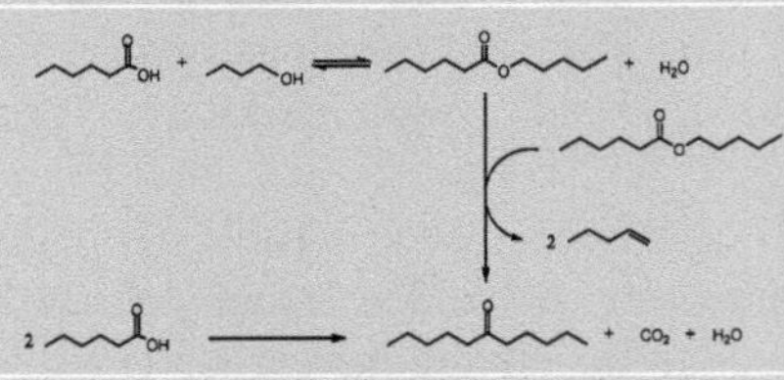 Ketonization of hexanoic acid or butyl hexanoate[107]
Oligomerization	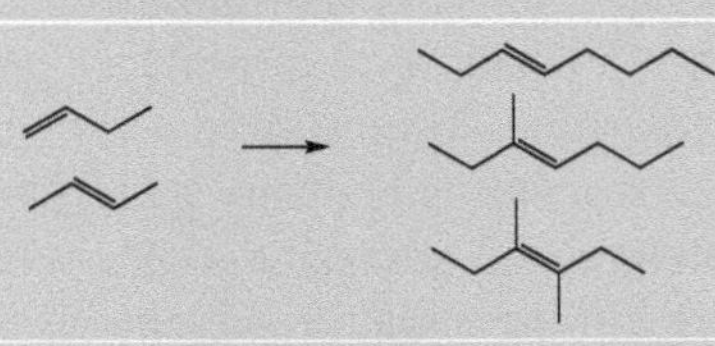 Oligomerization of 1-betene or 2-butene to C8 olefins[108]

HAA, hydroxyalkylation/alkylation.

reaction system includes furfural and acetone,[109] HMF and acetone,[110] furfural and methyl isobutyl ketone,[111] furfural and 3-pentanone,[112] levulinic acid and furfural,[105] and cyclopentanone self-aldol condensation.[113] Further hydrodeoxygenation of the precursors results in the formation of alkanes for hydrocarbon jet fuels.[114]

Cui et al.[115] reported a novel route for the production of renewable hydrocarbon biojet fuels from the biomass-derived platform chemicals furfural and 2,3-butanediol (BD). Furfural can be produced by the dehydration of xylose derived from hemicelluloses. BD can be produced by the microbial fermentation of cellulose-derived glucose. Carbonaceous solid acids bearing $-SO_3H$ groups can be prepared from biomass or its isolated polymeric components to catalyze the dehydration of BD to methyl ethyl ketone (MEK) in the liquid phase at low temperatures ($<170°C$). C9/C14 precursors can be obtained by carbon chain extension via the aldol condensation of furfural and MEK under alkaline conditions. Hydrodeoxygenation of the precursors resulted in the formation of hydrocarbon fuels, with C1–C4, C5–C8, and C9–C14 alkane products obtained in yields of 15.4%, 7.9%, and 73%, respectively. This process has a theoretical exergy efficiency similar to that of the bioethanol process, but higher than that of the biobutanol process. It can also achieve a flexible adjustment of the product profile with respect to chemicals and fuels.

Early success has been achieved for platform chemical routes to synthesize biojet fuels. Various possible reaction pathways can be chosen since the platform chemicals that are obtained from lignocellulosic biomass are diverse. The product profiles are also flexible, as proposed by Cui et al.[115] However, this route is only at the laboratory scale and is still far from commercialization. Many platform chemicals are even more valuable than the derived hydrocarbon fuels. This makes the production of jet-fuel from platform chemicals less attractive. Further design and optimization of the process to reduce the reaction steps and production cost should be considered.

5.2.2.6 Biological conversion to microbial lipids for hydroprocessed esters and fatty acids or direct conversion to hydrocarbon

Lignocellulose-derived sugars can also be converted into oil or lipid, known as microbial lipid or single cell oil, through microbial fermentation for HEFA to produce biojet fuel. This is because some microorganisms, for example, oleaginous fungi, algae, cyanobacteria, and bacteria, can produce lipids as storage compounds with triacylglycerols (TAGs) as their main constituents under certain conditions, such as an excessive carbon source with nitrogen limitation. Microorganisms that can accumulate intracellular lipids accounting for over 20% of the dry biomass weight in the form of lipid vesicles are classified as oleaginous microorganisms.[116] Currently, many oleaginous microorganisms have been developed, primarily including *Cryptococcus* sp., *Lipomyces* sp., *Rhodosporidium* sp., *Rhodotorula* sp., *Trichosporon* sp., *Yarrowia* sp., *Aspergillus* sp., *Mortierrela* sp., *Thamnidium* sp., *Candida* sp., *Zygosacchromyces* sp., *Zygorhynchus* sp., *Mucor* sp., *Torulopsis* sp., and *Pichia* sp.[117] The reported intercellular lipid content can be as high as 70% of the dry cell weight of *Rhodosporidium toruloides*.[118] However, when lignocellulose hydrolysate is used as a carbon source for lipid accumulation, the presence of inhibitors, such as organic acid, furfural,

HMF, and phenols, may show significantly negative effects on cell growth and lipid accumulation. As found by Zhao et al.,[119] when inhibitors were present singly in the culture medium, to obtain a high cell biomass accumulation of *R. toruloides*, the concentrations of formic acid, acetic acid, furfural, and vanillin should be lower than 2, 5, 0.5, and 1.5 g/L, respectively; however, the synergistic effects of the inhibitors could dramatically decrease the minimum critical inhibitory concentrations leading to significant inhibition in growth and lipid production. The fatty acid contents of microbial lipids are different depending on the microorganism species. However, the major fatty acid compositions are palmitic acid (C16:0), stearic acid (C18:0), oleic acid (C18:1), linoleic acid (C18:2), and linolenic acid (C18:3), which is in the range of various vegetable oils and animal tallows.[120] After the cell wall is disrupted by various mechanical, chemical, or biological approaches,[121] the intercellular lipid can be released, which can be used as a feedstock for HEFA production. The conversion of lignocellulose-derived sugars into lipid provides a way to produce biojet fuel, such as HEFA. However, this route is of high cost primarily due to the relatively low yield of sugar-to-lipid conversion. Typically, the theoretical yield of lipid from glucose is about 30% by weight (30 g lipid/100 g glucose), while the actual lipid yield is usually lower than 25%. Even if there is no consideration of the carbon loss in the subsequent processing to produce jet-fuel, this low yield makes it difficult to achieve large-scale application of this route in the short term.

Other biological ways to convert sugars to jet-fuel involve the metabolism of sugars by microbial organisms, typically genetically modified organisms (GMOs), to obtain jet-fuel precursors that are usually long carbon chain alkenes. An example of this route is the biological synthesis of farnesene through a process developed by Amyris Inc. using the mevalonate pathway in yeast cells. A maximal farnesene yield of 16.8 g/100 g sugar at a productivity of 16.9 g/L/d has been reported, with a carbon efficiency of 60%. Both C5 and C6 sugars can be converted. After recovery and purification, a 95% farnesene recovery and a 97% concentration can be obtained. However, although farnesene has been approved as a jet blend-stock and meets all ASTM standards for Jet A/A1, this process is still in its early stages and the cost needs to be further reduced for large-scale application.[65,68]

5.2.3 From sugar feedstock to biojet fuel

Sugars are important feedstocks for biofuel production. Besides being produced from the hydrolysis of lignocellulosic biomass, two other important sources to obtain sugars are starch and sugar crops, such as sugarcane and beet. Sugars can be converted into biojet fuels by chemical and biological processes as shown in Fig. 5.1. The processes relating to ATJ, platform chemicals, and biological conversion to microbial lipids have already been described. In addition, sugars can also be converted by APR to a mixture of water, hydrogen, and chemical intermediates, such as alcohols, ketones, acids, furans, paraffins, and other oxygenated hydrocarbons.[6] Compared with the HEFA route, the APR of sugars can form aromatics, which are key components of jet fuel. As reported by Wang et al.,[122] through the ARP of several sugars or polyols, a liquid fuel yield of 31.8% with a sorbitol conversion of 98% was obtained under the conditions: 280°C, weight hourly space velocity (WHSV) of 1.25 $hour^{-1}$, gas hourly space velocity (GHSV) of 2500 $hour^{-1}$, and 4.0 MPa hydrogen pressure, and the content of aromatics and naphthenes was above 72 wt.%. However, most reported APR technologies are still at the research and development stage, though a process (BioFormin Process) developed by Virent Energy Systems, Inc. seems to be in the pilot stage. The BioForming process can achieve a reduction in the oxygen content of carbohydrate feedstock with heterogeneous catalysts at moderate temperature (450°C−575°C) and pressure (10−90 bar) by a number of series and parallel reactions, including: (1) the reforming of sugars to generate hydrogen; (2) the dehydrogenation of alcohols/hydrogenation of carbonyls; (3) deoxygenation reactions;(4) hydrogenolysis; and (5) cyclization using in situ generated hydrogen for the defunctionalization of the highly reactive carbohydrate to a less reactive mono-oxygenated species.[123]

Compared to the sugar-to-alcohol biofuel (ethanol) process, the APR process of sugars showed some advantages, such as a higher volumetric energy density of the obtained hydrocarbon mixtures, no requirement for the energy-intensive dewatering steps, and being compatible with the current petroleum infrastructure and pipelines. However, being similar to other sugar-to-jet fuel routes, the current research progress has just demonstrated the technical feasibility of producing biojet fuel from sugars. Optimization, evaluation, and integration should be conducted to make it economically feasible.

5.3 Conclusion

Various routes have been developed to produce biojet fuels from renewable biomass feedstocks. Depending on the biomass used, the conversion routes are different. However, since biomass has a relatively high oxygen content, deoxygenation processing is necessary to finally obtain hydrocarbon liquid jet-fuels with a high volumetric energy density. Hydrogen is commonly used for deoxygenation in all of these routes. Oils or lipids are the most frequently used feedstock for biojet fuel production with a relatively simple and mature technical route. The oil-to-jet route refers to converting triglycerides or fatty acid into hydrocarbon liquid fuels in the jet fuel range by catalytic hydrogenation, decarboxylation, hydrodeoxygenation, hydrodecarbonylation, and/or isomerization under metal-based catalysts. This route is the closest to realizing commercialization. However, the high cost and unstable supply of oil or lipid-containing feedstocks has become an obstacle for the commercialization of the technology. Lignocellulosic biomass, such as forestry and agricultural residues, are abundant with wide geological distribution. However, due to the complex structure of the plant cell wall, the conversion of lignocellulose into biojet fuel is usually more complicated with multistep processing. After pretreatment, the lignocellulose can be pyrolyzed or liquefied through a thermochemical process to obtain bio-oil, which is further upgraded to obtain liquid hydrocarbon fuels. Another pathway is to gasify the lignocellulosic biomass to obtain syngas, which is further used as a feedstock for FT synthesis to obtain synthetic liquid hydrocarbon fuels. These two routes are relatively mature and they represent the most promising potential for commercialization. Lignocellulosic biomass can also be first converted into sugars for further conversion into biojet fuels either by chemical or biochemical processing or a combination. For sugar conversion into biojet fuel, one of the most mature routes is to convert sugar into alcohols, such as ethanol or butanol, followed by dehydration to form the corresponding olefins, which can be oligomerized for carbon chain extension to meet the chain length requirements for jet fuel. Sugars can also be converted chemically or biologically into platform chemicals, such as furfural or 5-HMF, followed by carbon−carbon coupling, hydrogenation, and hydrodeoxygenation to form hydrocarbon fuels. Sugars can also be used as carbon sources for microbial fermentation to accumulate microbial lipids, which can be further converted into jet fuel by hydroprocessing. Sugars can also be converted by APR into a mixture of water, hydrogen, and chemical intermediates, and finally into jet fuel with a high content of aromatics. However, most of these routes are only technically feasible at the laboratory stage or in pilot demonstrations. The economic feasibility still needs to be improved in terms of the technical update, catalyst development, feedstock logistics, energy efficiency, as well as process integration. For lignocellulose-to-jet conversion, the transportation of this low-density biomass may lead to extra costs and the consumption of fossil-based gasoline or diesel, which is also a key issue faced by cellulosic ethanol production. Thus a life cycle analysis should be well performed to trace the carbon footprint and evaluate the carbon efficiency of the biomass-to-jet process. Another important aspect of biomass-to-jet production lies in the availability of cheap hydrogen. Maybe it is no problem in a small-scale operation; however, for large-scale production, the source of hydrogen should be considered.

Acknowledgment

The authors are grateful for the support of this work by the National Natural Science Foundation of China (No. 21878176) and National Energy Administration Project (No. NY20130402).

References

1. Deane P. Biofuels for aviation. *Rapid Response Energy Brief* 2015;**5**:1–8.
2. Hari TK, Yaakob Z, Binitha NN. Aviation biofuel from renewable resources: Routes, opportunities and challenges. *Renew Sustain Energy Rev* 2015;**42**:1234–44.
3. Zhao XB, Qi F, Yuan CL, Du W, Liu DH. Lipase-catalyzed process for biodiesel production: enzyme immobilization, process simulation and optimization. *Renew Sustain Energy Rev* 2015;**44**:182–97.
4. Daggett D, Hadaller O, Hendricks R. Alternative fuels and their potential impact on aviation. In: *NASA/TM—2006-214365. NASA; 2006. Report No.: E-15568, NASA/TM-2006-214365, ICAS-2006-5.8.2.*
5. Zhang C, Hui X, Lin YZ, Sung CJ. Recent development in studies of alternative jet fuel combustion: progress, challenges, and opportunities. *Renew Sustain Energy Rev* 2016;**54**:120–38.
6. Mawhood R, Gazis E, de Jong S, Hoefnagels R, Slade R. Production pathways for renewable jet fuel: a review of commercialization status and future prospects. *Biofuel Bioprod Biorefin* 2016;**10**(4):462–84.
7. Wang WC, Tao L. Bio-jet fuel conversion technologies. *Renew Sust Energy Rev* 2016;**53**:801–22.
8. Vasquez MC, Silva EE, Castillo EF. Hydrotreatment of vegetable oils: a review of the technologies and its developments for jet biofuel production. *Biomass Bioenergy* 2017;**105**:197–206.
9. Serrano-Ruiz J, Dumesic J. Catalytic routes for the conversion of biomass into liquid hydrocarbon transportation fuels. *Energy Environ Sci* 2011;**4**(1):83–99.
10. Veriansyah B, Han JY, Kim SK, Hong SA, Kim YJ, Lim JS, et al. Production of renewable diesel by hydroprocessing of soybean oil: Effect of catalysts. *Fuel* 2012;**94**(1):578–85.
11. Zhang C, Nie H, Gao X, Qu H, Chu Y. Hydrodeoxygenation rules of palm oil. *Acta Petrolei Sin—Petroleum Process Sect* 2014;**30**(4):587–94.
12. Morgan T, Santillan-Jimenez E, Harman-Ware A, Ji Y, Grubb D, Crocker M. Catalytic deoxygenation of triglycerides to hydrocarbons over supported nickel catalysts. *Chem Eng J* 2012;**189**:346–55.
13. Wang H. *Biofuels production from hydrotreating of vegetable oil using supported noble metals, and transition metal carbide and nitride*. Wayne State University; 2012.
14. Snare M, Kubickova I, Maki-Arvela P, Eranen K, Murzin D. Heterogeneous catalytic deoxygenation of stearic acid for production of biodiesel. *Ind Eng Chem Res* 2006;**45**(16):5708–15.
15. Galadima A, Muraza O. Catalytic upgrading of vegetable oils into jet fuels range hydrocarbons using heterogeneous catalysts: a review. *J Ind Eng Chem* 2015;**29**:12–23.
16. Hileman JI, Stratton RW. Alternative jet fuel feasibility. *Transp Policy* 2014;**34**:52–62.
17. Liu G, Yan B, Chen G. Technical review on jet fuel production. *Renew Sust Energy Rev* 2013;**25**:59–70.
18. Widjaya ER, Chen GN, Bowtell L, Hills C. Gasification of non-woody biomass: a literature review. *Renew Sust Energy Rev* 2018;**89**:184–93.
19. Molino A, Larocca V, Chianese S, Musmarra D. Biofuels production by biomass gasification: a review. *Energies* 2018;**11**(4):811 Available from: https://doi.org/10.3390/en11040811.
20. Zhao X, Liu W, Deng Y, Zhu JY. *Lignocellulosic biomass-energized fuel cells: cases of high-temperature conversion*. Momentum Press; 2015.

21. Ciferno J, Marano J. *Benchmarking biomass gasification technologies for fuels, chemicals and hydrogen production.* U.S. Department of Energy, National Energy Technology Laboratory; 2002.
22. Demirbas A. Converting biomass derived synthetic gas to fuels via Fisher-Tropsch synthesis. *Energy Sour Part A* 2007;**29**(16):1507–12.
23. Schobert H. In: Varma A, editor. *Chemistry of fossil fuels and biofuels.* Cambridge University Press; 2013.
24. de Klerk A. Can Fischer–Tropsch Syncrude be refined to on-specification diesel fuel. *Energy Fuel* 2009;**23**(9):4593–604.
25. Schulz H. Short history and present trends of Fischer-Tropsch synthesis. *Appl Catal A-Gen* 1999;**186**(1–2):3–12.
26. Liu FX, Hao QL, Wang H, Yang Y, Bai L, Zhu YL, et al. Effect of potassium promoter on reaction performance of iron-based catalyst for Fischer-Tropsch synthesis in slurry reactor. *Chin J Catal* 2004;**25**(11):878–86.
27. Zhao G, Zhang C, Qin S, Xiang H, Li Y. Effect of interaction between potassium and structural promoters on Fischer-Tropsch performance in iron-based catalysts. *J Mol Catal A-Chem* 2008;**286**(1-2):137–42.
28. Wenjuan HOU, Baoshan WU, Xia AN, Tingzhen LI, Hongwei X, Yongwang LI. Effect of K on slurry Fischer-Tropsch synthesis over Fe-Cu-K-Si_O2catalysts. *J Fuel Chem Technol* 2008;**36**(2):186–91.
29. Zhang CH, Yang Y, Teng BT, Li TZ, Zheng HY, Xiang HW, et al. Study of an iron-manganese Fischer-Tropsch synthesis catalyst promoted with copper. *J Catal* 2006;**237**(2):405–15.
30. Wang Y, Kang J, Zhang Q. Research advances in catalysts for Fischer-Tropsch synthesis. *Petrochem Technol* 2009;**38**(12):1255–63.
31. Sun Y, Chen J, Wang J, Jia L, Hou B, Li D, et al. The development of cobalt-based catalysts for Fischer-Tropsch synthesis. *Chin J Catal* 2010;**31**(8):919–27.
32. Dhyani V, Bhaskar T. A comprehensive review on the pyrolysis of lignocellulosic biomass. *Renew Energy* 2018;**129**:695–716.
33. Wang S, Dai G, Yang H, Luo Z. Lignocellulosic biomass pyrolysis mechanism: a state-of-the-art review. *Prog Energy Combust Sci* 2017;**62**:33–86.
34. Sensoz S, Can M. Pyrolysis of pine (*Pinus brutia* Ten.) chips: 1. Effect of pyrolysis temperature and heating rate on the product yields. *Energy Sour* 2002;**24**(4):347–55.
35. Biswas B, Pandey N, Bisht Y, Singh R, Kumar J, Bhaskar T. Pyrolysis of agricultural biomass residues: comparative study of corn cob, wheat straw, rice straw and rice husk. *Bioresour Technol* 2017;**237**:57–63.
36. Huang H-J, Yuan X-Z. Recent progress in the direct liquefaction of typical biomass. *Prog Energy Combust Sci* 2015;**49**:59–80.
37. Toor SS, Rosendahl L, Rudolf A. Hydrothermal liquefaction of biomass: a review of subcritical water technologies. *Energy* 2011;**36**(5):2328–42.
38. Demirbas A. Mechanisms of liquefaction and pyrolysis reactions of biomass. *Energy Convers Manage* 2000;**41**(6):633–46.
39. Guo Y, Yeh T, Song W, Xu D, Wang S. A review of bio-oil production from hydrothermal liquefaction of algae. *Renew Sust Energy Rev* 2015;**48**:776–90.
40. *Alternative fuels from biomass sources* (Online course resource). Available from: <https://www.e-education.psu.edu/egee439/>; 2018.
41. Lian X, Xue Y, Zhao Z, Xu G, Han S, Yu H. Progress on upgrading methods of bio-oil: a review. *Int J Energy Res* 2017;**41**(13):1798–816.

42. Mortensen PM, Grunwaldt JD, Jensen PA, Knudsen KG, Jensen AD. A review of catalytic upgrading of bio-oil to engine fuels. *Appl Catal A—Gen* 2011;**407**(1-2):1–19.
43. Wildschut J, Mahfud FH, Venderbosch RH, Heeres HJ. Hydrotreatment of fast pyrolysis oil using heterogeneous noble-metal catalysts. *Ind Eng Chem Res* 2009;**48**(23):10324–34.
44. Zhang SP, Yan YJ, Li TC, Ren ZW. Upgrading of liquid fuel from the pyrolysis of biomass. *Bioresour Technol* 2005;**96**(5):545–50.
45. Zhao C, Kou Y, Lemonidou AA, Li X, Lercher JA. Highly selective catalytic conversion of phenolic bio-oil to alkanes. *Angew Chem Int Ed* 2009;**48**(22):3987–90.
46. Peng J, Chen P, Lou H, Zheng X. Catalytic upgrading of bio-oil by HZSM-5 in sub- and super-critical ethanol. *Bioresour Technol* 2009;**100**(13):3415–18.
47. Vichaphund S, Aht-Ong D, Sricharoenchaikul V, Atong D. Catalytic upgrading of jatropha waste fast pyrolysis vapors over synthesized HZSM-5 using analytical Py-GC/MS. *J Biobased Mater Bioenergy* 2013;**7**(2):252–8.
48. Vitolo S, Seggiani M, Frediani P, Ambrosini G, Politi L. Catalytic upgrading of pyrolytic oils to fuel over different zeolites. *Fuel* 1999;**78**(10):1147–59.
49. Xiu S, Shahbazi A. Bio-oil production and upgrading research: a review. *Renew Sust Energy Rev* 2012;**16**(7):4406–14.
50. Wang H, Tucker M, Ji Y. Recent development in chemical depolymerization of lignin: a review. *J Appl Chem* 2013;**2013** Available from: https://doi.org/10.1155/2013/838645 Article ID 838645.
51. Xu C, Arancon RAD, Labidi J, Luque R. Lignin depolymerisation strategies: towards valuable chemicals and fuels. *Chem Soc Rev* 2014;**43**(22):7485–500.
52. Zhao XB, Li SM, Wu RC, Liu DH. Organosolv fractionating pre-treatment of lignocellulosic biomass for efficient enzymatic saccharification: chemistry, kinetics, and substrate structures. *Biofuel Bioprod Biorefin* 2017;**11**(3):567–90.
53. Cheng F, Brewer CE. Producing jet fuel from biomass lignin: potential pathways to alkyl benzenes and cycloalkanes. *Renew Sust Energy Rev* 2017;**72**:673–722.
54. Wright ME, Harvey BG, Quintana RL. Highly efficient zirconium-catalyzed batch conversion of 1-butene: a new route to jet fuels. *Energy Fuel* 2008;**22**(5):3299–302.
55. Lenihan P, Orozco A, O'Neill E, Ahmad M, Rooney D, Walker G. Dilute acid hydrolysis of lignocellulosic biomass. *Chem Eng J* 2010;**156**(2):395–403.
56. Harmer MA, Fan A, Liauw A, Kumar RK. A new route to high yield sugars from biomass: phosphoric-sulfuric acid. *Chem Commun* 2009;(43):6610–12.
57. Heinonen J, Tamminen A, Uusitalo J, Sainio T. Ethanol production from wood via concentrated acid hydrolysis, chromatographic separation, and fermentation. *J Chem Technol Biotechnol* 2012;**87**(5):689–96.
58. Taherzadeh M, Karimi K. Acid-based hydrolysis processes for ethanol from lignocellulosic materials: a review. *Bioresources* 2007;**2**(3):472–99.
59. Maki-Arvela P, Salmi T, Holmbom B, Willfor S, Murzin D. Synthesis of sugars by hydrolysis of hemicelluloses—a review. *Chem Rev* 2011;**111**(9):5638–66.
60. Zhang YHP, Lynd LR. Toward an aggregated understanding of enzymatic hydrolysis of cellulose: noncomplexed cellulase systems. *Biotechnol Bioeng* 2004;**88**(7):797–824.
61. Zhao XB, Zhang LH, Liu DH. Biomass recalcitrance. Part I: the chemical compositions and physical structures affecting the enzymatic hydrolysis of lignocellulose. *Biofuel Bioprod Biorefin* 2012;**6**(4):465–82.
62. Zhao XB, Zhang LH, Liu DH. Biomass recalcitrance. Part II: Fundamentals of different pre-treatments to increase the enzymatic digestibility of lignocellulose. *Biofuel Bioprod Biorefin* 2012;**6**(5):561–79.

63. Sun Y, Cheng J. Hydrolysis of lignocellulosic materials for ethanol production: a review. *Bioresour Technol* 2002;**83**(1):1–11.
64. Rastogi M, Shrivastava S. Recent advances in second generation bioethanol production: an insight to pretreatment, saccharification and fermentation processes. *Renew Sust Energy Rev* 2017;**80**:330–40.
65. Chandel AK, Garlapati VK, Singh AK, Antunes FAF, da Silva SS. The path forward for lignocellulose biorefineries: bottlenecks, solutions, and perspective on commercialization. *Bioresour Technol* 2018;**264**:370–81.
66. Taha M, Foda M, Shahsavari E, Aburto-Medina A, Adetutu E, Ball A. Commercial feasibility of lignocellulose biodegradation: possibilities and challenges. *Curr Opin Biotechnol* 2016;**38**:190–7.
67. Ibrahim MF, Kim SW, Abd-Aziz S. Advanced bioprocessing strategies for biobutanol production from biomass. *Renew Sust Energy Rev* 2018;**91**:1192–204.
68. Wang W, Tao L, Markham J, Zhang Y, Tan E, Batan L, et al. *Review of Biojet Fuel Conversion Technologies. National Renew*. USA: Energy Laboratory; 2016. Contract No.: NREL/TP-5100-66291.
69. Zhang MH, Yu YZ. Dehydration of ethanol to ethylene. *Ind Eng Chem Res* 2013;**52**(28):9505–14.
70. Mascal M. Chemicals from biobutanol: technologies and markets. *Biofuel Bioprod Biorefin* 2012;**6**(4):483–93.
71. Taylor JD, Jenni MM, Peters MW. Dehydration of fermented isobutanol for the production of renewable chemicals and fuels. *Topics Catal* 2010;**53**(15–18):1224–30.
72. Wright M. *Process for the dehydration of aqueous bio-derived terminal alcohols to terminal alkenes (patent) US2014 2014-12-16.*
73. Kim H, Kim D, Park YK, Jeon JK. Synthesis of jet fuel through the oligomerization of butenes on zeolite catalysts. *Res Chem Intermediates* 2018;**44**(6):3823–33.
74. Krivan E, Tomasek S, Hancsok J. Application possibilities of zeolite catalysts in oligomerization of light olefins. *Periodica Polytechnica—Chem Eng* 2014;**58**(2):149–56.
75. Anbarasan P, Baer ZC, Sreekumar S, Gross E, Binder JB, Blanch HW, et al. Integration of chemical catalysis with extractive fermentation to produce fuels. *Nature* 2012;**491**(7423):235–9.
76. Chheda J, Roman-Leshkov Y, Dumesic J. Production of 5-hydroxymethylfurfural and furfural by dehydration of biomass-derived mono- and poly-saccharides. *Green Chem* 2007;**9**(4):342–50.
77. Sheldon R. Green and sustainable manufacture of chemicals from biomass: state of the art. *Green Chem* 2014;**16**(3):950–63.
78. Zheng HY, Zhu YL, Teng BT, Bai ZQ, Zhang CH, Xiang HW, et al. Towards understanding the reaction pathway in vapour phase hydrogenation of furfural to 2-methylfuran. *J Mol Catal A-Chem* 2006;**246**(1-2):18–23.
79. Zhang ZH. Synthesis of gamma-valerolactone from carbohydrates and its applications. *ChemSusChem* 2016;**9**(2):156–71.
80. Wu LP, Moteki T, Gokhale AA, Flaherty DW, Toste FD. Production of fuels and chemicals from biomass: condensation reactions and beyond. *Chem* 2016;**1**(1):32–58.
81. Bond JQ, Alonso DM, Wang D, West RM, Dumesic JA. Integrated catalytic conversion of gamma-valerolactone to liquid alkenes for transportation fuels. *Science* 2010;**327**(5969):1110–14.

82. Rosatella AA, Simeonov SP, Frade RFM, Afonso CAM. 5-Hydroxymethylfurfural (HMF) as a building block platform: biological properties, synthesis and synthetic applications. *Green Chem* 2011;**13**(4):754–93.
83. Shi N, Liu QY, Wang TJ, Ma LL, Zhang Q. One-pot degradation of cellulose into furfural compounds in hot compressed steam with dihydric phosphates. *ACS Sust Chem Eng* 2014;**2**(4):637–42.
84. Zhang L, Mao JB, Li SM, Yin JM, Sun XD, Guo XW, et al. Hydrogenation of levulinic acid into gamma-valerolactone over in situ reduced CuAg bimetallic catalyst: Strategy and mechanism of preventing Cu leaching. *Appl Catal B—Environ* 2018;**232**:1–10.
85. Mizugaki T, Nagatsu Y, Togo K, Maeno Z, Mitsudome T, Jitsukawa K, et al. Selective hydrogenation of levulinic acid to 1,4-pentanediol in water using a hydroxyapatite-supported Pt-Mo bimetallic catalyst. *Green Chem* 2015;**17**(12):5136–9.
86. Yang PP, Xia QN, Liu XH, Wang YQ. Catalytic transfer hydrogenation/hydrogenolysis of 5-hydroxymethylfurfural to 2,5-dimethylfuran over Ni-Co/C catalyst. *Fuel* 2017;**187**:159–66.
87. Huang YB, Chen MY, Yan L, Guo QX, Fu Y. Nickel-tungsten carbide catalysts for the production of 2,5-dimethylfuran from biomass-derived molecules. *ChemSusChem* 2014;**7**(4):1068–72.
88. Zu YH, Yang PP, Wang JJ, Liu XH, Ren JW, Lu GZ, et al. Efficient production of the liquid fuel 2,5-dimethylfuran from 5-hydroxymethylfurfural over Ru/Co3O4 catalyst. *Appl Catal B—Environ* 2014;**146**:244–8.
89. Gilkey MJ, Panagiotopoulou P, Mironenko AV, Jenness GR, Vlachos DG, Xu BJ. Mechanistic insights into metal lewis acid-mediated catalytic transfer hydrogenation of furfural to 2-methylfuran. *ACS Catal* 2015;**5**(7):3988–94.
90. Yan K, Chen A. Efficient hydrogenation of biomass-derived furfural and levulinic acid on the facilely synthesized noble-metal-free Cu-Cr catalyst. *Energy* 2013;**58**:357–63.
91. Dong F, Zhu YL, Zheng HY, Zhu YF, Li XQ, Li YW. Cr-free Cu-catalysts for the selective hydrogenation of biomass-derived furfural to 2-methylfuran: the synergistic effect of metal and acid sites. *J Mol Catal A-Chem* 2015;**398**:140–8.
92. Fu ZL, Wang Z, Lin WG, Song WL, Li SG. High efficient conversion of furfural to 2-methylfuran over Ni-Cu/Al_2O_3 catalyst with formic acid as a hydrogen donor. *Appl Catal A—Gen* 2017;**547**:248–55.
93. Zhou MH, Li J, Wang K, Xia HH, Xu JM, Jiang JC. Selective conversion of furfural to cyclopentanone over CNT-supported Cu based catalysts: model reaction for upgrading of bio-oil. *Fuel* 2017;**202**:1–11.
94. Li XL, Deng J, Shi J, Pan T, Yu CG, Xu HJ, et al. Selective conversion of furfural to cyclopentanone or cyclopentanol using different preparation methods of Cu-Co catalysts. *Green Chem* 2015;**17**(2):1038–46.
95. Serrano-Ruiz JC, Luque R, Sepulveda-Escribano A. Transformations of biomass-derived platform molecules: from high added-value chemicals to fuels via aqueous-phase processing. *Chem Soc Rev* 2011;**40**(11):5266–81.
96. Kikhtyanin O, Kelbichova V, Vitvarova D, Kubu M, Kubicka D. Aldol condensation of furfural and acetone on zeolites. *Catal Today* 2014;**227**:154–62.
97. Xing R, Subrahmanyam AV, Olcay H, Qi W, van Walsum GP, Pendse H, et al. Production of jet and diesel fuel range alkanes from waste hemicellulose-derived aqueous solutions. *Green Chem* 2010;**12**(11):1933–46.

98. Bui TV, Sooknoi T, Resasco DE. Simultaneous upgrading of furanics and phenolics through hydroxyalkylation/aldol condensation reactions. *ChemSusChem* 2017;**10**(7):1631–9.
99. Dutta S, Bohre A, Zheng WQ, Jenness GR, Nunez M, Saha B, et al. Solventless C-C coupling of low carbon furanics to high carbon fuel precursors using an improved graphene oxide carbocatalyst. *ACS Catal* 2017;**7**(6):3905–15.
100. Li G, Li N, Li S, Wang A, Cong Y, Wang X, et al. Synthesis of renewable diesel with hydroxyacetone and 2-methyl-furan. *Chem Commun* 2013;**49**(51):5727–9.
101. Gaertner CA, Serrano-Ruiz JC, Braden DJ, Dumesic JA. Ketonization reactions of carboxylic acids and esters over ceria-zirconia as biomass-upgrading processes. *Ind Eng Chem Res* 2010;**49**(13):6027–33.
102. Huang YB, Yang Z, Dai JJ, Guo QX, Fu Y. Production of high quality fuels from lignocellulose-derived chemicals: a convenient C-C bond formation of furfural, 5-methylfurfural and aromatic aldehyde. *RSC Adv* 2012;**2**(30):11211–14.
103. Byun J, Han J. Catalytic production of biofuels (butene oligomers) and biochemicals (tetrahydrofurfuryl alcohol) from corn stover. *Bioresour Technol* 2016;**211**:360–6.
104. Zhu CJ, Shen T, Liu D, Wu JL, Chen Y, Wang LF, et al. Production of liquid hydrocarbon fuels with acetoin and platform molecules derived from lignocellulose. *Green Chem* 2016;**18**(7):2165–74.
105. Wegenhart BL, Yang LN, Kwan SC, Harris R, Kenttamaa HI, Abu-Omar MM. From furfural to fuel: synthesis of furoins by organocatalysis and their hydrodeoxygenation by cascade catalysis. *ChemSusChem* 2014;**7**(9):2742–7.
106. Li GY, Li N, Wang ZQ, Li CZ, Wang AQ, Wang XD, et al. Synthesis of high-quality diesel with furfural and 2-methylfuran from hemicellulose. *ChemSusChem* 2012;**5**(10):1958–66.
107. Gaertner C, Serrano-Ruiz J, Braden D, Dumesic J. Catalytic coupling of carboxylic acids by ketonization as a processing step in biomass conversion. *J Catal* 2009;**266**(1):71–8.
108. Han J, Sen S, Luterbacher J, Alonso D, Dumesic J, Maravelias C. Process systems engineering studies for the synthesis of catalytic biomass-to-fuels strategies. *Comput Chem Eng* 2015;**81**:57–69.
109. Su MX, Li WZ, Zhang TW, Xin HS, Li S, Fan W, et al. Production of liquid fuel intermediates from furfural via aldol condensation over Lewis acid zeolite catalysts. *Catal Sci Technol* 2017;**7**(16):3555–61.
110. Cueto J, Faba L, Diaz E, Ordonez S. Performance of basic mixed oxides for aqueous-phase 5-hydroxymethylfurfural-acetone aldol condensation. *Appl Catal B—Environ* 2017;**201**:221–31.
111. Yang JF, Li N, Li GY, Wang WT, Wang AQ, Wang XD, et al. Solvent-free synthesis of C-10 and C-11 branched alkanes from furfural and methyl isobutyl ketone. *ChemSusChem* 2013;**6**(7):1149–52.
112. Chen F, Li N, Li SS, Yang JF, Liu F, Wang WT, et al. Solvent-free synthesis of C-9 and C-10 branched alkanes with furfural and 3-pentanone from lignocellulose. *Catal Commun* 2015;**59**:229–32.
113. Liang DC, Li GZ, Liu YH, Wu JM, Zhang XW. Controllable self-aldol condensation of cyclopentanone over MgO-ZrO_2 mixed oxides: origin of activity & selectivity. *Catal Commun* 2016;**81**:33–6.
114. Chheda JN, Dumesic JA. An overview of dehydration, aldol-condensation and hydrogenation processes for production of liquid alkanes from biomass-derived carbohydrates. *Catal Today* 2007;**123**(1–4):59–70.

115. Cui X, Zhao X, Liu D. A novel route for the flexible preparation of hydrocarbon jet fuels from biomass-based platform chemicals: a case of using furfural and 2,3-butanediol as feedstocks. *Green Chem* 2018;**20**(9):2018–26.
116. Liang M, Jiang J. Advancing oleaginous microorganisms to produce lipid via metabolic engineering technology. *Progr Lipid Res* 2013;**52**(4):395–408.
117. Yellapu SK, Bharti, Kaur R, Kumar LR, Tiwari B, Zhang X, et al. Recent developments of downstream processing for microbial lipids and conversion to biodiesel. *Bioresour Technol* 2018;**256**:515–28.
118. Xu JY, Zhao XB, Wang WC, Du W, Liu DH. Microbial conversion of biodiesel byproduct glycerol to triacylglycerols by oleaginous yeast *Rhodosporidium toruloides* and the individual effect of some impurities on lipid production. *Biochem Eng J* 2012;**65**:30–6.
119. Zhao XB, Peng F, Du W, Liu CM, Liu DH. Effects of some inhibitors on the growth and lipid accumulation of oleaginous yeast *Rhodosporidium toruloides* and preparation of biodiesel by enzymatic transesterification of the lipid. *Bioproc Biosyst Eng* 2012;**35**(6):993–1004.
120. Xu JY, Du W, Zhao XB, Zhang GL, Liu DH. Microbial oil production from various carbon sources and its use for biodiesel preparation. *Biofuel Bioprod Biorefin* 2013;**7**(1):65–77.
121. Lee A, Lewis D, Ashman P. Disruption of microalgal cells for the extraction of lipids for biofuels: processes and specific energy requirements. *Biomass Bioenergy* 2012;**46**:89–101.
122. Wang T, Qiu S, Weng Y, Chen L, Liu Q, Long J, et al. Liquid fuel production by aqueous phase catalytic transformation of biomass for aviation. *Appl Energy* 2015;**160**:329–35.
123. *Production of conventional liquid fuels from sugars.* Virent Energy Systems, Inc. Available from: <http://www.etipbioenergy.eu/images/Virent_Technology_Whitepaper.pdf>; 2008.

Chapter 6

Advances in bio-oil production and upgrading technologies

Atthapon Srifa[1], Weerawut Chaiwat[2], Pawnprapa Pitakjakpipop[3], Wipark Anutrasakda[4] and Kajornsak Faungnawakij[5]

[1]Department of Chemical Engineering, Faculty of Engineering, Mahidol University, Nakhon Pathom, Thailand, [2]Environmental Engineering and Disaster Management Program, School of Interdisciplinary Studies, Mahidol University, Kanchanaburi, Thailand, [3]National Metal and Materials Technology Center (MTEC), Pathum Thani, Thailand, [4]Department of Chemistry, Faculty of Science, Chulalongkorn University, Bangkok, Thailand, [5]National Nanotechnology Center (NANOTEC), National Science and Technology Development Agency (NSTDA), Pathum Thani, Thailand

6.1 Introduction

It is a known fact that the global energy demand has been increasing dramatically to serve the economy and population growth with heavy dependence on nonrenewable, fossil-based energy. The development and utilization of renewable energy have, therefore, become an urgent agenda, not only to provide a sustainable energy source but also to slow down the global warming effect. Bio-oil is one of the most important energy sources and can be produced from renewable feedstocks, such as lignocellulosic biomass, seaweed, and microalgae.[1] The production of bio-oil is conventionally obtained via pyrolysis and therefore, the liquid product is called pyrolysis oil. Simultaneously, biochar and gas products are produced via pyrolysis and the yield of each fraction depends strongly on the pyrolysis conditions and reactor design. High temperatures, above 300°C, without the presence of oxygen or extremely-low oxygen conditions are required for the effective conversion of feedstocks into liquid fuels, which basically contains lignin-derived oligomers, acids, alcohols, and esters.[1,2]

In the past few years, torrefaction has been proved to be an effective biomass pretreatment to enhance the efficiency of the pyrolysis process.[3] Nonetheless, the direct use of biocrude is still highly limited since it has high acidity (pH 2–5), high oxygen content, high viscosity, and low thermal/chemical stability as compared to conventional fuels (gasoline, diesel,

Sustainable Bioenergy. DOI: https://doi.org/10.1016/B978-0-12-817654-2.00006-X

biodiesel, bioethanol). In addition, hydrothermal liquefaction (HTL) has been developed to produce bio-oil. With a lower operating temperature under aqueous-phase conditions, bio-oils obtained from liquefaction are mainly composed of hydrophobic compounds with lower viscosity and higher solubility, compared to pyrolysis oil, yet it is not ready for conventional engines. Therefore bio-oil upgrading technology has been extensively developed in order to improve the fuel properties for practical uses. The hydrotreatment or hydrotreating process is a promising approach to upgrade bio-oil via deoxygenation, cracking, and/or isomerization using hydrogen gas.[4] Esterification is also applicable for upgrading with mild reaction conditions, while supercritical fluids can be employed with high pressure and/or temperature to fractionate the mixture species in bio-oil.

Fig. 6.1 shows patent and publication trends in the research and development of bio-oil technology during 2008–17. The number of publications and patents on the pyrolysis of various feedstocks into bio-oil typically increase each year between 2008 and 2017 and the numbers increase significantly to c. 1000 publications and c. 600 patents in 2017. There was, however, a small decline in publication numbers in 2017. The number of publications and patents on other related bio-oil technologies, including torrefaction, HTL, hydrotreatment, and esterification typically increase although there is a small fluctuation in some years. Since 2008, the number of publications and

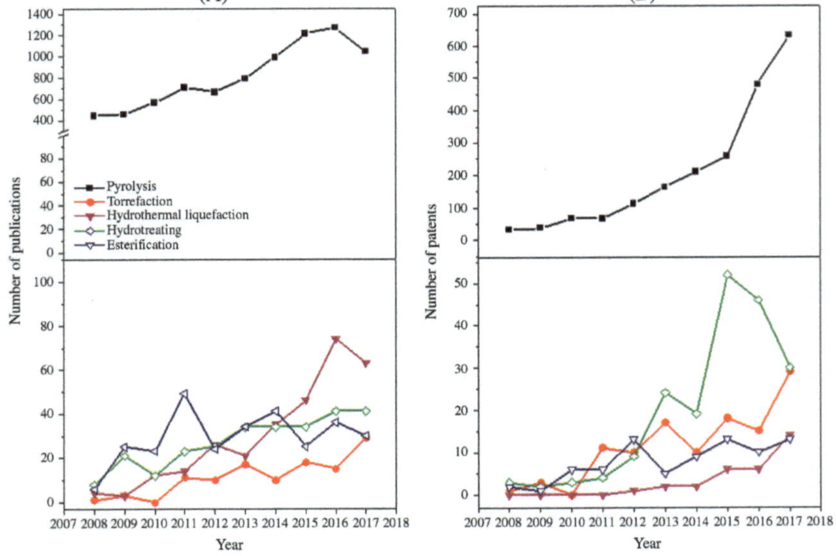

FIGURE 6.1 (A) Publication and (B) patent trends of bio-oil production and upgrading technology analyzed by Scopus (Publications) and Derwent–Clarivate Analytics (Patents) between 2008 and 2017. Analysis terms are [Biooil OR Bio-oil OR Pyrolysis oil] AND [Pyrolysis/Torrefaction/Hydrothermal liquefaction/Hydrotreatment OR Hydrotreating/Esterification].

patents on torrefaction keep increasing, and in 2017 they were recorded at c. 25–30 for each technology. HTL has become a promising candidate for bio-oil synthesis, especially in the past few years, as suggested by the growth rate of the publications. The upgrading of bio-oil via hydrotreatment or esterification has become more and more important for realizing the practical use of bio-oil. The number of publications and patents rose by around 20–40 per year between 2013 and 2017. However, during 2008–12, major effort was put into the esterification process, as suggested by the increased amount of papers and patents. It was also found that, at present, the supercritical fluid process for bio-oil production and upgrading has not been widely studied as no significant numbers of publications and patents were observed.

This chapter presents an overview of current advances in bio-oil production and upgrading technologies. Pyrolysis, torrefaction, and HTL are discussed as major bio-oil production processes. In addition, hydrotreating and esterification are reviewed as practical upgrading processes. In particular, the development of heterogeneous catalysts for improving the efficiency of these processes has been intensively discussed.

6.2 Bio-oil production technology

6.2.1 Pyrolysis and torrefaction

Pyrolysis is a thermochemically endothermic reaction under an inert atmosphere to decompose organic raw materials, such as biomass into solid (char), liquid (bio-oil), and gaseous products. Pyrolysis product distribution and quantitative and qualitative properties, particularly of bio-oil as a major product, are dependent on several factors, including: (1) type and particle size of biomass feedstock; (2) pyrolysis operating conditions, such as heating rate, feeding rate, vapor residence time, temperature, reaction time; and (3) reactor design technologies.[5–10] Fig. 6.2 summarizes the process of biomass pyrolysis with additional pretreatment methods for crude bio-oil production, including typical operating conditions and analyses of feedstock and bio-oil product properties.

6.2.1.1 Conventional pyrolysis of biomass

Biomass pyrolysis is conventionally complex due to the variations in the composition of its three major constituent ratios; cellulose, hemicellulose, and lignin. Each component exhibits different degradation pathways mainly dependent on pyrolysis temperature. During biomass pyrolysis, various thermo-chemical reactions, including dehydration, depolymerization, decarboxylation, and charring or crosslinking reactions, take place simultaneously and continuously with increases in reaction temperature. According to thermogravimetric analysis, hemicellulose, which is a branched heteropolysaccharide, is primarily decomposed at 250°C–350°C. The decomposition of

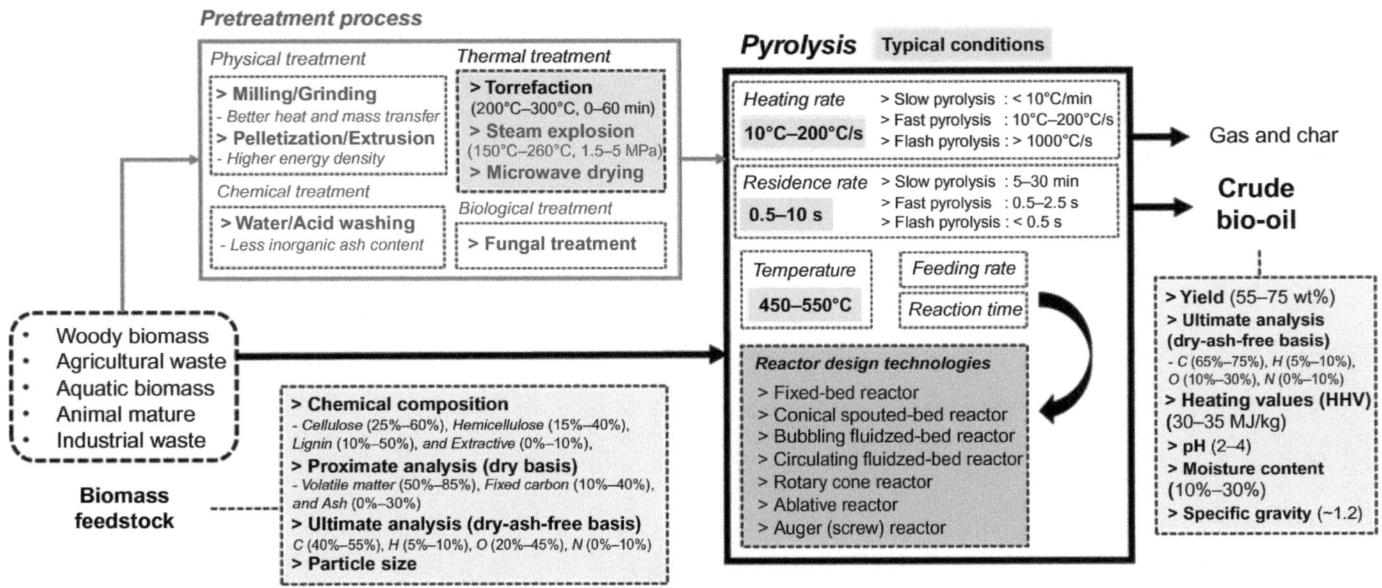

FIGURE 6.2 Overview of the process of biomass pyrolysis and its pretreatment methods for bio-oil production.

cellulose, a crystalline polysaccharide with a glucose monomer, mainly occurs at 300°C–400°C. A broad range, between 200°C and 600°C, can be observed for the gradual decomposition of lignin, which is the most stable three-dimensional structure and highly branched polyphenolic substance.[6–8] Some significant char (fixed-carbon) contents about 30–40 wt.% still remain after the pyrolysis of lignin at 900°C, while cellulose and hemicellulose are completely converted into volatile and gaseous products. Proximate and ultimate analyses to characterize the chemical properties of biomass feedstock, therefore, can be significantly varied with these different composition ratios in biomass structure as summarized in Fig. 6.2. Due to these reasons, biomasses containing high amounts of cellulose and hemicellulose with higher volatility and carbon contents would be preferable for bio-oil production via conventional pyrolysis.

In terms of the differences in heating rate and vapor residence times, the pyrolysis of biomass can be simply classified into: (1) slow pyrolysis (<10°C/min, 10–30 min); (2) fast pyrolysis (10–200°C/s, 0.5–2.5 seconds); and (4) flash pyrolysis (>1000°C/s, <0.5 seconds),[5,6] as indicated in Fig. 6.2. The increase in heating rate up to the final pyrolysis temperature, can enhance the yield of bio-oil product because higher heat transfer rates with shorter vapor residence times may minimize secondary reactions, such as the cracking and repolymerization of volatile tar. Various researchers have also concluded that a shorter vapor residence time by increasing the inert gas flow rate improves the bio-oil yields. However, the yield could be reduced if the gas flow is increased beyond certain limits since biomass solids and pyrolyzed volatiles might be partially entrained out of the reactor.[11,12]

To be effective in bio-oil production from biomass with relatively higher yields at 55–75 wt.%, fast pyrolysis has mostly been used as a typical thermo-chemical process with a temperature range of 450°C–550°C. The yield of bio-oil could be improved with an increase in temperature, but the opposite results were observed at temperatures that were too high (mostly above 500°C) due to the secondary cracking of volatile tar leading to an increase in gaseous yields.[13,14] Moreover, Ateş and Işıkdağ[15] and Ly et al.[16] reported that high-quality bio-oil with higher contents of carbon and hydrogen could be obtained when the pyrolysis temperature is increased. Not only the effect of temperature, but also other operating parameters, including type of biomass feedstock, also cause changes in product yields and qualities. Table 6.1 shows the optimum temperatures used in the pyrolysis of various biomasses to achieve the maximum yields of bio-oil products.[17–25]

6.2.1.2 Thermal pretreatment of biomass via torrefaction

To enhance pyrolysis efficiency in terms of heat and mass transfer and also to improve bio-oil yield and quality, the pretreatment of biomass feedstock is

TABLE 6.1 Maximum yields of bio-oil products at optimum temperatures in pyrolysis of various biomasses.

Biomass type (sample size)	Pyrolysis reactors/operating conditions	Optimum pyrolysis temperature (°C)	Maximum yield of bio-oil (wt.%)	Reference
Rice husk (0.63–1 mm)	Conical spouted bed reactor with a feeding rate of 1 g/min	450	70.0	17
Oil palm empty fruit bunch (355–50 μm)	Fluidized-bed reactor at a vapor residence time of 1.01–1.04 s with a feeding rate of 150 g/h	500	72.4	18
Neem deoiled cake (n/a)	Semi-batch reactor (30 g sample) at the heating rate of 25°C/min	400	40.2	19
Sugarcane bagasse (600–1180 μm)	Fixed-bed fire-tube heating reactor (200 g sample) with a vapor residence time of 5 s	475	56.0	20
Cassava rhizome 1 (250–425 μm)	Fluidized-bed reactor with a feeding rate of 300–500 g/h	472	63.2	21
Cassava rhizome 2 (355–500 μm)	Fluidized-bed reactor with a feeding rate of 150 g/h	510	65.0	22
Cassava stalk 1 (250–425 μm)	Fluidized-bed reactor with a feeding rate of 300–500 g/h	469	61.4	21
Cassava stalk 2 (355–500 μm)	Fluidized-bed reactor with a feeding rate of 150 g/h	475	62.0	22
Jatropha seed shell (0.125–1.40 mm)	Fluidized-bed pyrolyzer with a feeding rate of 0.94 kg/h and vapor residence time of 3.3–4.9 s	470	48.0	23
Pistachio shell (1.82 mm in average)	Fixed-bed reactor (20 g sample) with a heating rate of 7°C/min	550	20.5	24
Poplar (50 μm–3 mm)	Conical spouted bed reactor with a feeding rate of 25 kg/h for 4–6 h	455	69.0	25

necessary. Different individual or co-pretreatment methods, such as (1) physical treatment (grinding and milling); (2) thermal treatment (drying and torrefaction); (3) chemical treatment (water and acid washing); and (4) biological treatment (fungal and enzymes), can potentially be applied to obtain a uniform size of feedstock, remove moisture and oxygen content, and partially leach inorganic ash content from biomass as shown in Fig. 6.2. The particle size of feedstock is one of the most important factors affecting pyrolysis behavior. The grinding or milling of raw materials into more uniform and smaller sizes is the conventional physical treatment method to promote heat and mass transfer during fast pyrolysis.[7,10] Chemical treatments, like water and acid washing, have also been studied to reduce the content of inorganic ash, which could possibly affect pyrolysis product distribution with an improved bio-oil yield. The pyrolysis of lignocellulosic biomass biologically treated by fungus was found to be more efficient due to the selective decomposition of lignin and lower temperature requirement.[7] Moreover, drying using various types of dryers is recognized as the simplest way to remove water content thermally from biomass. Thermal treatments at low-temperature (200°C–300°C) for minutes to hours under inert conditions, that is, torrefaction, has been extensively used because it not only removes moisture, but also partially reduces the oxygen content in the original biomass, leading to a higher energy density of the feedstock.[26]

In the past few years, many researchers have focused on and studied the pyrolysis of torrefied biomass. Various biomass samples, commonly agricultural and woody residues, were conventionally torrefied in a range of 200°C–330°C for seconds to 30 minutes mostly using fixed-bed reactors prior to pyrolysis at 350°C–800°C to investigate the composition and properties of bio-oil products using advanced analytical techniques, such as Py-GCMS.[26] It can be briefly summarized that high-quality bio-oil could be produced with relatively low moisture and high carbon content, leading to a significant decrease in the O/C ratios of bio-oil generated by the pyrolysis of torrefied biomass.[27] Moreover, bio-oil obtained from the pyrolysis of torrefied biomass exhibited lower organic acid content according to the lower oxygen content in torrefied biomass, causing the bio-oil to be less corrosive, particularly for more severe torrefaction of original samples. Furthermore, aromatic hydrocarbon and phenol contents could be enriched when torrefied biomass was used to produce bio-oil due to the decomposition of lignin and sugars.[28–31] A combination of demineralization using acid treatment and the torrefaction of biomass has also been studied to upgrade both the quantitative and qualitative properties of bio-oil products[32,33] as shown in Fig. 6.3. Wigley et al.[33] used a 1% acetic acid washing at 30°C for 4 hours and torrefaction at 270°C for 20 minutes in series as an optimum pretreatment condition to increase the bio-oil yield from 55.3 to 57.8 wt.% and improve the quality and stability of the bio-oil products with much lower organic acid and moisture contents. However, various biomass types with different

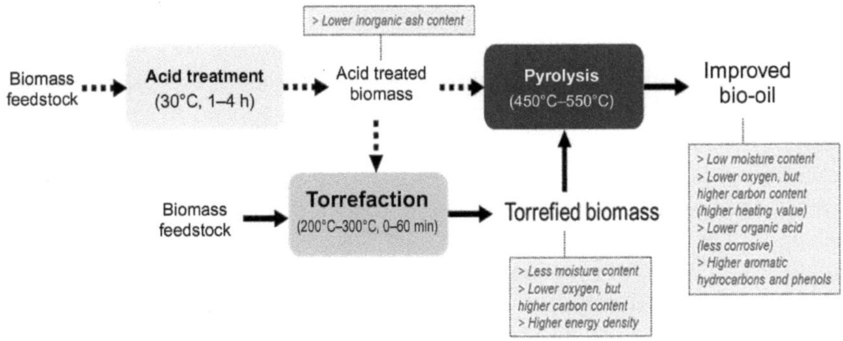

FIGURE 6.3 A combination pretreatment process of acid washing and torrefaction of biomass feedstock to improve bio-oil yield and qualities after pyrolysis.

chemical compositions and continuous reactors in larger scale tests for use in both torrefaction and subsequent pyrolysis should be further investigated in order to achieve a high pyrolysis efficiency for industrial scale-up.

6.2.1.3 *Design technologies of pyrolysis reactors*

For upgrading pyrolysis performance via fast pyrolysis, design reactor technology is a key factor for achieving a high heat transfer rate, controlled reaction temperature, and rapid cooling of pyrolysis volatile. Several types of semicontinuous and continuous reactors have been developed to improve heating rates and residence times. Table 6.2 presents a comparative overview of six available technologies of fast/flash pyrolysis reactors for bio-oil production from biomass; that is, (1) bubbling fluidized-bed reactor (BFB), (2) circulating BFB (CFB), (3) conical spouted bed reactor (CSBR), (4) rotating cone reactor (RCR), (5) ablative reactor, and (6) screw or auger reactor.[34–39] Each reactor design provides unique operating conditions and phenomenon with different advantages and disadvantages, as briefly reviewed in Table 6.2, in order to potentially simplify process operations with more compact sizing, leading to an economic feasibility for scale-up.[6,10]

BFB was implemented for the first time in fast pyrolysis for the semicontinuous production of bio-oil with a consistent high yield of 70%–75% due to the simplicity in the reactor design and operation. During inert gas fluidization with sand, small biomass particles (0.5–2 mm) could achieve good heat transfer during a short residence time (2–3 seconds).[34] CFB reduces the vapor residence time to 0.5–1 seconds and utilizes char and noncondensable gases for recirculation, leading to an increase in bio-oil yield and higher energy recovery. It is, therefore, feasible and currently commercialized in various industries. However, high gas velocities are necessary in fluidization systems, particularly in CFB, causing these systems to be more expensive in terms of the power requirement and the construction of the complex reactor

TABLE 6.2 Overview of pyrolysis reactor technologies for continuous bio-oil production.

Reactor type	Bio-oil yield (wt.%)	Feed particle size (mm)	Operating conditions and phenomena (pros/cons)	Reactor size and design	Scale-up	Reference
Bubbling fluidized bed	50–75	Fine/small (0.5–2)	High heat transfer rates with short residence times (2–3 s), *but high gas flow required with char accumulation problem*	Medium/less complex	Easy	34
Circulating fluidized bed	75–80	Fine/small (<2)	Higher heat transfer rates with short residence times (0.5–1 s), *but higher gas velocities required for recirculation of solids*	Medium/ complex	Easy/ feasible	35
Conical spouted bed	60–80	Coarse/small (1–3)	A substitute for fluidized bed with high mass and heat transfer, *but complex design with draft tube*	Medium/ complex	Difficult	36
Rotating cone	70	Small (0.1–4)	No carrier gas requirement, *but mechanical power through centrifugation required*	Small/ complex	Medium	37
Ablative	50–75	Small (0.5–3) to large (>5)	No inert gas required with high heat transfer on reactor surface, *but mechanical power and pressure required with possible heat losses*	Compact/ complex	Difficult	38
Screw/auger	65–75	Medium to pellet (>15)	Lower process temperature (400°C), *but moving parts in hot zone causing difficult heat transfer for scaling-up*	Compact/ simple	Not proven	39

systems.[35] CSBR is considered to be an alternative to fluidized beds. It can handle coarse particles (> 1 mm) based on conical geometry, indicating less pressure drop and grinding energy. However, it has some limitations for scale-up due to its complex design and difficult operation.[36] RCR is mostly based on mechanical centrifugation without the requirement for a carrier gas. RCR technologies are currently commercialized to industrial scales, typically producing bio-oil yields in the range of 60–70 wt.%.[37] The ablative reactor, with high heat transfer from the reactor surface plate using a mechanical press shows its advantage for handling large particles of biomass without gas flow. It is still difficult for larger scale upgrading due to its complexity in mechanical design in terms of reactor size ratio and the problems of possible heat loss and high-pressure requirement. Finally, a compact screw or auger reactor could lower the operating temperature to about 400°C, but scalability is still doubtful because there are moving parts are in the heating area, making suitable heat transfer in larger reactors difficult.[6,10]

According to a brief summary of previous research studies, the pyrolysis of biomass to provide good quality crude bio-oil with a relatively higher yield, various aspects of operating conditions and feedstock, appropriate pretreatment methods, and reactor designs should be entirely concerned with further developing feasible combined technologies in larger scales for commercialized bio-oil production from available biomasses.

6.2.2 Hydrothermal liquefaction

The HTL of biomass and microalgae as feedstocks is a thermal depolymerization process of wet lignocellulosic materials into liquid fuels, gas, and solids under pressurized hot water conditions in order to break down the biomass structure into shorter components. Generally, this process can be operated in the temperature and pressure ranges of 250°C–350°C and 40–220 bar, respectively,[1,40–43] which are lower than those of the conventional pyrolysis process. It should be noted that the moisture content of biomass is generally as high as 50%–90%, and therefore HTL provides the advantage of energy efficiency by converting biomass without drying pretreatment. Typically, the reaction proceeds using homogeneous and/or heterogeneous catalysts with or without reducing gas in order to improve the quality of liquid products. Hydrocarbon materials, biomass, and microalgae are thermochemically converted into water-insoluble molecules with low viscosities. The produced bio-oils can be directly utilized for heavy engines or further upgraded into liquid transportation fuels, such as green diesel, biogasoline, and bio-jet fuels. The composition of the produced liquid fuels is dependent on the process conditions.

Besides the three major components of biomass, minor components include ash and extractive organic compounds, such as uronic acids, triglycerides, alkaloids, pigments, terpenes, and waxes. As mentioned earlier, hemicellulose is bound to lignin, and cellulose strands are interlaced with hemicellulose. In the process of HTL, carbon–hydrogen structured lignocellulosic materials are depolymerized into hydrocarbon liquid fuels, gas, char, and water-soluble compounds in the presence of pressurized water under the optimized temperature and pressure.[44] The liquid fuels, after water separation, can be directly utilized as energy sources in burners, combustion engines, and boilers. Furthermore, the produced bio-oil can be upgraded into a variety of products, such as advanced transportation fuels and value-added chemicals and materials, including aromatics, biopolymers, and biolubricants through hydrotreating or esterification processes.[45]

The Pacific Northwest National Laboratory demonstrated a laboratory-scale reactor for HTL using loblolly pine and corn stovers as agricultural feedstocks[42] (Fig. 6.4). They investigated the effect of process conditions, such as temperature and time, in particular, on the recycling of water-soluble product in order to enhance the oil yields and suppress a large amount of wastewater generation. The experiments were conducted in a continuous stirred-tank reactor (CSTR) with a size of 1 L integrated with plug flow

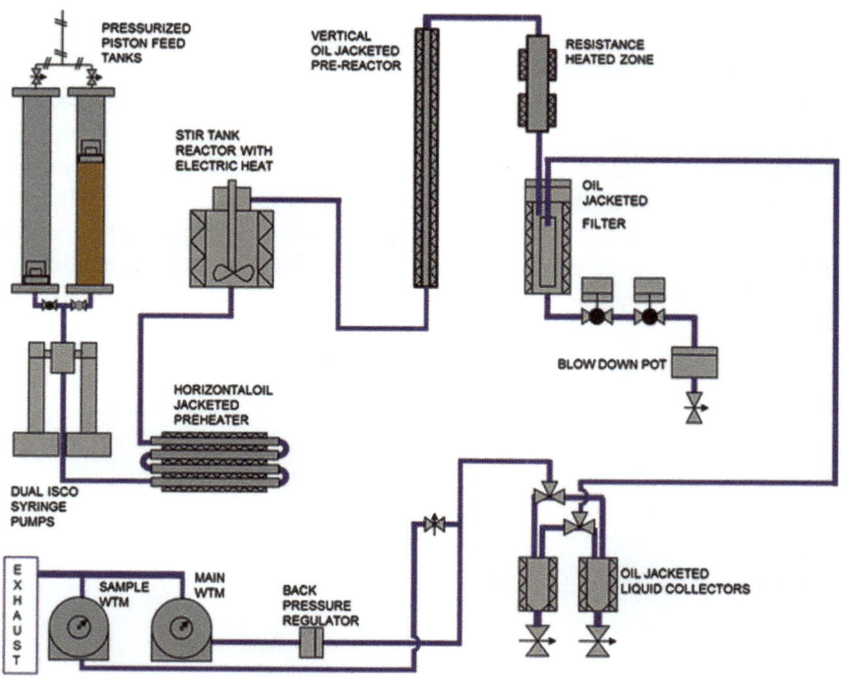

FIGURE 6.4 Laboratory reactor system for hydrothermal liquefaction process developed by the PNNL.[42] *PNNL*, Pacific Northwest National Laboratory. *Adapted from Elliott DC, Biller P, Ross AB, Schmidt AJ, Jones SB. Hydrothermal liquefaction of biomass: developments from batch to continuous process.* Bioresour Technol *2015;**178**:147–56, no need of permission; article is available under the terms of the Creative Commons Attribution License.*

compartments and operated in fully continuous mode. Interestingly, the CSTR succeeded in operating under the temperature range of 160°C–300°C with good mixing in the design system.

6.2.2.1 Hydrothermal liquefaction mechanism

The HTL pathway and mechanisms have not been clearly demonstrated so far due to the complexity of the process. The proposed pathway of the process comprises of three major reaction networks, including depolymerization, decomposition, and recombination.[46] Briefly, carbon–hydrogen structured lignocellulosic materials are firstly decomposed and then depolymerized into smaller active molecules. Subsequently, the polymerization of the produced compounds proceeds to generate liquid hydrocarbon-range alkanes, gas, biochar, and other solid compounds in the presence of water under moderate temperature and pressure. Interestingly, the operating conditions, including temperature, pressure, and time strongly contribute to the degree of each reaction (repolymerization, condensation, and decomposition). Due to the complexity of lignocellulosic biomass, the reaction pathway of the HTL of biomass and microalgae are consequently complicated.[40] Fig. 6.5 shows a process overview of the HTL from carbon–hydrogen structured lignocellulosic materials and the details of each reaction.

In the first step, the depolymerization of biomass into macromolecules first occurs by dissolving it in water, while the physical and chemical properties of biomass feedstocks strongly affect the solubility of the generated macromolecules. In addition, the composition and structure of hemicellulose and cellulose in the various types of biomass and microalgae influence the thermal stability of the produced bio-oils.[47] Moreover, the transformation of

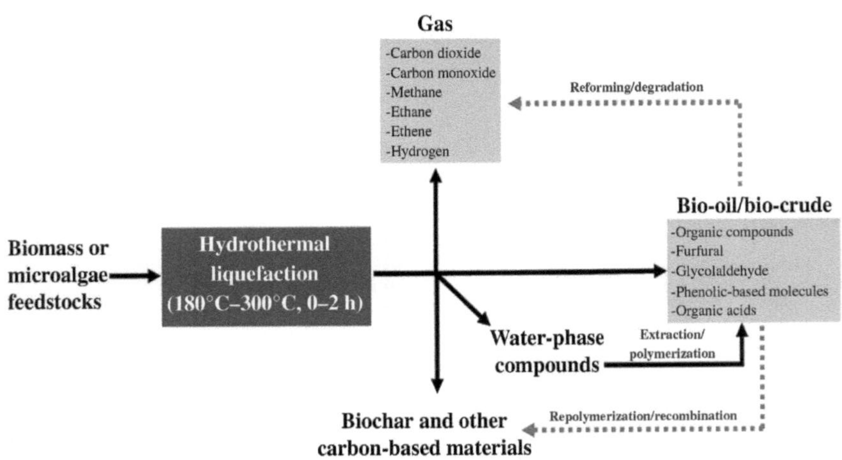

FIGURE 6.5 Proposed pathway of the hydrothermal liquefaction process from carbon–hydrogen structured lignocellulosic materials.

long chain polymeric compounds with major elements of hydrogen, oxygen, and carbon into shorter compounds can be optimized by changing the process conditions (residence time, temperature, and pressure).[47]

In the second step, the decomposition of the shorter compounds produced from the first step subsequently proceeds through cleavage, dehydration, decarboxylation, and deamination processes. Simultaneously, the degradation of biomass takes place to generate water molecules, carbon dioxide molecules, and amino acid through dehydration, decarboxylation, and deamination reactions, respectively.[48] The dehydration and decarboxylation reactions involve the removal of oxygen in the form of water and carbon dioxide, respectively, to generate insoluble and soluble compounds. Macromolecules in the biomass are further transformed into soluble oligomers and monomers by the hydration reaction.[40] Furthermore, the structure of cellulose could cleave into glucose monomers in the presence of water under moderate temperature and pressure. Fructose is simultaneously generated with a higher reactivity as compared to glucose. Accordingly, fructose could further degrade into different types of product forms via isomerization, hydrolysis, dehydration, and reverse-aldol defragmentation reactions in addition to the rearrangement and recombination processes to generate other value-added compounds.[3] The main decomposed products are water-soluble compounds, such as small organic compounds, furfural, glycolaldehyde, phenolic-based molecules, and organic acids.[40]

In the third step, the repolymerization and recombination of high reactivity fragments finally take place.[40] It should be deduced that the reverse reaction continuously proceeds to the initial processing steps if the operation occurs in the absence of hydrogen.[46] In the presence of hydrogen, the generated free radicals can be directly transformed into the desired stable product species, thus retaining the product yield. In contrast, without hydrogen and with high free radicals contained in the process, highly reactive fragments with small molecular weight can be recombined or repolymerized to generate larger molecular weight compounds and other carbon species.[46]

6.2.2.2 *Catalysts for hydrothermal liquefaction*

In the HTL of lignocellulosic biomass, the addition of small amounts of catalysts could suppress the side reactions, in particular, to increase the reaction rate at low reaction pressure and temperature conditions. In addition, the formation of solid residues could be retarded by the addition of catalysts, thus improving the liquid hydrocarbon yield and quality.[40] Both homogeneous and heterogeneous catalysts have been suggested for the HTL process.

The most common homogeneous catalysts reported in the literature are alkali and acid catalysts. Phosphoric acid, hydrochloric acid, acetic acid, perchloric acid, formic acid, and sulfuric acid are commonly used as acid catalysts with high activity and selectivity. However, the HTL process with the use of acid catalysts can produce high oxygen content bio-oil. This strongly affects the quality of bio-oil and must be further upgraded under high temperature and pressure conditions through a hydrotreating process. The strong corrosive effects of the acids on HTL equipment limit their large-scale industrial application. Therefore alkali-based catalysts, such as K_2CO_3, Na_2CO_3, $MgMnO_2$, KOH, and NaOH, have been extensively utilized in the literature.[1,49–51]

Several researchers have investigated alkali-based catalysts for HTL. Jindal et al.[49] screened NaOH, Na_2CO_3, KOH, and K_2CO_3-based alkali catalysts for the HTL of various types of waste woods under pressurized water at 280°C for 15 minutes in a batch reactor. A 34.9 wt.% bio-oil yield and a 6.8 wt.% solid residue yield were achieved using 1 M K_2CO_3. Typically, K_2CO_3 and KOH would provide a higher bio-oil yield than that achieved with Na_2CO_3 and NaOH. Interestingly, it was found that bio-oils produced using K_2CO_3 were mainly composed of carbon and hydrogen with low oxygen content compared to the original feedstock. Singh et al.[51] reported on the utilization of KOH and K_2CO_3 as alkali catalysts for the HTL of water hyacinth. The experiments were conducted in a temperature range of 250°C–300°C under different water hyacinth to water ratios of 1:3, 1:6, and 1:12 with residence times in the range of 15–60 minutes. It was reported that the bio-oil yield was significantly increased, while the highest yield of 23 wt.% and a conversion of 89% were achieved in 1 N KOH solution. This could be due to the fact that alkali catalysts promote decarboxylation, while biological molecule dehydration is inhibited. However, the water–gas shift reaction that favors H_2 and CO_2 formation in the process can be promoted by raising the pH in the solution. In addition, Long et al.[50] investigated the solid alkali catalysts: $MgMnO_x$, $MgNiO_x$, $MgFeO_x$, $MgCrO_x$, $MgZnO_x$, and $MgAlO_x$, for the HTL of bagasse in subcritical water. The results confirmed that solid alkali catalysts substantially promote bagasse liquefaction. A 93.7% bagasse conversion and 59.5% water-soluble fraction yield was obtained using $MgMnO_2$ under HTL conditions at 250°C in 15 minutes. It should be noted that major components in the HTL products are phenolic compounds derived from lignin, when the solid alkali catalysts were used. In addition, it can be deduced that the $MgMnO_2$ catalyst exhibited a good recyclability with no significant degradation after the stability test.

Heterogeneous catalysts, particularly metal catalysts, have been reported in the literature. Many precious and nonprecious metals, including Ru, Pt, Pd, Co, Mo, Ni, and Pt, supported on SiO_2, Al_2O_3, TiO_2, and zeolites have been utilized for the process.[40,52] However, sintering, dissolution, catalyst poisoning, external and internal diffusion limitations, and inactivation are key issues in the HTL process when using heterogeneous catalysts. Introducing hydrogen, a reducing gas, can prevent catalyst deactivation under high temperature and pressure conditions.

Wang et al.[52] studied the HTL of microalgae *Nannochloropsis* under iron, cobalt, nickel, molybdenum, and manganese supported TiO_2 catalysts. It was found that the highest yield, conversion, and quality of produced bio-oil were obtained using Ni/TiO_2 as a catalyst. They found that a 48.23% bio-crude yield and 89.28% conversion rate were achieved at a temperature of 300°C under Ni/TiO_2 catalysts. Furthermore, the viscosity and total acid numbers were improved due to the lighter hydrocarbon formation. Meryemoglu et al.[53] evaluated the catalytic performance of the carbon-supported Ru catalysts in HTL of dried kenaf hydrolysate. The experiments were conducted in the presence of tetralin under H_2 at a temperature of 300°C. In the absence of a catalyst, they reported that an oil and gas yield of 48.6% was obtained. Similarly, 77.2% and 72.7% of oil and gas yields were achieved using the commercial Ru/C and self-prepared Ru/C, respectively. It suggested that the prepared Ru/activated carbon showed a similar performance to the commercial one. Biller et al.[54] tested Co/Mo, Ni/Al, and Pt/Al catalysts for the HTL of biomass. They found that the calorific value and oxygen content in the produced bio-oil were improved significantly using these catalysts. It indicated that heterogeneous catalysts play an important role in reducing the nitrogen and oxygen contents in the liquid product, which can improve bio-oil quality. Younas et al.[55] studied the HTL of rice straw in the presence and absence of a NiO nanocatalyst aiming at producing bio-oil in a temperature range of 200°C–300°C. Light and heavy oils were generated under optimized conditions. At a temperature of 300°C, the bio-oil yields of light and heavy oils were increased to 13.2% and 17.2%, respectively, under the NiO nanocatalyst. The carbon recovery of bio-oil was increased to 52.8% with a 46.6% energy recovery under catalytic liquid fraction conditions. However, the product compositions and heating values of the bio-oil did not improve with the use of the NiO nanocatalyst. Phenols, alcohols, and ketones were found in the light oils, while the heavy oils contained alkanes and organic acids. Table 6.3 summarizes the findings of a survey on the HTL of biomass and microalgae.[56–61] In addition, as reported in the literature, the utilization of different types of catalysts can change the reaction pathways, biomass conversion, and composition and yields of gas, solid, and bio-oil. The selection of catalyst for the HTL process depends on the feedstock used and the variety of the desired products. The catalyst selection is important because they can show positive or negative effects on the process. Therefore the relationship between the reaction pathways and types of catalyst are crucial to generating the desired products in order to accelerate the industrialization of HTL.

6.2.2.3 *Limitations and prospects of hydrothermal liquefaction*

The HTL of biomass and microalgae in the presence or absence of a catalyst can be implemented to convert the abundant bioresources into bio-oils for an

TABLE 6.3 Summary of process conditions and product compositions from hydrothermal liquefaction of biomass and microalgae.

Feedstock	Reactor	Catalyst	Process conditions	Products	Reference
Microalgae (*Spirulina platensis*)	Autoclave reactor with ultrasonic pretreatments	–	Liquefaction temperature: 260°C, 300°C, and 340°C Pressure: 6.5, 9, and 11 MPa	Acids (1.39%), alkenes (7.24%), ketones (2.27%), esters (19.85%), and nitrogen-containing heterocycles (37.4%)	56
Microalgae (*Spirulina*)	Stainless steel batch reactor	Ni/TiO_2	Liquefaction temperature: 230°C–290°C Time: 10–120 min	Bio-oil yield (amides, esters, nitriles, hydroperoxide, and alkanes) (43.05 wt.%) and energy recovery value (64.62%)	57
Microalgae (*Chlorella*)	Stainless steel batch reactor	CuO/SBA-15	Liquefaction temperature: 300°C Time: 30 min Pressure: 6 MPa	Aliphatic hydrocarbons (64.64 wt.%), aromatic hydrocarbons (1.07 wt.%), oxygen-containing compounds (5.71 wt.%), nitrogen-containing compounds (1.84 wt.%), and nitrogen- and oxygen-containing compounds (23.94 wt.%)	58
Corn Stover and Pine Forest	Continuous processing bench scale system	–	Liquefaction temperature: 350°C Pressure: 20.7 MPa	Glycolic and acetic acid, alcohols, acetone, cyclopenta-ones, and inorganic sodium silicon sulfur	59
Waste furniture sawdust	Stainless steel autoclave	–	Liquefaction temperature: 180°C–300°C Time: 0–60 min Pressure: 0–2 MPa	Highest sawdust conversion (61.6 wt.%) and bio-oil yield (alcohols, acids, aldehydes, ketones, esters, phenols, and their derivatives) (12.7 wt.%)	60
Kraft lignin	Small, continuous, pilot unit equipped with a fixed-bed reactor	ZrO_2, K_2CO_3, and KOH	Liquefaction temperature: 350°C Pressure: 25 MPa	Bio-oil yield (phenolic monomers) (61.0 ± 2.7 wt.%)	61

efficient energy/fuel production technology. This technology is expected to provide more advantages compared to other bio-oil production technologies, such as pyrolysis or torrefaction processes.[62,63] However, several drawbacks, such as inability to scale-up, need to be considered.

The requirement for highly advanced apparatus, such as reactors, is one of the limitations of this process because the reaction is operated under high temperature and pressure. In particular, operating under an atmosphere of reducing gases, such as hydrogen, can significantly stabilize the catalyst life span and increase the production yield. Nonetheless, the product cost will be increased due to the additional cost of the hydrogen itself and the related facilities. In addition, the formation of coke, tar, and solid residue during the process may block the functioning of the equipment. The bio-oils, after water separation, still contain several complex components and low quality. Thus intensive development in bio-oil upgrading or purification techniques is needed for the practical use of HTL bio-oils as fuels and chemicals. Process intensification and optimization are key factors for developing HTL of raw or pretreated biomass from a bench- or a pilot-scale process to industrial-scale applications. Thus it is crucial to develop multifunctional catalysts with excellent performance and long lifespans in order to enhance the energy efficiency, product quality, and minimize both the operating and capital costs of the HTL process.

6.3 Bio-oil upgrading technology

6.3.1 Hydrotreating

One of the most promising methods for the removal of oxygenated compounds from bio-oil is hydrodeoxygenation (HDO), in which the oxygen-based functional group in bio-oil is hydrotreated through a reaction with H_2-forming water or/and methanol. Possible reactions that may occur in a HDO process include decarboxylation, hydrogenation, hydrogenolysis, hydrocracking, isomerization, and dehydration. Significant research has been carried out to develop catalysts for the process aiming at improving catalyst activities and upgrading the fuel properties as well as prolonging the catalyst life span. Herein, the types of catalysts and catalyst modifications for improving catalytic performance and stability are summarized. There are three main categories of catalyst employed in HDO, which are transition metal sulfide (TMS), noble metals, and transition metal phosphides and carbides.

6.3.1.1 Transition metal sulfide catalysts

TMS is a conventional class of catalyst in the hydrodesulfurization (HDS) process used in oil refineries for removing sulfur compounds by reacting with hydrogen gas and releasing sulfur as hydrogen sulfide (H_2S). Sulfided

$CoMo/Al_2O_3$ and $NiMo/Al_2O_3$ catalysts are commercial catalysts commonly used in HDS. It is generally accepted that sulfur-containing compounds can adsorb through sulfur atoms on sulfur vacancy sites of metal sulfide catalysts. Consequently, the hydrotreating reaction of the adsorbed compounds with H_2 molecules can effectively take place on the catalyst surface. Romero et al.[64] have proposed that the sulfur vacancy sites located on the edge of MoS_2 could be active sites for the adsorption of oxygen-containing species. These vacancy sites are active for different pathways depending on the adsorption mode of the oxygenated molecule on the active sites. It has been proposed that the possible adsorption mode of 2-ethyl phenol on the vacancy site of the Mo-based catalyst could bind through its oxygen atom and aromatic rings. The oxygen surface bond is either associatively or dissociatively adsorbed onto vacancy sites, thus activating the C–O bond(s) for scission. C–O bond(s) scission takes place after the partial hydrogenation of the aromatic rings at the ortho position or full hydrogenation of the ring for HDO. The Mo-based sulfide phase can be stabilized in the HDS of petroleum oil because of the presence of H_2 and H_2S in the reaction environment as a reactant and a forming gas, respectively. Bio-oil contains oxygen-containing compounds with a low sulfur content; therefore, sulfided catalysts could be converted into their oxide forms and then deactivated. To maintain the sulfide form of sulfided catalysts, a sulfur source, such as H_2S, carbon disulfide (CS_2), or dimethyl sulfide $(CH_3)_2S$ needs to be added into the feed.[64,65] Sulfur contamination in the upgraded fuel, which is undesirable in terms of environmental concern, is a drawback when this type of catalyst is used for HDO.[66] Water and salt impurities (i.e., K, Cl, and S) in bio-oil have been reported as causes of the deactivation of the Mo-based sulfide catalyst.[65,67]

6.3.1.2 *Noble metal catalysts*

Noble metal catalysts are a promising class of catalyst for HDO as they are active for the dissociative absorption of H_2. Noble metal catalysts are superior to sulfided catalysts because they are more tolerant to water degradation and a sulfur cofeed is not required. In addition, carbon deposition on Rh, Pd, and Pt catalysts was also lower than on the conventional sulfided $CoMo/Al_2O_3$ catalyst.[66] Huber et al. explored[4] the HDO of the aqueous fraction of bio-oil under Ru/C and Pt/C at low temperature (125°C) and high temperature. The product distributions are tunable by pressure and temperature. The high price of materials and the nature of noble metal catalysts have motivated researchers to search for alternative catalytic materials. It is advisable to keep the process away from the complete hydrogenation of aromatics in bio-oils as it leads to a lower octane number and higher H_2 consumption.

6.3.1.3 Metal phosphides and metal carbides

Metal phosphides are an alternative class of catalyst that have attracted great attention for the HDO process because they provide better activity in HDS and hydrodenitrogenation (HDN) compared to commercial Mo-based catalysts and are comparatively inexpensive compared to noble metals. Metal phosphides have been extensively investigated for hydroprocessing by Oyama et al.[68,69] Transition metal phosphides, including Ni_2P, Co_2P, Fe_2P, WP, and MoP were applied for guaiacol HDO. The most active phosphide was Ni_2P, and the major products from the reaction were benzene, methoxybenzene, and phenol. Sulfided $CoMo/Al_2O_3$ showed lower activity than that of the metal phosphides in guaiacol HDO. Although 5% Pd/Al_2O_3 had the highest performance among the catalysts after a short contact time, the undesirable compound, catechol, which is an oxygen-containing compound, was obtained.[68] Bui et al.[70] investigated the catalytic activity of phosphide catalysts based on the same amount of active sites loaded in a reactor for HDO of 2-methyltetrahydrofuran (2-MTHF) at 300°C and atmospheric pressure. The catalytic activity of the silica-supported metal phosphide catalysts followed the order of $Ni_2P > WP > MoP > CoP > FeP$. The catalytic performances of the metal phosphides were also compared to a commercial Pd/Al_2O_3 catalyst. The results showed that all the phosphides provided better activity in the HDO of 2-MTHF. Boullosa-Eiras et al.[71] conducted the HDO of phenol at 350°C–450°C and an H_2 pressure of 25 bar under TiO_2-supported molybdenum carbide, nitride, phosphide, and oxide. Selectivity toward hydrogenolysis-forming benzene as the main product was obtained for all catalysts prepared in this work. MoC_2 provided the highest phenol conversion with a moderate deactivation. The catalytic activity of Mo_2C was also observed for the vapor phase HDO under ambient pressure of a mixture of phenolic model compounds consisting of anisole, *m*-cresol, guaiacol, and 1,2-dimethoxybenzene.[72] The reaction proceeded until the complete deoxygenated of benzene and toluene for all the reactions studied irrespective of how high the molar ratio of H_2 to phenolic compounds (up to 3300) was. Favorable aromatic products were correlated to the strong adsorption of Ar–O bond on Mo_2C and its in situ surface modification by oxygenates during the reaction. XRD results showed the stability of the bulk structure of Mo_2C after the reaction.

6.3.1.4 The promotional effects by catalyst promoters, supports, and support modification

The promotional effects of catalytic promoters, supports, as well as support modification have been studied to improve catalytic performance. Wang et al.[73] demonstrated that the sulfonated carbon support of ruthenium catalyst ($Ru/AC\text{-}SO_3H$) was an effective modification for catalyst stability and activity in the HDO of acetophenone. The modified support benefited the HDO reactivity and selectivity of deoxygenated products because it provided better metal dispersion and strong acid site numbers compared to Ru/AC and Ru/SiO_2. Lin[74] illustrated the cobalt-modified molybdenum carbide (Co-Mo_2C) catalyst for the selective cleavage of the C=O bond in furfural HDO, thus producing 2-methylfuran (2-MF). The elongation of the C=O bond during adsorption on the modified molybdenum carbide was evidenced by density-functional theory (DFT) calculation. The change in the adsorption configuration of furfural facilitated C=O scission, hence promoting selectivity to 2-MF. The stability of the catalyst was also improved with the modification of the cobalt. Wu[75] investigated the effects of support in the atmospheric HDO of guaiacol under Ni_2P supported on SiO_2, Al_2O_3, and ZrO_2. The different product distributions indicated that the supports had a great impact on the catalytic behavior of the same Ni_2P active phase. It has been proposed that guaiacol initially undergoes direct deoxygenation, demethylation, and demethoxylation producing consequent products of anisole, catechol, and phenol, respectively. The product distributions revealed that Ni_2P/SiO_2 favored direct demethoxylation and direct deoxygenation pathways, whereas Ni_2P/ZrO_2 and Ni_2P/Al_2O_3, which have higher surface acidity, prefer demethylation. Yang[76] studied Ce-promoted Ni/SBA-15 catalysts for anisole HDO under mild conditions. The cerium-containing catalyst promoted nickel dispersion and catalytic activity. Benzene, which is a direct HDO product, is favorable for anisole HDO with increasing cerium content in Ce-Ni/SBA-15. H_2-TPR results suggested that the addition of cerium affected the reduction temperature of NiO to Ni toward a lower temperature. Wang et al.[77] compared the use of Al_2O_3 and its modified composites with CeO_2, ZrO_2, and CeO_2-ZrO_2 as supports for Pt catalysts in *p*-cresol HDO. It was proposed that hydrogenation and hydrogenolysis (direct deoxygenation) occurred during *p*-cresol HDO. Hydrogenation was a predominant reaction for all Pt catalysts. The addition of ZrO_2 to Al_2O_3 enhanced the hydrogenolysis of the C–O bond (direct deoxygenation) without ring saturation in *p*-cresol HDO, thus producing toluene. The catalyst activities and selectivity to deoxygenated product followed the order: $Pt/CeO_2-ZrO_2-Al_2O_3 > Pt/CeO_2-ZrO_2-Al_2O_3 > Pt/Al_2O_3 > Pt/ZrO_2-Al_2O_3$.

6.3.2 Esterification

The pyrolysis of biomass feedstocks leads to the formation of crude bio-oil that contains a complex mixture of hundreds of organic compounds, including acids, alcohols, aldehydes, ketones, ethers, furans, and phenols.[78,79] These contents give rise to several undesirable properties of crude bio-oil, such as strong corrosiveness, low heating value, immiscibility with petroleum-derived fuels primarily due to the large presence of oxygen (30–45 wt.%) and water (15–30 wt.%) as well as high acidity (pH = 2–4).[80] Other undesirable properties of crude bio-oil include instability, high viscosity, and poor combustion. These shortcomings are mainly caused by the fact that crude bio-oil can be deteriorated by the presence of organic acids since the latter can catalyze the polymerization of sugars and sugar derivatives (furfural, hydroxyl aldehyde, and hydroxyl acetone) in crude bio-oil during storage and transportation.[81–83]

Due to the mentioned limitations, the stability and physicochemical properties of crude bio-oil need to be enhanced so that it can be practically used as biofuel.[84] One way to achieve this goal is via esterification, which is a simple and low-cost technique that can convert the organic acids in crude bio-oil into their ester analogs. By eliminating the presence of acids in crude bio-oil, such as 1–10 wt.% of acetic acid and trace amounts of formic and propanoic acids, the major cause of the corrosiveness of the reactors and the instability of the bio-oil can be eliminated.[85,86] Nonetheless, despite of the simplicity of the esterification process of crude bio-oil, where high energy consumption and complicated operating processes are not required, the use of a catalyst is still necessary for the process and this will be discussed next.

6.3.2.1 Catalysts

An esterification reaction can generally be catalyzed by an acid or a base. However, since the use of base catalysts can lead to undesired reactions, such as neutralization and saponification, the use of acid catalysts is more common for this purpose.[87] Various acids are present in crude bio-oil; however, their acidity, in general, is relatively weak and therefore these acids are not suitable for use as intrinsic catalysts for the esterification of bio-oil.[88] As for external catalysts, the acid-catalyzed esterification of crude bio-oil may be performed using homogeneous acid catalysts, such as sulfuric acid and hydrochloric acid. In a study using sulfuric acid as the catalyst for the reaction of bio-oil and methanol,[88] it was found that the esterification of carboxylic acids, the hydrolysis of sugars, and the conversion of phenolic and large aromatic compounds were efficiently catalyzed due to the homogeneous dispersion of hydrogen ions of sulfuric acid. While homogeneous acid catalysts offer some advantages, such as high product yield, they also come with some disadvantages, such as high corrosion and difficulty in catalyst recovery. In particular, poor recovery can affect the cost-efficiency of the esterification process through the consumption of time and energy. For the esterification of bio-oil, therefore, the development of a more efficient heterogeneous acid catalyst system is necessary to address these drawbacks.

For bio-oil upgradation via esterification, a heterogeneous acid catalyst system containing zeolites is among the most widely used catalysts due to its sufficiently strong acidic sites as well as high surface area and the high thermal stability of the latter. Examples of zeolites used for this purpose include HZSM-5, HY, Hbeta, faujasite, mordenite, and ferrierite.[82,89–91] One study investigated the performance of the majority of these zeolites in catalyzing the esterification of acetic acid and *o*-cresol in crude bio-oil at 180°C for 1 hour[92]; it was found that a key factor affecting the yield of cresol acetate was the pore size of the catalysts. Specifically, the use of Hbeta and faujasite, which have larger pores, gave significantly higher yields as compared to the use of HZSM-5, ferrierite, and mordenite, which have smaller pores. These findings highlight the problem with the mass transfer effect, which generally occurs with the use of microporous zeolites (pore size <2 nm). Also, since there are high molecular weight compounds present in crude bio-oil, pore blockage is likely to occur with the use of microporous zeolites.

To alleviate the problem of pore size, attempts have been made to develop zeolites with hierarchical porosity.[82,89,91,92] There is a report that investigates the influence of the hierarchical porosity of faujasite (Si/Al = 30) on the catalytic properties in the esterification of bio-oil[82]; and it was found that introducing mesoporosity into the structure of faujasite can enhance catalytic activity. Specifically, the turnover frequency (TOF) for the esterification of acetic acid with two types of alcohol (methanol and benzyl alcohol) was increased by 30%–50% as compared to the unmodified faujasite. The enhanced activity of the hierarchically porous faujasite was most likely due to the increase in both acid site accessibility and hydrophobicity.

For bio-oil upgrading via esterification, heterogeneous acid catalyst systems other than those based on zeolites have also been reported. Examples include acid-functionalized porous silica,[78,93] sulfated metal oxide,[94,95] sulfated mixed oxide,[95] ion-exchange resin,[88,96] and acid-functionalized clay.[97]

6.3.2.2 *Types of alcohol*

Esterification is a reaction between a carboxylic acid and an alcohol, and the type of the latter is an important factor affecting the results of the esterification process. While methanol is among the most common choices for the purpose of reacting with carboxylic acids in bio-oil esterification,[88,94,97] investigations on the influence of other types of alcohol for a similar purpose have also been reported.[82,98] For instance, among five types of alcohol (methanol, anisyl alcohol, benzyl alcohol, *p*-cresol, and *n*-butanol) reacting with acetic acid catalyzed by faujasite,[82] the experimental results showed that methanol was the most reactive alcohol. As described by the Langmuir–Hinshelwood bimolecular surface reaction model, the use of a longer-chain aliphatic alcohol (*n*-butanol) or aromatic alcohols (anisyl alcohol, benzyl alcohol, and *p*-cresol) was accompanied by an increase in electron-donating properties, therefore resulting in a less stable protonated transition state and a slower reaction rate. The latter can also be affected by the size and the solubility of the alcohol used.[98]

In spite of the finding that methanol shows the highest reactivity with organic acids in bio-oil, it should be noted that methanol is an external alcohol that is not initially present in bio-oil. In other words, due to the fact that methanol is an external alcohol, its higher reactivity might not be sufficient to increase the overall efficiency of the catalytic system. Thus it is more desirable to develop an effective catalytic esterification process based on alcohols already present in bio-oil, such as guaiacol, cresols, acetol, and levoglucosan.[99] Such a process would lower both the acidity and the oxygen content in bio-oil. A challenge, however, is that high temperatures are required for this type of catalytic system in order to obtain high yields of ester products. Particularly, the use of high temperatures in this type of system would also lead to the formation of coke and the latter can retard the reaction and deactivate the catalyst.

6.3.2.3 Formation of coke

The formation of coke is highly likely during the esterification of bio-oil with alcohols at high temperatures; such formation is inevitable, even in cases where the reaction is set up at atmospheric pressure with temperatures limited to lower than 100°C. In fact, the formation of coke has a complex mechanism and can occur at various temperatures and heating rates.[100] In a study that investigated the formation of coke during the esterification of bio-oil using Amberlyst-70 as the catalyst[88,101]; it was found that a small amount of coke was formed at room temperature, however, further increases in reaction temperature (from room temperature to 170°C) also increases the yield of coke. Moreover, coke formation can be expedited by incrementing catalyst loading and reaction time, where an increase in catalyst loading can lead to an increase in the number of acidic sites available for coking. Nonetheless, the effect of reaction time on the formation of coke is not linear; where coke was mostly formed during the first 30 minutes of the reaction (11.5%) and only a 3.9% increase in coking was found after 120 minutes of reaction. Different types of alcohols, such as methanol, ethanol, 1-propanol, 1-butanol, and phenol, did not strongly influence the degree of coke formation. However, the coke formation rate was significantly higher under the use of ethylene glycol, presumably due to its multiple hydroxyl groups.

To mitigate the formation of coke, a technique was proposed which involved the simultaneous esterification and acetalation of bio-oil with online solvent extraction in butanol to upgrade the bio-oil.[102] This technique resulted in a conversion of 97% of acids and 100% of aldehydes into the corresponding esters, hemiacetals, and acetals. The result was considered as a significant improvement of a similar system that operated without extraction. Importantly, the proposed system could also effectively decrease the occurrence of side reactions that would cause the formation of coke.

6.3.2.4 Integrated two-stage upgrading process

Due to the complex composition of crude bio-oil, where more than a few hundred organic compounds with different functional groups and reactivities are present, a single esterification process is not sufficient for upgrading the model bio-oil. Therefore the use of hydrogenation–esterification, which involves a two-stage upgrading process, has been reported.[79,90,103] The catalysts used for this integrated process include Cu/SBA-15, Raney Ni, Ni/HZSM-5, and Ru/HZSM-5. In a study that investigates the use of hydrogenation–esterification for crude bio-oil upgrading using Cu-SBA-15 as the catalyst and methanol as the solvent and coreactant at 300°C[104]; it was found that acetic acid could efficiently be converted into ethanol and methyl acetate, and that furfural, hydroxyacetone, and guaiacol could also be efficiently hydrogenated. Moreover, the hydrogenation–esterification process presented in this work allows for the acid content in crude bio-oil to decrease substantially from 25.0 to 0.4 wt.%.

In another study, an integrated two-stage upgrading process involving esterification–alkylation was reported, where Amberlyst-36 was used as the catalyst.[104,105] It was found that by using aliphatic alcohols (methanol, ethanol, and *n*-butanol) as the coreactant for esterification and 2-MF as the alkylating reagent for alkylation, the integrated process was highly efficient for converting reactive compositions in crude bio-oil, such as acids, aldehydes, and ketones, into their corresponding esters and ethers. The converted products were highly stable and the solubility of the upgraded bio-oil was also improved. In summary, an integrated two-stage upgrading process is effective for upgrading crude bio-oil for use as high-grade liquid fuels.

6.3.2.5 Characteristic properties of bio-oil before and after upgrading via esterification

A comparison between the physicochemical properties (acid value, viscosity, high heating value, and water content) of crude bio-oil and upgraded bio-oil via esterification or an integrated two-stage upgrading process is shown in Table 6.4.[94,95,104–106] It can be observed that after the upgrading process, the acid value, viscosity, and water content decrease, while the high heating value increases. These changes result in the enhanced stability and higher overall quality of the upgraded bio-oil so that the latter is more suitable for use as biofuel. It should be noted that since water is a major component of crude bio-oil and it is also a coproduct from the esterification reaction, separation techniques for water removal are crucial. Particularly, techniques performed during esterification, such as an azeotropic distillation system[105] and online solvent extraction,[102] not only lower the water content in the reaction, but also help drive the reaction equilibrium toward the formation of ester products.

TABLE 6.4 Comparison of properties of bio-oil before and after upgrading via esterification or integrated two-stage upgrading process.

Reaction	Reaction parameters	Acid value (mg NaOH/g)		Viscosity (mm^2/s)		High heating value (MJ/kg)		Water content (wt.%)		Reference
		Before	After	Before	After	Before	After	Before	After	
Esterification	Reactants: bio-oil, MeOH									
	Catalyst: SO_4^{2-}/ZrO_2	70–110	0.27	4.21	4.17	38.08	42.03	0.37	0.19	94
	Conditions: 100°C, 6 h									
Esterification	Reactants: bio-oil, EtOH									
	Catalyst: ZrO_2/TiO_2	pH 2.5	pH 5.3	–	–	8.08	22.7	40.1	16.8	95
	Conditions: 80°C, 6 h									
Esterification–alkylation	Reactants: bio-oil, *n*-butanol, 2-methylfuran									
	Catalyst: Amberlyst-36	41.12	6.17	18.49	5.64	14.46	30.85	27.82	3.21	105
	Conditions: 100°C, 4 h									
Esterification–alkylation	Reactants: bio-oil, MeOH, 2-methylfuran									
	Catalyst: Amberlyst-36 Conditions: 150°C, 2 h	40.7	6.20	18.49	2.21	14.46	17.28	29.78	20.87	104
Oxidative pretreatment and esterification	Reactants: bio-oil, ozone/H_2O_2, *n*-butanol									
	Catalyst: Ru/γ-Al_2O_3	90.3	34.3	12.0	10.1	16.0	35.3	30.4	3.6	106
	Conditions: 340°C, 1.5 h									

6.4 Conclusion

The pyrolysis process (slow, fast, and flash) along with torrefaction have been technically described in relation to advancements in reactor and process design, operating conditions, and catalysts, aiming at improving bio-oil quality and process efficiency. Among the different pyrolysis processes, fast pyrolysis operated in a typical temperature range of 450°C–550°C would give a bio-oil yield above 50%. Biomass pretreatment by physical grinding and chemical torrefaction (200°C–330°C) would provide more suitable feedstocks for pyrolysis with improved bio-oil yields above 70%. The fast pyrolysis of torrefied biomass produces bio-oil with a lower organic acid content and higher aromatic hydrocarbon and phenol contents. In addition, the HTL process, at relatively low temperatures of 250°C–350°C and high pressure of 40–220 bar, has been alternatively developed for bio-oil production; this process takes place in water and the drying of biomass is not required.

Further upgrading the bio-oil products produced from these processes is indispensable for the practical use of bio-oil in various applications. The catalytic upgrading of bio-oil can be implemented via several approaches, including hydrotreating and esterification, while the catalytic materials play an important role in the upgrading processes. TMSs, noble metals, and transition metal phosphides and carbides have been extensively developed for the hydrotreating process in a typical temperature range of 300°C–400°C and a pressure range of 10–50 bar; and the final products can be used as gasoline, jet, and diesel (bio)fuels. Acid catalysts, especially hierarchical zeolites, have been developed for the esterification process in a typical temperature range of 60°C–200°C. The obtained products are alkyl ester-type biodiesels, which can be efficiently utilized as diesel blends. The developed technology paves the way for the present and future utilization of biomass via bio-oil platforms.

Acknowledgment

The authors acknowledge the financial support of the Thailand Research Fund, NANOTEC, and the Synchrotron Light Research Institute (BRG6080015). This work was also supported by the National Nanotechnology Center (NANOTEC, project number: P1750381). The authors are grateful to Dr. Thanitporn Narkkun for patent analysis data.

References

1. Zhou CH, Xia X, Lin CX, Tong DS, Beltramini J. Catalytic conversion of lignocellulosic biomass to fine chemicals and fuels. *Chem Soc Rev* 2011;**40**(11):5588–617.
2. Rahman MM, Liu R, Cai J. Catalytic fast pyrolysis of biomass over zeolites for high quality bio-oil—a review. *Fuel Process Technol* 2018;**180**:32–46.

3. Zhang X, Wilson K, Lee AF. Heterogeneously catalyzed hydrothermal processing of C5–C6 sugars. *Chem Rev* 2016;**116**(19):12328–68.
4. Sanna A, Vispute TP, Huber GW. Hydrodeoxygenation of the aqueous fraction of bio-oil with Ru/C and Pt/Ccatalysts. *Appl Catal B Environ* 2015;**165**:446–56.
5. Bridgwater AV. Review of fast pyrolysis of biomass and product upgrading. *Biomass Bioenergy* 2012;**38**:68–94.
6. Guda VK, Steele PH, Penmetsa VK, Li Q. *Fast pyrolysis of biomass: recent advances in fast pyrolysis technology. Recent advances in thermo-chemical conversion of biomass.* Elsevier; 2015. p. 177–211. https://doi.org/10.1016/B978-0-444-63289-0.00007-7.
7. Kan T, Strezov V, Evans TJ. Lignocellulosic biomass pyrolysis: a review of product properties and effects of pyrolysis parameters. *Renew Sustain Energy Rev* 2016;**57**:1126–40.
8. Guedes RE, Luna AS, Torres AR. Operating parameters for bio-oil production in biomass pyrolysis: a review. *J Anal Appl Pyrolysis* 2018;**129**:134–49.
9. Baloch HA, et al. Recent advances in production and upgrading of bio-oil from biomass: a critical overview. *J Environ Chem Eng* 2018;**6**(4):5101–18.
10. Qureshi KM, Kay Lup AN, Khan S, Abnisa F, Wan Daud WMA. A technical review on semi-continuous and continuous pyrolysis process of biomass to bio-oil. *J Anal Appl Pyrolysis* 2018;**131**:52–75.
11. Biradar CH, Subramanian KA, Dastidar MG. Production and fuel quality upgradation of pyrolytic bio-oil from Jatropha Curcas de-oiled seed cake. *Fuel* 2014;**119**:81–9.
12. Sulaiman F, Abdullah N. Optimum conditions for maximising pyrolysis liquids of oil palm empty fruit bunches. *Energy* 2011;**36**(5):2352–9.
13. Isahak WNRW, Hisham MWM, Yarmo MA, Yun Hin T. A review on bio-oil production from biomass by using pyrolysis method. *Renew Sustain Energy Rev* 2012;**16**(8):5910–23.
14. Jung SH, Kang BS, Kim JS. Production of bio-oil from rice straw and bamboo sawdust under various reaction conditions in a fast pyrolysis plant equipped with a fluidized bed and a char separation system. *J Anal Appl Pyrolysis* 2008;**82**(2):240–7.
15. Ateş F, Işıkdağ MA. Evaluation of the role of the pyrolysis temperature in straw biomass samples and characterization of the oils by GC/MS. *Energy Fuels* 2008;**22**(3):1936–43.
16. Ly HV, Kim SS, Choi JH, Woo HC, Kim J. Fast pyrolysis of *Saccharina japonica* alga in a fixed-bed reactor for bio-oil production. *Energy Convers Manag* 2016;**122**:526–34.
17. Alvarez J, Lopez G, Amutio M, Bilbao J, Olazar M. Bio-oil production from rice husk fast pyrolysis in a conical spouted bed reactor. *Fuel* 2014;**128**:162–9.
18. Abdullah N, Gerhauser H. Bio-oil derived from empty fruit bunches. *Fuel* 2008;**87**(12):2606–13.
19. Volli V, Singh RK. Production of bio-oil from de-oiled cakes by thermal pyrolysis. *Fuel* 2012;**96**:579–85.
20. Islam MR, Parveen M, Haniu H. Properties of sugarcane waste-derived bio-oils obtained by fixed-bed fire-tube heating pyrolysis. *Bioresour Technol* 2010;**101**(11):4162–8.
21. Pattiya A, Suttibak S. Production of bio-oil via fast pyrolysis of agricultural residues from cassava plantations in a fluidised-bed reactor with a hot vapour filtration unit. *J Anal Appl Pyrolysis* 2012;**95**:227–35.
22. Pattiya A. Bio-oil production via fast pyrolysis of biomass residues from cassava plants in a fluidised-bed reactor. *Bioresour Technol* 2011;**102**(2):1959–67.
23. Kim SW, Koo BS, Ryu JW, Lee JS, Kim CJ, Lee DH, et al. Bio-oil from the pyrolysis of palm and Jatropha wastes in a fluidized bed. *Fuel Process Technol* 2013;**108**:118–24.
24. Apaydin-Varol E, Pütün E, Pütün AE. Slow pyrolysis of pistachio shell. *Fuel* 2007;**86**(12):1892–9.

25. Makibar J, Fernandez-Akarregi AR, Amutio M, Lopez G, Olazar M. Performance of a conical spouted bed pilot plant for bio-oil production by poplar flash pyrolysis. *Fuel Process Technol* 2015;**137**:283–9.
26. Chen Z, Wang M, Jiang E, Wang D, Zhang K, Ren Y, et al. Pyrolysis of torrefied biomass. *Trends Biotechnol* 2018;**36**:1287–98 Available from: https://doi.org/10.1016/j.tibtech.2018.07.005.
27. Meng J, Park J, Tilotta D, Park S. The effect of torrefaction on the chemistry of fast-pyrolysis bio-oil. *Bioresour Technol* 2012;**111**:439–46.
28. Chen D, Li Y, Deng M, Wang J, Chen M, Yan B, et al. Effect of torrefaction pretreatment and catalytic pyrolysis on the pyrolysis poly-generation of pine wood. *Bioresour Technol* 2016;**214**:615–22.
29. Ren S, Lei H, Wang L, Yadavalli G, Liu Y, Julson J. The integrated process of microwave torrefaction and pyrolysis of corn stover for biofuel production. *J Anal Appl Pyrolysis* 2014;**108**:248–53.
30. Chen D, Zheng Z, Fu K, Zeng Z, Wang J, Lu M. Torrefaction of biomass stalk and its effect on the yield and quality of pyrolysis products. *Fuel* 2015;**159**:27–32.
31. Ren S, Lei H, Wang L, Bu Q, Chen S, Wu J, et al. The effects of torrefaction on compositions of bio-oil and syngas from biomass pyrolysis by microwave heating. *Bioresour Technol* 2013;**135**:659–64.
32. Ukaew S, Schoenborn J, Klemetsrud B, Shonnard DR. Effects of torrefaction temperature and acid pretreatment on the yield and quality of fast pyrolysis bio-oil from rice straw. *J Anal Appl Pyrolysis* 2018;**129**:112–22.
33. Wigley T, Yip ACK, Pang S. A detailed product analysis of bio-oil from fast pyrolysis of demineralised and torrefied biomass. *J Anal Appl Pyrolysis* 2017;**123**:194–203.
34. Carrier M, et al. Impact of the lignocellulosic material on fast pyrolysis yields and product quality. *Bioresour Technol* 2013;**150**:129–38.
35. Duanguppama K, Suwapaet N, Pattiya A. Fast pyrolysis of contaminated sawdust in a circulating fluidised bed reactor. *J Anal Appl Pyrolysis* 2016;**118**:63–74.
36. Fernandez-Akarregi AR, Makibar J, Lopez G, Amutio M, Olazar M. Design and operation of a conical spouted bed reactor pilot plant (25kg/h) for biomass fast pyrolysis. *Fuel Process Technol* 2013;**112**:48–56.
37. Wagenaar BM, Venderbosch RH, Carrasco J, Strenziok R, van der Aa BJ. *Rotating cone bio-oil production and applications. Progress in thermochemical biomass conversion.* Wiley; 2008. p. 1268–80 Available from: http://dx.doi.org/10.1002/9780470694954.ch105.
38. Luo G, Chandler DS, Anjos LCA, Eng RJ, Jia P, Resende FLP. Pyrolysis of whole wood chips and rods in a novel ablative reactor. *Fuel* 2017;**194**:229–38.
39. Puy N, Murillo R, Navarro MV, López JM, Rieradevall J, Fowler G, et al. Valorisation of forestry waste by pyrolysis in an auger reactor. *Waste Manag* 2011;**31**(6):1339–49.
40. Cao L, Zhang C, Chen H, Tsang DCW, Luo G, Zhang S, et al. Hydrothermal liquefaction of agricultural and forestry wastes: state-of-the-art review and future prospects. *Bioresour Technol* 2017;**245**(Pt A):1184–93.
41. de Caprariis B, De Filippis P, Petrullo A, Scarsella M. Hydrothermal liquefaction of biomass: influence of temperature and biomass composition on the bio-oil production. *Fuel* 2017;**208**:618–25.
42. Elliott DC, Biller P, Ross AB, Schmidt AJ, Jones SB. Hydrothermal liquefaction of biomass: developments from batch to continuous process. *Bioresour Technol* 2015;**178**:147–56.

43. Gollakota ARK, Kishore N, Gu S. A review on hydrothermal liquefaction of biomass. *Renew Sustain Energy Rev* 2018;**81**:1378–92.
44. Magdeldin M, Kohl T, Järvinen M. Techno-economic assessment of the by-products contribution from non-catalytic hydrothermal liquefaction of lignocellulose residues. *Energy* 2017;**137**:679–95.
45. Duan P, Zhang C, Wang F, Fu J, Lü X, Xu Y, et al. Activated carbons for the hydrothermal upgrading of crude duckweed bio-oil. *Catal Today* 2016;**274**:73–81.
46. Toor SS, Rosendahl L, Rudolf A. Hydrothermal liquefaction of biomass: a review of subcritical water technologies. *Energy* 2011;**36**(5):2328–42.
47. Zein M, Winter R. Effect of temperature, pressure and lipid acyl chain length on the structure and phase behaviour of phospholipid–gramicidin bilayers. *Phys Chem Chem Phys* 2000;**2**(20):4545–51.
48. Jena U, McCurdy AT, Warren A, Summers H, Ledbetter RN, Hoekman SK, et al. Oleaginous yeast platform for producing biofuels via co-solvent hydrothermal liquefaction. *Biotechnol Biofuels* 2015;**8**(1):167.
49. Jindal MK, Jha MK. Catalytic hydrothermal liquefaction of waste furniture sawdust to bio-oil. *Indian Chem Eng* 2016;**58**(2):157–71.
50. Long J, Li Y, Zhang X, Tang L, Song C, Wang F. Comparative investigation on hydrothermal and alkali catalytic liquefaction of bagasse: process efficiency and product properties. *Fuel* 2016;**186**:685–93.
51. Singh R, Balagurumurthy B, Prakash A, Bhaskar T. Catalytic hydrothermal liquefaction of water hyacinth. *Bioresour Technol* 2015;**178**:157–65.
52. Wang W, Xu Y, Wang X, Zhang B, Tian W, Zhang J. Hydrothermal liquefaction of microalgae over transition metal supported TiO_2 catalyst. *Bioresour Technol* 2018;**250**:474–80.
53. Meryemoglu B, Hasanoglu A, Irmak S, Erbatur O. Biofuel production by liquefaction of kenaf (*Hibiscus cannabinus* L.) biomass. *Bioresour Technol* 2014;**151**:278–83.
54. Biller P, Riley R, Ross AB. Catalytic hydrothermal processing of microalgae: decomposition and upgrading of lipids. *Bioresour Technol* 2011;**102**(7):4841–8.
55. Younas R, Hao S, Zhang L, Zhang S. Hydrothermal liquefaction of rice straw with NiO nanocatalyst for bio-oil production. *Renew Energy* 2017;**113**:532–45.
56. Zhang B, Feng H, He Z, Wang S, Chen H. Bio-oil production from hydrothermal liquefaction of ultrasonic pre-treated *Spirulina platensis*. *Energy Convers Manag* 2018;**159**:204–12.
57. Tian W, Liu R, Wang W, Yin Z, Yi X. Effect of operating conditions on hydrothermal liquefaction of *Spirulina* over Ni/TiO_2 catalyst. *Bioresour Technol* 2018;**263**:569–75.
58. Li J, Fang X, Bian J, Guo Y, Li C. Microalgae hydrothermal liquefaction and derived biocrude upgrading with modified SBA-15 catalysts. *Bioresour Technol* 2018;**266**:541–7.
59. Panisko E, Wietsma T, Lemmon T, Albrecht K, Howe D. Characterization of the aqueous fractions from hydrotreatment and hydrothermal liquefaction of lignocellulosic feedstocks. *Biomass Bioenergy* 2015;**74**:162–71.
60. Jindal MK, Jha MK. Effect of process parameters on hydrothermal liquefaction of waste furniture sawdust for bio-oil production. *RSC Adv* 2016;**6**(48):41772–80.
61. Belkheiri T, Andersson SI, Mattsson C, Olausson L, Theliander H, Vamling L. Hydrothermal liquefaction of kraft lignin in subcritical water: influence of phenol as capping agent. *Energy Fuels* 2018;**32**(5):5923–32.
62. Gerber Van Doren L, Posmanik R, Bicalho FA, Tester JW, Sills DL. Prospects for energy recovery during hydrothermal and biological processing of waste biomass. *Bioresour Technol* 2017;**225**:67–74.

63. Tekin K, Karagöz S, Bektaş S. A review of hydrothermal biomass processing. *Renew Sustain Energy Rev* 2014;**40**:673–87.
64. Romero Y, Richard F, Brunet S. Hydrodeoxygenation of 2-ethylphenol as a model compound of bio-crude over sulfided Mo-based catalysts: promoting effect and reaction mechanism. *Appl Catal B Environ* 2010;**98**(3–4):213–23.
65. Mortensen PM, Gardini D, Damsgaard CD, Grunwaldt JD, Jensen PA, Wagner JB, et al. Deactivation of Ni-MoS_2 by bio-oil impurities during hydrodeoxygenation of phenol and octanol. *Appl Catal A Gen* 2016;**523**:159–70.
66. Gutierrez A, Kaila RK, Honkela ML, Slioor R, Krause AOI. Hydrodeoxygenation of guaiacol on noble metal catalysts. *Catal Today* 2009;**147**(3–4):239–46.
67. Badawi M, Paul JF, Cristol S, Payen E, Romero Y, Richard F, et al. Effect of water on the stability of Mo and CoMo hydrodeoxygenation catalysts: a combined experimental and DFT study. *J Catal* 2011;**282**(1):155–64.
68. Zhao HY, Li D, Bui P, Oyama ST. Hydrodeoxygenation of guaiacol as model compound for pyrolysis oil on transition metal phosphide hydroprocessing catalysts. *Appl Catal A Gen* 2011;**391**(1–2):305–10.
69. Zhang J, Matsubara K, Yun GN, Zheng H, Takagaki A, Kikuchi R, et al. Comparison of phosphide catalysts prepared by temperature-programmed reduction and liquid-phase methods in the hydrodeoxygenation of 2-methylfuran. *Appl Catal A Gen* 2017;**548** (June):39–46.
70. Bui P, Cecilia JA, Oyama ST, Takagaki A, Infantes-Molina A, Zhao H, et al. Studies of the synthesis of transition metal phosphides and their activity in the hydrodeoxygenation of a biofuel model compound. *J Catal* 2012;**294**:184–98.
71. Boullosa-Eiras S, Lødeng R, Bergem H, Stöcker M, Hannevold L, Blekkan EA. Catalytic hydrodeoxygenation (HDO) of phenol over supported molybdenum carbide, nitride, phosphide and oxide catalysts. *Catal Today* 2014;**223**:44–53.
72. Chen CJ, Lee WS, Bhan A. Mo_2C catalyzed vapor phase hydrodeoxygenation of lignin-derived phenolic compound mixtures to aromatics under ambient pressure. *Appl Catal A Gen* 2016;**510**:42–8.
73. Wang J, Zhang Y, Zhang M, Wang Z, Zhang M. Hydrodeoxygenation of acetophenone over a sulfonated carbon-supported ruthenium catalyst. *Catal Today* 2018;**314**(December 2017):164–9.
74. Lin Z, Wan W, Yao S, Chen JG. Cobalt-modified molybdenum carbide as a selective catalyst for hydrodeoxygenation of furfural. *Appl Catal B Environ* 2018;**233**(March):160–6.
75. Wu SK, Lai PC, Lin YC, Wan HP, Lee HT, Chang YS. Atmospheric hydrodeoxygenation of guaiacol over alumina-, zirconia-, and silica-supported nickel phosphide catalysts. *ACS Sustain Chem Eng* 2013;**1**(3):349–58.
76. Yang Y, Ochoa-Hernández C, Pizarro P, O'shea VAP, Coronado JM, Serrano DP. Ce-promoted Ni/SBA-15 catalysts for anisole hydrotreating under mild conditions. *Appl Catal B Environ* 2016;**197**:206–13.
77. Wang W, Wu K, Liu P, Li L, Yang Y, Wang Y. Hydrodeoxygenation of p-cresol over Pt/ Al_2O_3 catalyst promoted by ZrO_2, CeO_2, and CeO_2-ZrO + . *Ind Eng Chem Res* 2016;**55** (28):7598–603.
78. Zhang P, Sun Y, Zhang Q, Guo Y, Song D. Upgrading of pyrolysis biofuel via esterification of acetic acid with benzyl alcohol catalyzed by Brønsted acidic ionic liquid functionalized ethyl-bridged organosilica hollow nanospheres. *Fuel* 2018;**228**:175–86.

79. Xu Y, Limin Z, Jiamin C, Xinghua Z, Longlong M, Tiejun W, et al. One step hydrogenation–esterification of model compounds and bio-oil to alcohols and esters over Raney Ni catalysts. *Energy Convers Manag* 2016;**108**:78–84.
80. Valle B, Remiro A, García-Gómez N, Gayubo AG, Bilbao J. Recent research progress on bio-oil conversion into bio-fuels and raw chemicals: a review. *J Chem Technol Biotechnol* 2018. Available from: https://doi.org/10.1002/jctb.5758.
81. Ciddor L, Bennett JA, Hunns JA, Wilson K, Lee AF. Catalytic upgrading of bio-oils by esterification. *J Chem Technol Biotechnol* 2015;**90**(5):780–95.
82. Osatiashtiani A, Puértolas B, Oliveira CCS, Manayil JC, Barbero B, Isaacs M, et al. On the influence of Si:Al ratio and hierarchical porosity of FAU zeolites in solid acid catalysed esterification pretreatment of bio-oil. *Biomass Convers Biorefinery* 2017;**7**(3):331–42.
83. Ortega JV, Renehan AM, Liberatore MW, Herring AM. Physical and chemical characteristics of aging pyrolysis oils produced from hardwood and softwood feedstocks. *J Anal Appl Pyrolysis* 2011;**91**(1):190–8.
84. Fermoso J, Pizarro P, Coronado JM, Serrano DP. Advanced biofuels production by upgrading of pyrolysis bio-oil. *Wiley Interdiscip Rev Energy Environ* 2017;**6**(4):e245.
85. Oasmaa A, Kuoppala E. Fast pyrolysis of forestry residue. 3. Storage stability of liquid fuel. *Energy Fuels* 2003;**17**(4):1075–84.
86. Mullen CA, Boateng AA, Goldberg NM, Lima IM, Laird DA, Hicks KB. Bio-oil and biochar production from corn cobs and stover by fast pyrolysis. *Biomass Bioenergy* 2010;**34**(1):67–74.
87. Lee AF, Bennett JA, Manayil JC, Wilson K. Heterogeneous catalysis for sustainable biodiesel production via esterification and transesterification. *Chem Soc Rev* 2014;**43**(22):7887–916.
88. Wu L, Hu X, Wang S, Hasan MDM, Jiang S, Li T, et al. Acid-treatment of bio-oil in methanol: the distinct catalytic behaviours of a mineral acid catalyst and a solid acid catalyst. *Fuel* 2018;**212**:412–21.
89. Rodponthukwaji K, Wattanakit C, Yutthalekha T, Assavapanumat S, Warakulwit C, Wannapakdeeb W, et al. Catalytic upgrading of carboxylic acids as bio-oil models over hierarchical ZSM-5 obtained via an organosilane approach. *RSC Adv* 2017;**7**(57):35581–9.
90. Hilten R, Weber J, Kastner JR. Continuous upgrading of fast pyrolysis oil by simultaneous esterification and hydrogenation. *Energy Fuels* 2016;**30**(10):8357–68.
91. Milina M, Mitchell S, Pérez-Ramírez J. Prospectives for bio-oil upgrading via esterification over zeolite catalysts. *Catal Today* 2014;**235**:176–83.
92. Milina M, Mitchell S, Michels NL, Kenvin J, Pérez-Ramírez J. Interdependence between porosity, acidity, and catalytic performance in hierarchical ZSM-5 zeolites prepared by post-synthetic modification. *J Catal* 2013;**308**:398–407.
93. Manayil JC, Inocencio CVM, Lee AF, Wilson K. Mesoporous sulfonic acid silicas for pyrolysis bio-oil upgrading via acetic acid esterification. *Green Chem* 2016;**18**(5):1387–94.
94. Li L, Yan B, Li H, Yu S, Li S, Yu H, et al. SO_4^{2-}/ZrO_2 as catalyst for upgrading of pyrolysis oil by esterification. *Fuel* 2018;**226**:190–4.
95. Liu Y, Li Z, Leahy JJ, Kwapinski W. Catalytically upgrading bio-oil via esterification. *Energy Fuels* 2015;**29**(6):3691–8.
96. Lee Y, Shafaghat H, Kim JK, Jeon JK, Jung SC, Lee IG, et al. Upgrading of pyrolysis bio-oil using WO_3/ZrO_2 and Amberlyst catalysts: evaluation of acid number and viscosity. *Korean J Chem Eng* 2017;**34**(8):2180–7.

97. Wang A, Wang J, Lu C, Xu M, Lv J, Wu X. Esterification for biofuel synthesis over an eco-friendly and efficient kaolinite-supported $SO_4^{2-}/ZnAl_2O_4$ macroporous solid acid catalyst. *Fuel* 2018;**234**:430–40.
98. Weerachanchai P, Tangsathitkulchai C, Tangsathitkulchai M. Effect of reaction conditions on the catalytic esterification of bio-oil. *Korean J Chem Eng* 2011;**29**(2):182–9.
99. Mullen CA, Boateng AA. Chemical composition of bio-oils produced by fast pyrolysis of two energy crops. *Energy Fuels* 2008;**22**(3):2104–9.
100. Xiong Z, Syed-Hassan SSA, Xu J, Wang Y, Hu S, Su S, et al. Evolution of coke structures during the pyrolysis of bio-oil at various temperatures and heating rates. *J Anal Appl Pyrolysis* 2018;**134**:336–42.
101. Wu L, Hu X, Wang S, Mourant D, Song Y, Li T, et al. Formation of coke during the esterification of pyrolysis bio-oil. *RSC Adv* 2016;**6**(89):86485–93.
102. Qin F, Cui H, Yi W, Wang C. Upgrading the water-soluble fraction of bio-oil by simultaneous esterification and acetalation with online extraction. *Energy Fuels* 2014;**28**(4):2544–53.
103. Chen J, Cai Q, Lu L, Leng F, Wang S. Upgrading of the acid-rich fraction of bio-oil by catalytic hydrogenation-esterification. *ACS Sustain Chem Eng* 2016;**5**(1):1073–81.
104. Ye J, Jiang J, Xu J. Effect of alcohols on simultaneous bio-oil upgrading and separation of high value-added chemicals. *Waste Biomass Valorization* 2017;**9**(10):1779–85.
105. Lu J, Guo S, Fu Y, Chang J. Catalytic upgrading of bio-oil by simultaneous esterification and alkylation with azeotropic water removal. *Fuel Process Technol* 2017;**161**:193–8.
106. Tanneru SK, Parapati DR, Steele PH. Pretreatment of bio-oil followed by upgrading via esterification to boiler fuel. *Energy* 2014;**73**:214–20.

Chapter 7

New trends in biogas production and utilization

Luana Cardoso Grangeiro[1], Sâmilla Gabriella Coêlho de Almeida[1], Bruna Sampaio de Mello[1], Lucas Tadeu Fuess[2], Arnaldo Sarti[1,3] and Kelly J. Dussán[1,3]

[1]Department of Biochemistry and Chemical Technology, Institute of Chemistry, São Paulo State University-Unesp, Araraquara, Brazil, [2]Biological Processes Laboratory (LPB), São Carlos School of Engineering (EESC), University of São Paulo (USP), São Carlos, Brazil, [3]São Paulo State University (Unesp), Bioenergy Research Institute (IPBEN), Araraquara, Brazil

7.1 Introduction

Challenging issues associated with the availability and trade of fossil energy have stimulated the search for alternative energy sources. Moreover, various attempts have been made during the few past decades to both diversify the energy matrix on a global scale and mitigate cases of pollution resulting from the production and use of fossil fuels. The exploitation of a wide variety of renewable energy sources was consequently encouraged, leading to the development and further improvement of conversion systems to efficiently use wind, solar radiation, biomass, and other bioenergy sources.[1] Classical examples of this trend include the production of automotive biofuels, for example, bioethanol and biodiesel, using different feedstocks such as sugarcane, corn, and cassava for ethanol and soybean and palm oils for biodiesel. Biogas from the anaerobic processing of organic compounds is also characterized as another important bioenergy source, which is usually associated with the reduction of the polluting load of residual agroindustrial streams and solid wastes.[1]

Anaerobic biodigestion comprises a natural process in which different microbial groups are syntrophically associated with the conversion of complex organic compounds primarily into biogas,[2,3] namely a methane-rich gaseous blend, which also contains carbon dioxide as a relevant component. The conversion of simple organic compounds into organic acids and hydrogen and their further conversion into methane, are the most relevant steps

Sustainable Bioenergy. DOI: https://doi.org/10.1016/B978-0-12-817654-2.00007-1

in anaerobic processes as they are key steps for producing a wide range of bioproducts. Operating anaerobic systems under acidogenic conditions targets the production of biohydrogen and/or organic acids as final metabolites,[4,5] whereas maintaining adequate environmental conditions for methanogens will lead to biogas production (methane-rich) as the sole product. Anaerobic biodigestion may be efficiently coupled to the management of wastewaters and waste in a wide range of industrial processes, thus positively impacting the energy balance of the processes by decreasing bioenergy losses. Although producing biohydrogen and soluble phase metabolites from anaerobic digestion (AD) still require significant fundamental and applied research, the versatility of the process fits completely into the concept of biorefinering, in which a broad range of value-added bioproducts may be obtained from a single type of raw material.[5]

Currently, the production of methane from anaerobic systems may be considered a well-established technology considering the development of specific reactor configurations and operating conditions for different types of wastewater. Understanding the role of microbial populations has also enabled a better control of anaerobic systems. In particular, the energy potential of biogas may be exploited by different approaches, which may include the direct production of electricity and/or thermal energy in prime movers[6,7] and the production of biomethane, that is, purified biogas, to replace natural gas in industrial applications and vehicles.[7–9] In particular, biogas purification for biomethane production may be also achieved through microbial metabolism,[10] enabling the recovery of elemental sulfur from sulfide-rich biogas streams. Apart from its conventional use, studies have also investigated the use of methane (from dilute emissions) as a substrate to produce value-added products, such as polyhydroxyalkanoates, exopolysaccharides, and methanol,[11,12] which characterizes a more noble biogas application, although massive research is still required to efficiently be able to implement full-scale plants for such purposes.

Despite the positive aspects of using biogas as a reliable bioenergy source, some challenges must still be overcome worldwide. Although, there are successful examples regarding biogas use, especially in Germany,[13] in some cases, the implementation of anaerobic plants to manage residual streams is still far from reality, and therefore well-defined regulations for the use of biogas are still required. Given the technological level associated with the anaerobic processing of a wide range of organic wastewaters, implementing full-scale anaerobic plants still depends on a more favorable economic encouragement in several countries, as well as requiring more restrictive environmental regulations to define proper management approaches for residual streams.[8,14] Highlighting the relevance of biogas energy based on consistent literature reports is also imperative to promoting a better acceptance of its potential to be exploited as a reliable, renewable energy source in full-scale applications.

In this context, this chapter aims to discuss detailed aspects of biogas production and use in an effort to present an overview of the available applications for an efficient bioenergy recovery from anaerobic systems. Aspects regarding renewable energy sources are initially introduced, followed by a comprehensive presentation of anaerobic processes and biogas production, including some characteristics of biogas purification and upgrading. Both biohydrogen and methane production are considered based on the operation of acidogenic and methanogenic systems. Finally, trends and perspectives concerning the exploitation of biogas from anaerobic biodigestion as a bioenergy source are highlighted. Overall, this text compiles critical information to better understand the potential of anaerobic biodigestion as a robust processing step for bioenergy production.

7.2 Renewable energy

The natural mechanism responsible for maintaining the terrestrial temperature in acceptable ranges for survival is called the greenhouse effect. The responsible gases are mainly: carbon dioxide (CO_2), methane (CH_4), nitrous oxide (N_2O), halocarbons, and water vapor; which absorb a fraction of the sun's rays and distribute them in the form of infrared radiation into the atmosphere, hence warming the planet.[15] About 70% of the sun's rays are absorbed by these gases and about 30% cannot overcome this barrier and are reflected back into space. However, considering the increase in their emissions, this barrier in the atmosphere offers a greater resistance to heat dissipation, thus causing an imbalance in the greenhouse effect.[16]

This effect has been intensified due to anthropogenic activities; mainly the burning of fossil fuels, industrial activities, accelerated growth of the economy, and population increase in emerging countries.[17] The main consequence of this is global warming as the temperature of the planet tries to offset this energy imbalance, and so increasing its average. Thus the main challenge of the economy in the coming years is to ensure a certain quality of life for humanity that will have a low environmental impact. Currently emissions already have a substantial impact on climate change. Possible ways of facing these environmental issues have been considered, such as reducing dependence on nonrenewable energy sources, decreasing the emission of harmful gases into the atmosphere, increasing the use of renewable energies, and creating policies that incentivize renewable energies.[18]

Environmental impacts caused by greenhouse gas emissions are evident all over the world, therefore, the search has increased for cleaner energy sources as a way to mitigate these consequences.[19] Based on this premise, renewable energy is any energy source, in addition to not being derived from a finite source, where its entire production results in a real reduction of greenhouse gas emissions.[17] However, the energy matrix was diversified, reducing dependence on fossil fuels. Some of the main sources of renewable energy have already been in use for a long time, such as solar energy[20–22] and hydroelectric power plants.[23] However, modern bioenergy sources derived from biomass have received much attention in many countries in the past few years. This chapter mainly focuses on biomethane and biohydrogen.

Biodiesel, bioethanol, and biogas are examples of the three major biofuels derived from biomass. The production of renewable energy from energy crops and agricultural or industrial waste or wastewater is a viable solution because it links two strategies of interest: being an appropriate destination for various waste streams and facilitating energy production. However, in the first instance a large capital investment is required to develop efficient bioenergy production, storage, and distribution technologies. One of the sectors of interest is transport as it accounts for 23% of the world's CO_2 emissions into the atmosphere.[24]

The incentive policy of each country is directly related to the transition of the energy system, which aims to reduce global greenhouse gas emissions. This transition of the energy supply chain includes changes in energy technologies, and social, economic, and regulatory policies.[25] Some countries are more prominent because they have policies that encourage the use of technologies that produce renewable energy and target the reduction of fossil fuels, as is the case in some countries of the European Union.[26] The diversification of energy sources in order to intensify the use of bioenergy is seen as one of the most effective ways to promote low-carbon energy solutions. In Finland, the use of forest industry waste accounts for 80% of the total renewable energy used in the region. By increasingly using these sources in a sustainable way and expanding the use of bioenergy in other regions, reductions in carbon emissions can be achieved more rapidly.[25,27]

When compared to conventional energy sources, some bioenergy sources have an unstable character, such as solar and wind energy.[20,21] The behavior of solar radiation and wind speed may vary throughout the season. The integration of several bioenergy sources has been explored since as a single source is not sufficient for continuous supply. The use of combined bioenergy sources was first noticed in remote areas, and an expansion of this concept has since been seen.[28,29]

7.2.1 Renewable energy sources

Methane (CH_4) is the most abundant organic gas in the atmosphere and it is the main gas (50%–90%) found in natural gas. On the other hand, biomethane is the main gas found in biogas produced by the anaerobic degradation of organic compounds by microorganisms. Biogas also contains carbon dioxide (CO_2) and nitrogen (N_2), and other trace species, such as hydrogen sulfide (H_2S) (between 0.2% and 0.4%), hydrogen (H_2), ammonia (NH_3), and water vapor.

Hydrogen (H) is the lightest element, with its most common isotope consisting of only one proton and one electron. This element is a major constituent of water and all organic matter and is distributed not only on the earth, but also throughout the Universe. It can form H_2 molecules, which are smaller when compared to most other molecules. The H_2 molecule is colorless, odorless, and tasteless (similar to methane) and diffuses faster than any other gas. Hydrogen has a density of 0.09 kg/m^3 and a greater electrical conductivity than any other solid element. Moreover, gaseous hydrogen has one of the highest heat capacities (14.4 kJ/kg K).[30] The interest in using hydrogen as a fuel is due to the fact that hydrogen combustion is considered a clean process because it leads to the formation of water as a unique product. This combustion releases 242 kJ/mol or 121 kJ/g of energy per unit mass higher than other available fuels. The released energy can be used in combustion engines or converted into electrical energy by means of a fuel cell.[31]

7.2.2 Renewable energy worldwide: trends and market

In recent years, global energy consumption has increased and the main source is fossil fuel.[32] However, it is a finite fuel source and using it on a large scale is associated with negative environmental impacts. Due to the fact that it is a centralized energy source, there has been an increase in the price of fuels, thus driving efforts toward renewable energy.[22] In 2010, renewable energy accounted for 19% of the world's energy supply, including traditional biomass sources, hydropower, and less usual energy sources; such as modern bioenergy, which has been studied in recent years. Biomethane and biohydrogen are found in these new technologies and represent a percentage of almost 2% worldwide. The energy sources of traditional biomasses have grown slowly and have even declined in some regions, while bioenergy has been growing faster in developed and some developing countries.[33]

The growth of renewable energies was more significant for some technologies between 2004 and 2009. Significant growth for biodiesel and ethanol, with an annual increase of 51% and 20%, respectively, was observed. Bioenergy such as hydroelectric and biomass energies are growing at rates of 3%–5%, where in some countries this rate exceeds the global average.[32] In Brazil, the primary energy matrix between 2006 and 2015 consisted of 54.2% nonrenewable sources and 45.8% renewable sources. The main renewable sources exploited in Brazil are hydroelectric and biomass; especially sugarcane, with ethanol production and the direct combustion of the lignocellulosic residues generated as a source of heat and electricity production.[34] In addition, Brazil has a highly diversified and solid agroindustry sector, where the residues of major crops, such as coffee,[35] soy,[36] and citrus,[37] can be used to produce heat through direct burning or the production of high value-added energy sources; for instance, the generation of bioelectricity from the conversion of thermal energy generated in boilers by steam into mechanical energy where turbines transform it into electric energy.

Countries in Europe have increased the use of biomass for energy production, along with China and India. This increase is justified by the development of plants that combine the generation of heat and electricity from biomass residues with high calorific value. Since 2000, Germany has enforced a law that incentivizes the production of bioenergy, especially focusing on wind energy production, solid biomass, and biogas. At the end of 2015, electricity consumption of renewable energy in the country was 31.6% and 7% for biomass, especially biogas.[38] Among the bioenergy types, biohydrogen can be obtained by two processes: photosynthesis and the dark fermentation process. Hydrogen has one of the highest production values when compared to conventional energy sources. It is of great interest because it does not cause the emission of harmful gases at the time of combustion; having as a byproduct only the release of water in fuel cells, besides being more energetic than hydrocarbons.[33] Although, biohydrogen production technology is still in the early stages, it aims for high market potential in the future. Countries such as Japan, the United States, and Taiwan have invested greatly in bioenergy and consequently have some patent ownership.[39]

In addition to the production cost, the production of biohydrogen is considered unstable in terms of large-scale yield. Several factors are related to this parameter and have been studied for an increased production, such as: type of feedstock used, configuration of bioreactor, organic loading rates (OLRs), hydraulic detention, pH of the medium, temperature, microorganisms used, and pretreatment steps to increase production potential.[40] In 2012, the approximate annual production of hydrogen was 368 TCM; however, almost half of it came from the steam reforming of natural gas, a process that demands high temperatures and pressures. This method must consider the price and availability of natural gas because to produce biohydrogen the burning of natural gas (consisting mainly of methane) is required. The percentage equivalent to the production from biomass is only 1%.[41] However, biohydrogen technology is a current field of research and plays an important role in terms of using renewable sources.[33]

On the other hand, biomethane, a coproduct of AD, has a better consolidated production in the bioenergy market. Several European Union countries have shown rapid progress as there exists energy policies with legal and financial solutions in such countries that support the reduction of fossil fuel consumption. In Germany, there was a rapid increase in the number of biogas plants, with an increase of almost eight thousand plants that produce biogas from 1990 to 2013.[26] In Brazil, landfills are still an alternative for the disposal of urban waste. The biogas generated in this practice can be collected and used as a renewable energy source. Even though there are a large number of landfills in the country, it only has 17 landfill biogas power plants with a production capacity of about 122 MW.[42] This is considered to be a low number, given the Brazilian potential for exploiting this bioenergy source. The United Kingdom, which has a smaller population, has 422 landfill biogas power plants corresponding to 1051 MW of capacity.

Therefore although the number of biogas industries is increasing each year, there is a variation in biogas production capacity and utilization routes. Political support and investment subsidies are the main factors determining which sector of biogas will be used. The leading producer of biogas is Germany with an annual production of 42 TW h, while Brazil accounts for an annual output of 0.6 TW h.[13] Most countries use biogas for heat and electricity cogeneration and for upgrading to biomethane.

7.2.3 Bioenergy technologies

Currently, there are many bioenergy production technologies in practice, which can be categorized into: chemical, biochemical, and thermal technologies.[43] Whichever technology is chosen, each has its advantages and disadvantages, as well as requirements for implementation.[44] Chemical technologies are characterized by the contact of the feedstock with at least one chemical agent that will act by breaking the natural resource for conversion into liquid fuel; for example, biodiesel production technology utilizes a transesterification reaction. However, biochemical technologies are characterized by the contact of the feedstock with biomolecules (e.g., enzymes) or microorganisms that act by metabolizing or catalyzing certain processes. Examples of this approach include fermentation and AD processes, such as is used for biogas production.[45] The main parameter of the thermal process is heat, where a transformation of the feedstock takes place through the oxidation of excess air, in which the heat generated can be used to produce electricity or as a direct heating source. An example is the direct combustion of lignocellulosic residues for heating boilers in an industrial plant.[43]

Some bioenergy production technologies can be adapted according to the available feedstock, resulting in different final products: electricity, heat, and liquid and gaseous fuels. However, some feedstocks offer technological limitations, such as animal manure and wastewater, which are characterized by a high moisture content and require AD as a technological route to produce biomethane or biohydrogen.[44]

7.3 Biogas production

AD is the main source of biogas. It is a biological process by which a complex mixture of symbiotic microorganisms transforms organic materials under oxygen-free conditions into biogas, nutrients, and additional cell matter. Furthermore, AD is a robust process that is widely used. The first industrial-scale digesters date back to the first half of the 20th century.[2] Nowadays, it is a reliable technology used in Europe and Asia, and it is used to treat more than 10% of organic waste in several European countries. In Germany, there are over 500 AD facilities to treat the organic fraction of municipal solid waste and over 4000 on-farm digesters.[46] In Europe, the total number of currently operating biogas plants is roughly 14,500 with an installed electric capacity of about 7.9 GW.[47] A study by Salomon et al.[48] shows that Brazil has the potential to generate electricity from biogas obtained from the AD of organic residues of about 1.05–1.13 GW, which corresponds to 1.16%–1.24% of the total electricity generated in the country.

AD is both a waste treatment technology, which enhances environmental quality, and a sustainable energy producing technology.[49] It has also been evaluated as one of the most energy-efficient and environmentally beneficial technologies for bioenergy production.[50]

7.3.1 Anaerobic digestion

The bioconversion of organic material into biogas is accomplished by a consortium of microorganisms. This consortium is composed of two biological kingdoms: bacteria and archaea. They carry out the degradation process of organic materials, mainly composed of carbohydrates, proteins, and lipids, into simpler compounds through four main steps: hydrolysis, acidogenesis, acetogenesis, and methanogenesis.[51] Complex polymeric materials, such as polysaccharides, proteins, and lipids (fat and grease), are hydrolyzed by the extracellular enzymes of hydrolytic bacteria to soluble products of a small enough size to allow them to be transported across the cell membrane. These relatively simple, soluble compounds are anaerobically oxidized by acidogenic bacteria to short-chain fatty acids, alcohols, carbon dioxide, hydrogen, and ammonia. Short-chain fatty acids (other than acetate) are converted by acetogenic bacteria into acetate, hydrogen gas, and carbon dioxide. Finally, aceticlastic methanogens convert the acetate into methane and carbon dioxide and hydrogenotrophic methanogens reduce the hydrogen and carbon dioxide into methane.[52,53]

Liquid AD handles feedstocks with solid concentrations of between 0.5% and 15%. Animal manure, sewage sludge, and food waste are generally treated by liquid AD. On the other hand, solid-state AD (SS-AD) generally occurs at solid concentrations higher than 15%. The organic fractions of municipal solid waste and lignocellulosic biomass, such as crop residues and energy crops, can be processed through SS-AD.[54] Since the early 1990s in Europe, SS-AD has been the dominant AD process used to treat municipal solid waste.[55]

The utilization of a single feedstock in the AD process sometimes causes a hindrance to the process due to its nutritional imbalance, lack of diversified microorganisms, and low biogas production. Therefore various studies are focused on the codigestion of organic materials. Anaerobic codigestion

(AcoD) refers to the simultaneous AD of two or more substrates, which is a promising potential option to overcome the disadvantages of mono-digestion and improve the economic viability of AD plants by facilitating a higher biomethane production.[56] The process has several advantages: improving the reliability of feedstock; diluting toxic substances; increasing the diversity of microorganism species due to the nutrition available from a wide variety of organic wastes enabling the consortia to be more resilient to environmental adversities; higher biomethane yield per unit of digester volume; and the economic scale can increase as the quantity of waste increases.[57] The numbers of codigestion plants are continuously increasing in many European countries and have become standard practice.[58] Astals et al.[59] demonstrated that the AcoD of pig manure and glycerol under mesophilic conditions showed an increase in biogas production of up to 400% over the mono-digestion of the same substrates.

Concerning hydrogen, choosing the best production method will depend on the amount of energy to be produced and its degree of purity. The literature reports various examples of technologies available for hydrogen production. There are conventional technologies and biological technologies.[60] Hydrogen production technologies using conventional energy sources include the steam reforming of natural gas and petroleum, the catalytic decomposition of natural gas, the partial oxidation of the heavy hydrocarbon fraction of petroleum, and the gasification of coal or coke.[61] These methods are extremely energy-intensive, require high temperatures ($>700°C$), and emit large amounts of carbon oxides, sulfur, and nitrogen, as well as ash containing radioactive substances and heavy metals into the atmosphere.[62,63]

Methods such as water electrolysis require electricity from nuclear or coal- or natural gas-fired power plants. This process results from a lack of carbon dioxide emission into the atmosphere.[64] Moreover, other approaches, such as the photocatalytic, plasmochemical, magnetolytic, or radiolytic splitting of water, as well as the high-temperature plasma gasification of hydrocarbons are being developed.[61] The decomposition of water yielding hydrogen and oxygen can also take place through mechanocatalytic splitting.[65] AD for hydrogen production is also referred to as dark fermentation and is carried out in the acidogenic phase from the inhibition of the methanogenic phase, mainly by applying proper OLRs and pH conditions.[51] This process is an attractive approach for bioenergy generation because it has the potential to achieve important advantages over conventional electro- and thermochemical approaches. Moreover, this type of biological process has the potential to reuse wastewater and residues as carbon sources as well as the potential to produce value-added metabolites, such as organic acids and biopolymers.[4]

Various microorganisms are involved in the steps of the AD process, such as hydrolytic and fermentative bacteria, and when the feedstocks used to produce bioenergy are rich in sulfate during the process, sulfate reduction to sulfide may occur by a specific group of sulfate-reducing bacteria.[66] In general, microorganisms that can be commonly found in the AD process include: *Clostridium* spp., *Peptococcus anaerobius, Bifidobacterium* spp., *Micrococcus* spp., *Bacillus* spp., *Desulfovibrio* spp., *Corynebacterium* spp., *Lactobacillus, Actinomyces, Staphylococcus* spp., *Escherichia coli, Methanobacterium, Methanobacillus, Methanococcus, Methanospirillum, Methanosaeta, Methanothrix,* and *Methanosarcina.*[31,67]

7.3.2 Feedstock

All types of biomass can be used as feedstock for biogas production if they contain carbohydrates, proteins, and fats as major components. In the SS-AD processes, a wide range of organic solids found in municipal, industrial, and agricultural waste can be used as substrates.[54] Typical municipal biomass wastes include municipal sewage waste, fruit/vegetal waste, and food waste. In fact, implementing AD has been one of the major developments in the treatment of municipal solid waste in recent years.[68]

Agricultural waste consists of animal manure, food crops (e.g., leaves and vegetable waste), and dedicated energy crops (e.g., maize, beets, sugarcane, and wheat). Maize silage has become a substrate of focus in Germany due to the fact that maize is one of the most used crops from which a high yield of biomethane can be obtained,[69] as shown in a study conducted by Pakarinen et al.,[70] which obtained a biomethane yield of 379 Nd m^3/kg/VS, without from maize silage.

Liquid AD substrates include domestic wastewater, industrial wastewater, and sewage sludge. Domestic wastewater has been used as a promising substrate for biogas (70%–80% biomethane) production due to the increasing number of upflow anaerobic sludge blanket (UASB) reactors in wastewater treatment plants.[71] Sewage sludge; food-processing wastewater, such as the effluent from meat-packing and sugar processing; and wastewater from pulp and paper mills are examples of other feedstocks that have been commonly used in the anaerobic process due to their high biomethane yields. These wastewaters contain readily degradable organics and sewage sludge, as well as a wide range of nutrients and alkalinity sources.[52]

Briefly, the literature describes the use of several types of wastewaters as substrates in fermentative processes, for examples: cheese whey and food processing,[72–75] soft-drink wastewater, vinasse from various feedstocks, and cassava processing wastewater.[4,76–80]

7.3.3 Factors affecting the anaerobic digestion process

According to Speece (1983),[52] performance losses in the AD process can be associated with operational conditions. This mainly includes inappropriate pH conditions and the application of organic under- or overloads. Research shows that the main parameters that need to be monitored due to their influence on the biological production of biogas in anaerobic reactors include: pH, hydraulic detention time (HDT), OLRs, temperature, and the inoculum and substrate required for the process.[60,81,82]

7.3.3.1 pH

The concentration of H^+ ions plays an important role in the AD processes because pH value has a direct impact on the activity of various strains found in a mixed culture, the metabolic pathways of microorganisms, as well as their morphology and cell structure.[83] Several studies have focused on the influence of the initial pH on AD production because without properly controlling the effluent, pH changes may occur in the metabolic pathways or in cell flocculation and adherence phenomena.[4,74,84] Studies on pH have shown that optimum pH values can be found in the range of approximately 5.0–7.0.[85–87]

7.3.3.2 Hydraulic detention time

HDT is an indispensable parameter and can be handled using flow control, with the intention of observing the behavior of biogas production.[88,89] This parameter is the time that a substrate remains in a reactor chamber. A minimal HDT reduces the reactor volume and, thus, reduces capital costs.[52] It has a large impact on biogas productivity during AD processes that are carried out in a continuous or semicontinuous mode.[61]

7.3.3.3 Organic loading rate

The OLR refers to the amount of organic material per unit reactor volume, which is subjected to the AD process in the reactor in a given unit time period.[90] Previous studies have proposed that the applied OLR acts as a key factor in the production of biogas.[4,91,92] The OLR may be influenced by the substrate concentration and the HDT, but different variations, apart from these parameters, affects the system differently. Studies show that increasing the concentration of substrate generates an increase in the OLR, and this may improve biogas production, but high concentrations of substrate may generate inhibitions in the system or even changes in the microbial pathways.[92,93]

7.3.3.4 Temperature

Temperature affects the growth rate of bacteria and the efficiency of the conversion of substrates into biogas in the AD process. Thus microorganisms can be classified into several temperature groups for which the maximum yield of the produced hydrogen falls into temperature ranges suitable for the growth conditions of psychrophiles (0°C–25°C), mesophiles (25°C–45°C), thermophiles (45°C–65°C), extreme thermophiles (65°C–80°C), and hyperthermophiles (above 80°C).[94] It is important to understand that the optimum temperature selection depends on the type of bacteria used during fermentation, both for pure cultures and consortia. Moreover, the activity of specific enzymes that are responsible for biogas production, depend on the temperature for which they can take an optimum value. Therefore the optimum fermentation temperature depends on the type of bacteria and the kind of substrate used.[61]

7.3.3.5 Substrate

The choice of an appropriate raw material should be one that has a source of organic compounds that serve as substrates in the AD process. Different sugars have been used as substrates, for example, monosaccharides (glucose), and disaccharides (lactose or sucrose) are the most preferred sources of carbon for metabolic conversions by microorganisms in fermentation processes.[61] Moreover, materials such as domestic wastewater, vinasse, food waste, corn, rice, paper, and other sources of organic matter may be used as substrates.[61]

7.3.4 Pretreatment

Pretreatment methods are commonly necessary in order to increase biomass digestion to reduce the retention time and prepare the inoculum for hydrogen production. It is a strategy to improve biogas production and digester performance. These methods can be divided into physicochemical and biological processes. Physicochemical pretreatments include: physical methods, which involve mechanical disruption; thermal energy disruption; and the use of chemical agents, such as acid, alkaline, and oxidants substances. Biological pretreatment of biomass comprises approaches such as, but not limited to, enzymatic hydrolysis, predigestion, as well as using fungi or biosurfactants.[95]

7.3.4.1 Substrate pretreatment

The choice of pretreatment method depends on the characteristics and structure of the biomass used. Substrate pretreatment increases the formation of biodegradable substrates, without the loss of matter during the process.[60] Sewage sludge, animal byproducts, lignocellulosic, and algae biomass are examples of feedstocks that usually need pretreatment. Mechanical pretreatment (grinding or milling) is applied to reduce the particulate size and increase the surface area, thus reducing the crystallinity and grade of polymerization.[60] Thermal pretreatment has been successfully applied at an industrial scale and it leads to pathogen removal and reduces the viscosity of the digestate, while the main objective is the disintegration of cell membranes, which results in the solubilization of organic compounds.[96]

The most commonly used alkalis are sodium, ammonium, calcium, and potassium hydroxides. On the other hand, acid pretreatment can be conducted using either a concentrated acid or a dilute acid, such as sulfuric acid, hydrochloric acid, nitric acid, and acetic acid.[97] Acidic pretreatments and oxidative methods, such as ozonation, are also used to enhance biogas production and improve the hydrolysis rate.[96] Organosolv is a chemical pretreatment that involves the contact of the feedstock with aqueous acetic acid, ethanol, or methanol at 190°C for 60 minutes leading to the solubilization of the lignin and hemicellulose fractions of the feedstock, thereby making the cellulose accessible to hydrolytic bacteria.[98]

Enzymes like cellulases, cellobiases, endoglucanase, xylanases, and pectinases, etc., are used to increase biogas yields, as well as to decrease the viscosity of the feedstock or the fermentation media.[99] Fungal pretreatments are used in the degradation of lignocellulose by brown-, white- and soft-rot fungi, of which white-rot fungi are the most effective because of their capacity to delignify biomass.[100–102]

7.3.4.2 *Inoculum pretreatment*

Some studies have suggested that acidogenic processes require treatment of the inoculum to minimize the growth of methanogenic bacteria, which consume hydrogen. However, if the interest is in biomethane production, this pretreatment is not necessary. Among the techniques adopted, the most important are acid, alkaline, thermal, and chloroform treatments, among others.[88,92,103,104]

7.3.4.3 *Reactors*

Technological advancements have led to improved AD reactor design and operations, enabling better efficiency in removing organic material and optimizing biogas recovery. The AD process is commonly carried out in standard rate digesters with segregated liquid layers and in high-rate digesters involving mixing, heating, and uniform feeding facilities.[98] Several types of bioreactors are currently in use, but the three main groups of bioreactors commonly include batch reactors, a one-stage continuously fed system, and a two-stage or multistage continuously fed system. In one-stage digestion, all the microbiological phases of AD occur in one tank, while multistage digesters allow for the growth of different microbial groups in different chambers.[58]

Many studies have demonstrated that the reactor configuration has a significant effect on hydrogen production and it is mainly related to microorganism retention in the reactor, effective contact between the microorganism and the substrate, and its hydraulic behavior. In general, one of the biggest challenges in hydrogen production is to manipulate the cell concentration in the reactor in such a way that it leads to a stable production in the long term.[4,61,81] Different types of reactor configurations are used for biohydrogen production, but the most frequently studied are batch reactors, continuously stirred tank reactors (CSTRs), anaerobic packed-bed reactors (APBR), and anaerobic fluidized-bed reactors.[61]

Anaerobic batch reactors are the least expensive to operate and the simplest reactors used primarily in laboratories for preliminary investigations to select optimum conditions for dark fermentation. It consists of a single-tank fill and draw unit in which all the phases of the anaerobic treatment occur in one tank.[61] Several authors concluded that batch digesters may be more suitable for organic waste digestion due to the dry solid contents and their fibrous characteristics. Reports proposed that a production of 579 mL H_2/g DQO using vinasse as substrate under mesophilic conditions is possible.[58,77]

CSTRs are one of the most commonly used reactors for full-scale biogas production to treat high-strength and high levels of suspended solids in wastewaters, such as animal manure and organic industrial wastes. CSTRs are cylindrical in shape and are equipped with a mechanical stirring system. A feedstock is fed into the reactor and the effluent is removed in a continuous or sequential mode. They are commonly used owing to their simple design, easy adjustment of operating conditions (pH, temperature), and their stirring action, which facilitates homogeneous conditions of the medium and provides a good contact between the microorganism and substrate.[61] Ekstrand et al.[105] studied the codigestion of paper and pulp wastewater sludge and activated sludge in two CSTRs in series with sludge recirculation. The group showed that the codigestion of the nutrient-poor fiber sludge with activated sludge demonstrated a higher stability of the AD and high biomethane yield of 230 N mL/gVS. While treating cassava stillage, Luo et al.[78–106] obtained a H_2 yield of between 14 mL/gVS, (under mesophilic conditions) and 69.6 mL/gVS (under thermophilic conditions) due to a decrease in the propionate concentration and the inhibition of homoacetogens.

UASB is a well-known reactor that has been used in anaerobic processes in many different types of plants for treating both industrial and municipal wastewaters. It is a high-rate digester and its design allows for high biomethane yields due to the high organic load rate applied. The main structure of the reactor consists a dense sludge bed located in the bottom in which all biological processes take place and that ensures good wastewater–biomass contact.[58,107] Cabral et al.[108] monitored five full-scale UASB reactors treating municipal wastewater that achieved a biomethane production ranging from 81 to 142 NL/kgCOD_{remov}, as well as biogas losses. The group also estimated the amount of electric energy to be generated by the biomethane, then concluded that it was possible to supply approximately 60% of the total consumption of the wastewater treatment plant.

In SS-AD processes, many reactor configurations have been used on an industrial scale, including the Valorga, Kompogas, and Dranco SS-AD systems, which are the most prevalent and have mainly been used for the commercial processing of feedstocks with solid contents ranging from 20% to 40%. The reactors are operated in a continuous single stage and the biogas yields of the systems range from 0.3 to 0.5 m^3/kg volatile solids.[55]

APBRs have been tested successfully for hydrogen production using different substrates because of their simple construction and increased cell retention time, which is essential for hydrogen production.[4] These reactors have a biofilm support structure (media), made of materials such as activated carbon, PVC (polyvinyl chloride) supports, hard rock particles, or ceramic rings for biomass immobilization. APBRs have many advantages compared to other types of reactors that use suspended cells, including the possibility of achieving a higher cell concentration and a longer cell retention time.[109] Moreover, Wu et al.[110] studied a packed-bed reactor filled with immobilized-cell beads for the dark fermentation of glucose, fructose, and sucrose. The hydrogen yield was 0.70 mol H_2/mol glucose.

There are many studies on two or multiple-stage reactors because the separation of the AD process into two stages, for instance, hydrolysis/acidogenesis and acetogenesis/methanation, can increase the conversion rate of organic material into biomethane. However, the cost of such a complex system is a significant drawback, and therefore few multiple-stage AD units currently operate to produce biogas fuel on a commercial scale.[111] An anaerobic baffled reactor (ABR) is a multiple-stage reactor that consists a series of upflow reactors connected by means of baffles that force wastewater to flow under and over them as it passes from the influent to the effluent.[112] Because of its characteristics, high organic rates can be applied and therefore high biomethane yields can be obtained. A full-scale (450 m^3) multiple-plate anaerobic reactor was studied by Guiot et al.[113] to treat wastewater from the cheese industry, and a high efficiency of organic removal was obtained along with a biomethane production rate averaging around 4 m^3/m^3/d due to its multistage configuration.

7.4 Biogas purification and upgrading

Biogas produced during the AD process is mainly composed of biomethane, carbon dioxide, and water vapor. To apply biogas in applications that require a high CH_4 concentration; CO_2, water, and other undesirable compounds, such as hydrogen sulfide and ammonia, must be removed by purification and upgrading processes.[114] In purification processes, trace components that are harmful to the natural gas grid, appliances, or end-users are removed; and these processes can be divided into two generic categories: physicochemical methods and biological processes.[115,116] Biogas upgrading is the removal of carbon dioxide from biogas to raise its calorific value. Biogas purification is a compulsory step while biogas upgrading is usually optional.[117]

Physicochemical purification methods include chemical and physical adsorption, chemical and physical absorption, water scrubbing, organic solvent scrubbing, amine scrubbing, pressure swing adsorption (PSA), and gas separation membranes. Chemical adsorption and absorption processes have been explored both scientifically and technically; many of them are commercially offered by companies such as Schmack–Biogas AGCarboTech; Eco-Tec Inc. BgPur BioGas Purification System Process; Guild Associates, Inc. Guild PSA Technology.[115]

7.4.1 Hydrogen sulfide removal

The removal of hydrogen sulfide (H_2S) can be achieved by different methods, such as biological processes and physicochemical processes, for instance, absorption in water, chemical absorption, using membranes, and adsorption in activated carbon.[118] However, physicochemical H_2S removal processes are expensive due to their high chemical requirements and energy and disposal costs. Biological processes are considered to be economically and environmentally friendly.[115] These processes consist of transforming H_2S into S^0 using chemotropic and phototrophic bacterial species in both aerobic and anaerobic pathways. There are mainly three different design approaches in biological desulfurization: biofilters, bioscrubbers, and biotrickling filters.[118] All these methods have various advantages and disadvantages and, depending on the resources available, the best one is the least expensive.

7.4.2 Carbon dioxide removal

Carbon dioxide (CO_2) removal is also known as biogas upgrading. It is a multistep gas separation process involving the removal of the bulk CO_2 component, gas drying, the removal of minor components, and compression.[119] The product is called biomethane with a CH_4 content typically exceeding 95% (v/v), and it can be utilized to produce SNG (substitute natural gas used for injecting into the grid) and/or CNG (compressed natural gas employed as a fuel in transportation).[120] Several technologies are commercially available to separate CO_2 from raw biogas or are currently being investigated, such as: absorption in liquids, adsorption on solid surfaces, membrane separation, cryogenic separation, and methanation.[120] Nevertheless, in terms of economic and environmental sustainability, an attractive alternative solution for biogas upgrading is the biological method of hydrogenotrophic methanogenesis. In this concept, CO_2 and H_2 can be biologically converted into methane by the action of autochthonous hydrogenotrophic methanogens without any additional energy.[121]

7.5 Biomethane and biohydrogen: applications and trends

Biogas can be used as a heat source, for power generation, or it can be upgraded to biomethane for vehicle fuel or to feed natural gas grids.[41]

The main constituent of biogas is biomethane and in smaller amounts CO_2, H_2S, nitrogen, hydrogen, oxygen, and water vapor. One of the factors that cause a low calorific value in biogas is the presence of an incombustible fraction, mainly CO_2; in addition to the disadvantages of other contaminating gases, such as H_2S, which causes corrosion in motors, boilers, and chimneys. For this reason, it is necessary to remove the contaminating gases, thus giving a greater value to the biofuel produced and a greater diversity of uses.[24]

Biomethane, obtained from biogas upgrading, reaches the same standard as natural gas through upgrading which removes CO_2 and cleaning which removes the impurities present. The use of biogas can be categorized into: raw and upgraded forms of biomethane. When raw biogas is used it includes the generation of electricity and heat and when upgraded to biomethane it can be used as fuel in internal combustion engines or injected into natural gas networks.[13,15] Some countries in the European Union stand out because they use biogas to produce electricity; however, in Sweden, new biogas plants have turned to the production of vehicular fuels.[122]

Biohydrogen can also be obtained from AD, through techniques and stages of pretreatment that intensify the production of biohydrogen instead of biomethane by inhibiting the methanogenic phase. It is considered a clean biofuel when compared to natural gas. It can be used in a blend to generate heat and energy, as well as vehicular fuel and transport.[123] Some recurring problems in terms of selling biohydrogen are scheduling, storage, and transportation. Some strategies to increase biohydrogen yields have been studied and developed, for example, during the phase of dark fermentation the biohydrogen generated can be combined with a fuel cell; when the formation of the biohydrogen begins, it is transferred directly to the fuel cell to generate electricity.[124]

7.6 Conclusion and future perspectives

The production of biomethane and biohydrogen has advanced in the past few years, with these being known as efficient substitutes for the most diverse areas, and being able to be used as a source of heating, power generation, and vehicular fuel. Research that investigates large-scale production technologies, transport, and storage are of fundamental importance for the intensification of the use of this from of bioenergy. Policymakers struggle with the question of how to become less dependent on gasoline and at the same time how to reduce greenhouse gas emissions. Bioenergy represents an alternative and future fuel. It offers the highest potential benefits in terms of diversified supply and decreases pollutant and greenhouse gas emissions. Moreover, all of these properties are being evaluated and promoted worldwide as a benign replacement for gasoline, heating oil, natural gas, and other fuels in transportation applications.

It is a well-known fact that biogas production is developing rapidly in regions where there are government policy incentives and rigorous environmental regulations. In addition, it is important to note that the use of lignocellulosic sources, such as cattle residues, fruit residues, and vegetal residues, will allow for the establishment of small family-oriented biogas industries. This will establish a lucrative chain of biogas production, and therefore, the creation of biogas training and development centers with the aim of creating future qualified professionals in this sector. Consequently, potential commercial uses of biogas in the context of developing economies, such as combustion in households, fuel for cooking, heating, and often generating electricity on a small scale as well as vehicle fuel, among others, must be included. However, it is necessary to establish technologies that result in low operating costs, low energy consumption, and high CO_2 and S capture efficiencies, that is, the proper purification of biogas.

Furthermore, for the dissemination of biogas plants, several improvements are necessary for process monitoring and process control. Likewise, the influence of the microbial community structure on process stability requires more efforts for improving biogas yield because strong variations in the community structures occur during the biological process, which influences the process efficiency. Molecular analyses have shown the presence of numerous recently unknown bacteria, which may have an important influence on AD. With the increasing number of biogas plants, an improvement in the effluent quality is necessary in order to avoid the contamination of groundwater.

Acknowledgements

The authors are grateful to FAPESP, São Paulo Research Foundation (grant number #2016/23209-0 and #2015/15880-1).

References

1. Chisti Y. The status of biofuels. *Biofuel Res J* 2015;**7**:253–7.
2. Mccarty PL, Smith DP. Anaerobic wastewater treatment. *Environ Sci Technol* 1986;**20**:1200–6.
3. Mosey FE. Mathematical modeling of the anaerobic digestion process: regulatory mechanisms for the formation of short-chain volatile acids from glucose. *Water Sci Technol* 1983;**15**(8–9):209–32.
4. Fuess LT, Kiyuna LSM, Garcia ML, Zaiat M. Operational strategies for long-term biohydrogen production from sugarcane stillage in a continuous acidogenic packed-bed reactor. *Int J Hydrogen Energy* 2016;**41**:8132–45.
5. Poggi-Varaldo HM, et al. Biohydrogen, biomethane and bioelectricity as crucial components of biorefinery of organic wastes: a review. *Waste Manage Res* 2014;**32**(5):353–65.
6. Fuess LT, Klein BC, Chagas MF, Rezende MCAF, Garcia ML, Bonomi A, et al. Diversifying the technological strategies for recovering bioenergy from the two-phase anaerobic digestion of sugarcane vinasse: an integrated techno-economic and environmental approach. *Renew Energy* 2018;**122**:674–87.

7. Moraes BS, Junqueira TL, Pavanello LG, Cavalett O, Mantelatto PE, Bonomi A, et al. Anaerobic digestion of vinasse from sugarcane biorefineries in brazil from energy, environmental, and economic perspectives: profit or expense? *Appl Energy* 2014;**113**:825–35.
8. Fuess LT, Zaiat M. Economics of anaerobic digestion for processing sugarcane vinasse: applying sensitivity analysis to increase process profitability in diversified biogas applications. *Process Saf Environ Prot* 2018;**115**:27–37.
9. Souza SNM, Borsoi A, Santos RF, Secco D, Frigo EP, Silva MJ. Production potential of biogas in sugar and ethanol plants for use in urban buses in Brazil. *J Food Agric Environ* 2012;**10**:908–10.
10. Muñoz R, Meier L, Dias I, Jeison D. A critical review on the state-of-the-art of physical/chemical and biological technologies for an integral biogas upgrading. *Rev Environ Sci Biotechnol* 2015;**14**:727–59.
11. Cantera S, Sánchez-Andrea I, Lebrero R, García-Encina PA, Stams AJM, Muñoz R. Multi-production of high-added market value metabolites from diluted methane emissions via methanotrophic extremophiles. *Bioresour Technol* 2018;**267**:401–7.
12. Lebrero R, Chandran K. Biological conversion and revalorization of waste methane streams. *Crit Rev Environ Sci Technol* 2018;**47**(22):2133–57.
13. Hakawati R, Smyth BM, Mccullough G, Rosa F, Rooney D. What is the most energy efficient route for biogas utilization: heat, electricity or transport? *Appl Energy* 2017;**206**:1076–87.
14. Fuess LT, Rodrigues IJ, Garcia ML. Fertirrigation with sugarcane vinasse: foreseeing potential impacts on soil and water resources through vinasse characterization. *J Environ Sci Health* 2017;**52**(11):1063–72.
15. Akitt JW. Some observations on the greenhouse effect at the earth's surface. *Spectrochim Acta - A: Mol Biomol Spectrosc* 2018;**188**:127–34.
16. Kirk-Davidoff D. *The greenhouse effect, aerosols, and climate change. Green chemistry.* Elsevier Inc.; 2018. p. 211–34.
17. Kang S, Selosse S, Maïzi N. Contribution of global GHG reduction pledges to bioenergy expansion. *Biomass Bioenergy* 2016;**111**:142–53.
18. Lee DH. Econometric assessment of bioenergy development. *Int J Hydrogen Energy* 2017;**42**(45):27701–17.
19. Lindoso DP, Rocha JD, Debortoli N, Parente ICI, Eiró F, Bursztyn M, et al. Climate change and vulnerability to drought in the semiarid: the case of smallholder farmers in the Brazilian northeast. *Clim Change Brazil: Econ Soc Regul Aspects* 2011;235–54.
20. Sansaniwal SK, Sharma V, Mathur J. Energy and exergy analyses of various typical solar energy applications: a comprehensive review. *Renew Sustain Energy Rev* 2018;**82**:1576–601.
21. Ishaq H, Dincer I, Naterer GF. Science direct performance investigation of an integrated wind energy system for co-generation of power and hydrogen. *Int J Hydrogen Energy* 2018;1–12.
22. Manzano-Agugliaro F, Alcayde A, Montoya FG, Zapata-Sierra A, Gil C. Scientific production of renewable energies worldwide: an overview. *Renew Sustain Energy Rev* 2013;**18**:134–43.
23. Apergis N, Chang T, Gupta R, Ziramba E. Hydroelectricity consumption and economic growth nexus: evidence from a panel of ten largest hydroelectricity consumers. *Renew Sustain Energy Rev* 2016;**62**:318–25.

24. Sahota S, Vijay VK, Subbarao SMP, Chandra R, Ghosh P, Shah G, et al. Characterization of leaf waste based biochar for cost effective hydrogen sulphide removal from biogas. *Bioresour Technol* 2018;**250**:635–41.
25. Dahal K, Juhola S, Niemelä J. The role of renewable energy policies for carbon neutrality in Helsinki metropolitan area. *Sustain Cities Soc* 2018;**40**:222–32.
26. Piwowar A, Dzikuć M, Adamczyk J. Agricultural biogas plants in Poland – selected technological, market and environmental aspects. *Renew Sustain Energy Rev* 2016;**58**:69–74.
27. Koponen K, Sokka L, Salminen O, Sievänen R, Pingoud K, Ilvesniemi H, et al. Sustainability of forest energy in northern Europe. Technology 2015; 237.
28. Guo S, Liu Q, Sun J, Jin H. A review on the utilization of hybrid renewable energy. *Renew Sustain Energy Rev* 2018;**91**:1121–47.
29. Welfle A. Balancing growing global bioenergy resource demands-Brazil's biomass potential and the availability of resource for trade. *Biomass Bioenergy* 2017;**105**:83–95.
30. Gupta RB. *Hydrogen fuel: production, transport, and storage.* 1st ed. CRC Press; 2008.
31. Chernicharo C.A.L. Reatores anaeróbios - princípios do tratamento biológico de águas residuárias. Belo horizonte: Departamento de Engenharia Sanitária e Ambiental. UFMG. 1997; 5: 2 edição.
32. Cruz-Peragon F, Palomar JM, Casanova PJ, Dorado MP, Manzano-Agugliaro F. Characterization of solar flat plate collectors. *Renew Sustain Energy Rev* 2012;**16**(3):1709–20.
33. Demirbas A. Global renewable energy projections. *Energy Sour B: Econom Plan Policy* 2009;**4**(2):212–24.
34. Da Silva CMS, Carneiro ACO, Vital BR, Figueiró CG, Fialho LF, Magalhães MA, et al. Biomass torrefaction for energy purposes – Definitions and an overview of challenges and opportunities in Brazil. *Renew Sustain Energy Rev* 2018;**82**:2426–32.
35. García CA, Pena A, Betancourt R, Cardona CA. Energetic and environmental assessment of thermochemical and biochemical ways for producing energy from agricultural solid residues: coffee cut-stems case. *J Environ Manage* 2017;**216**:160–8.
36. Huang X, Cao JP, Zhao XY, Wang JX, Fan X, Zhao YP, et al. Pyrolysis kinetics of soybean straw using thermogravimetric analysis. *Fuel* 2016;**169**:93–8.
37. Taghizadeh-Alisaraei A, Hosseini SH, Ghobadian B, Motevali A. Biofuel production from citrus wastes: a feasibility study in Iran. *Renew Sustain Energy Rev* 2017;**69**:1100–12.
38. Scheftelowitz M, Becker R, ThräN D. Improved power provision from biomass: a retrospective on the impacts of German energy policy. *Biomass Bioenergy* 2018;**111**:1–12.
39. Lai WH, Chen HY, Chang FY, Wu CC, Lin CH, Huang SR. Market and patent analysis of commercializing biohydrogen technology. *Int J Hydrogen Energy* 2011;**36**(21):14049–58.
40. Dinesh GK, Chauhan R, Chakma S. Influence and strategies for enhanced biohydrogen production from food waste. *Renew Sustain Energy Rev* 2018;**92**:807–22.
41. Karthic P, Joseph S. Comparison and limitations of biohydrogen production processes. *Res J Biotechnol* 2012;**7**(2):59–71.
42. Lima RM, Santos AHM, Pereira CRS, Flauzino BK, Pereira ACOS, Nogueira FJH, et al. Spatially distributed potential of landfill biogas production and electric power generation in Brazil. *Waste Manage* 2017;**74**:323–34.
43. Silva DAL, Delai I, Montes MLD, Ometto AR. Life cycle assessment of the sugarcane bagasse electricity generation in Brazil. *Renew Sustain Energy Rev* 2014;**32**:532–47.
44. Fiorese G, Catenacci M, Bosetti V, Verdolini E. The power of biomass: experts disclose the potential for success of bioenergy technologies. *Energy Policy* 2014;**65**:94–114.

45. Aslani A, Mazzuca-Sobczuk T, Eivazi S, Bekhrad K. Analysis of bioenergy technologies development based on life cycle and adaptation trends. *Renew Energy* 2018;**127**:1076–86.
46. Labatut RA, Largus TA, Norman RS. Biochemical methane potential and biodegradability of complex organic substrates. *Bioresour Technol* 2011;**102**(3):2255–64.
47. Wellinger A. Biomethane injection in Europe e an overview. In: *Egatec - European Gas Technology Conference, Vienna, Austria*, 2015.
48. Salomon KR, Lora EES. Estimate of the electric energy generating potential for different sources of biogas in Brazil. *Biomass Bioenergy* 2009;**33**(9):1101–7.
49. Panwar NL, Kothari R, Tyagi VV. Thermo chemical conversion of biomass ecofriendly energy routes. *Renew Sustain Energy Rev* 2012;**16**(4):1801–16.
50. Weiland P. Biogas production: current state and perspectives. *Appl Microbiol Biotechnol* 2010;**85**(4):849–60.
51. Metcalf & Eddy. Tratamento de efluentes e recuperação de recursos. 5 edição; 2016.
52. Speece RE. Anaerobic biotechnology for industrial wastewater treatment. *Environ Sci Technol* 1983;**17**(9):416a–27a.
53. Pavlostathis SG, Giraldo, Gomez E. Kinetics of anaerobic treatment: a critical review. *Crit Rev Environ Sci Technol* 1991;**21**(5-6):411–90.
54. Li Y, Stephen YP, Jiying Z. Solid-state anaerobic digestion for methane production from organic waste. *Renew Sustain Energy Rev* 2011;**15**(1):821–6.
55. Brown D, Yebo L. Solid state anaerobic co-digestion of yard waste and food waste for biogas production. *Bioresour Technol* 2013;**127**:275–80.
56. Hagos K, Zong J, Li D, Liu C, Lu X. Anaerobic co-digestion process for biogas production: progress, challenges and perspectives. *Renew Sustain Energy Rev* 2017;**76**:1485–96.
57. Shah FA, Mahmood Q, Rashid N, Pervez A, Raja IA, Shah MM. Co-digestion, pretreatment and digester design for enhanced methanogenesis. *Renew Sustain Energy Rev* 2015;**42**:627–42.
58. Braun R, Wellinger A. Potential of co-digestion. *Int Energy Agency (IEA) Bioenergy* 2009.
59. Astals S, Nolla-Ardèvol V, Mata-Alvarez J. Anaerobic co-digestion of pig manure and crude glycerol at mesophilic conditions: biogas and digestate. *Bioresour Technol* 2012;**110**:63–70.
60. Anzola-Rojas MP, Fonseca SG, Silva CC, Oliveira VM, Zaiat M. The use of the carbon/nitrogen ration and specific organic loading rate as tools for improving biohydrogen production in fixed-bed reactors. *Biotechnol Rep* 2015;**5**:46–54.
61. Łukajtis R, et al. Hydrogen production from biomass using dark fermentation. *Renew Sustain Energy Rev* 2018;**91**:665–94.
62. Kapdan IK, Kargi F. Bio-hydrogen production from waste materials. *Enzyme Microbial Technol* 2006;**38**:569–82.
63. Momirlan M, Veziroglu T. Current status of hydrogen energy. *Renew Sustain Energy Rev* 2002;**6**:141–79.
64. Sinha P, Pandey A. An evaluative report and challenges for fermentative biohydrogen production. *Int J Hydrogen Energy* 2011;**36**:7460–78.
65. Protti S, Albini A, Serpon N. Photocatalytic generation of solar fuels from the reduction of H_2O and CO_2: a look at the patent literature. *Phys Chem* 2014;**16**(37):19790–827.
66. Lens PN, Visser A, Janssen AJH, Hulshoff Pol LW, Lettinga G. Biotechnological treatment of sulfate-rich wastewaters. *Crit Rev Environ Sci Technol* 1998;**28**(1):41–88.
67. Tchobanoglous G, Burton FL, Stensel HD. *Metcalf & Eddy wastewater engineering: treatment and reuse*. 4th ed. New York: Mcgraw-Hill; 2003.

68. Appels L, Lauwers J, Degrève J, Helsen L, Lievens B, Willems K, et al. Anaerobic digestion in global bio-energy production: potential and research challenges. *Renew Sustain Energy Reviews* 2011;**15**(9):4295–301.
69. Martínez-Gutiérrez E. Biogas production from different lignocellulosic biomass sources: advances and perspectives. *3 Biotech* 2018;**8**(5):233.
70. Pakarinen A, et al. Evaluation of annual bioenergy crops in the boreal zone for biogas and ethanol production. *Biomass Bioenergy* 2011;**35**(7):3071–8.
71. Chernicharo CAL, van Lier JB, Noyola A, Ribeiro TB. Anaerobic sewage treatment: state of the art, constraints and challenges. *Rev Environ Sci Biotechnol* 2015;**14**(4):649–79.
72. Perna V, Castell OE, Wenzel J, Zampol C, Fontes Lima DM, Borzacconi L, et al. Hydrogen production in an Upflow anaerobic packed bed reactor used to treat cheese whey. *Int J Hydrogen Energy* 2013;**38**:54–62.
73. Scoma A, Bertin L, Fava F. Effect of hydraulic retention time on biohydrogen and volatile fatty acids production during acidogenic digestion of dephenolized olive mill wastewaters. *Biomass Bioenergy* 2013;**48**:51–8.
74. Dareioti MA, Vavouraki AI, Kornaros M. Effect of pH on the anaerobic acidogenesis of agro-industrial wastewaters for maximization of bio-hydrogen production: a lab-scale evaluation using batch tests. *Bioresour Technol* 2014;**162**:218–27.
75. Peixoto G, Saavedra NK, Varesche MB, Zaiat M. Hydrogen production from soft-drink wastewater in an Upflow anaerobic packed-bed reactor. *Int J Hydrogen Energy* 2011;**36**:8953–66.
76. Fernandes BS, Peixoto G, Albrecht FR, Del Aguila NKS, Zaiat M. Potential to produce biohydrogen from various wastewaters. *Energy Sustain Develop* 2010;**14**(2):143–8.
77. Luo G, Xie L, Zhou Q, Angelidaki I. Enhancement of bioenergy production from organic wastes by two-stage anaerobic hydrogen and methane production process. *Bioresour Technol* 2011;**102**:8700–6.
78. Rosa PRF, Santos SC, Sakamoto IK, Varesche MBA, Silva EL. The effects of seed sludge and hydraulic retention time on the production of hydrogen from a cassava processing wastewater and glucose mixture in an anaerobic fluidized bed reactor. *Int J Hydrogen Energy* 2014;**39**:13118–27.
79. Andreani CL, et al. Hydrogen production from cassava processing wastewater in an anaerobic fixed bed reactor with bamboo as a support material. *Engenharia Agrícola* 2015;**35**(3):578–87.
80. Gomes SD, Fuess LT, Mañunga T, Gomes PCFL, Zaiat M. Bacteriocins of lactic acid bacteria as a hindering factor for biohydrogen production from cassava flour wastewater in a continuous multiple tube reactor. *Int J Hydrogen Energy* 2016;**41**:8120–813.
81. Fernandes BS, Saavedra NK, Mintinguer SI, Sette LD, Oliveira VM, Varesche MBA, et al. The effect of biomass immobilization support material and bed porosity on hydrogen production in an Upflow anaerobic packed-bed bioreactor. *Appl Biochem Biotechnol* 2013;**170**:1348–66.
82. Gois, G.N.S.B. Produção de hidrogênio e metano a partir da vinhaça de cana-de-açúcar em reatores anaeróbios de leite. Dissertação (Mestrado em Engenharia Química) – Universidade Federal de Alagoas. Centro de Tecnologia. Programa de Pós-Graduação em Engenharia Química; 2017.
83. Lin CY, Lay CH, Sen B, Chu CY, Kumar G, Chen CC, et al. Fermentative hydrogen production from wastewaters: a review and prognosis. *Int J Hydrogen Energy* 2012;**37**:15632–42.

84. Antonopoulou G, Gavala HN, Skiadas IV, Lyberatos G. Influence of pH on fermentative hydrogen production from sweet sorghum extract. *Int J Hydrogen Energy* 2010;**35**:1921–8.
85. Mu Y, Yu HQ, Wang Y. Evaluation of three methods for enriching H_2-producing cultures from anaerobic sludge. *Enzyme Microbial Technol* 2007;**40**:947–53.
86. Liu I, Whang L, Ren W, Lin P. The effect of pH on the production of biohydrogen by clostridia: thermodynamic and metabolic considerations. *Int J Hydrogen Energy* 2011;**36**:439–49.
87. Amorim ELC, Sader LT, Silva EL. Effect of substrate concentration on dark fermentation hydrogen production using an anaerobic fluidized bed reactor. *Appl Biochem Biotechnol* 2012;**166**:1248–63.
88. Amorim ELC, Barros AR, Damianovic MHRZ, Silva EL. Anaerobic fluidized bed reactor with expanded clay as support for hydrogen production thorough dark fermentation of glucose. *Int J Hydrogen Energy* 2009;**34**:783–90.
89. Tenca A, Schievano A, Perazzolo F, Adani F, Oberti R. Biohydrogen from thermophilic co-fermentation of swine manure with fruit and vegetable waste: maximizing stable production without pH control. *Bioresour Technol* 2011;**102**:8582–8.
90. Arimi MM, Knode J, Kiprop A, Namango SS, Zhang Y, Geißena SU. Strategies for improvement of biohydrogen production from organic-rich wastewater: a review. *Biomass Bioenergy* 2015;**75**:101–18.
91. Kim DH, Han SK, Kim SH, Shin HS. Effect of gas spargin on continuous fermentative hydrogen production. *Int J Hydrogen Energy* 2006;**31**:2158–69.
92. Hafez H, Nakhla G, Naggar HE, Elbeshbishy E, Baghchehsaraee B. Effect of organic loading on a novel hydrogen bioreactor. *Int J Hydrogen Energy* 2010;**35**(1):81–92.
93. Shen L, Bagley DM, Liss SN. Effect of organic loading rate on fermentative hydrogen production from continuous stirred tank and membrane bioreactors. *Int J Hydrogen Energy* 2009;**34**:3689–96.
94. Levin DB, Pitt L. Biohydrogen production: prospects and limitations to practical application. *Int J Hydrogen Energy* 2004;**29**:173–85.
95. Zhen G, Lu X, Kato H, Zhao Y, Li Y. Overview of pretreatment strategies for enhancing sewage sludge disintegration and subsequent anaerobic digestion: current advances, full-scale application and future perspectives. *Renew Sustain Energy Rev* 2017;**69**:559–77.
96. Ariunbaatar J, Panico A, Esposito G, Pirozzi F. Lens PN. Pretreatment methods to enhance anaerobic digestion of organic solid waste. *Appl Energy* 2014;**123**:143–56.
97. Rodriguez C, Alaswad A, Benyounis KY, Olabi AG. Pretreatment techniques used in biogas production from grass. *Renew Sustain Energy Rev* 2017;**68**:1193–204.
98. Zhang Q, Hu J, Duu-Jong L. Biogas from anaerobic digestion processes: research updates. *Renew Energy* 2016;**98**:108–19.
99. Carrere H, et al. Review of feedstock pretreatment strategies for improved anaerobic digestion: from lab-scale research to full-scale application. *Bioresour Technol* 2016;**199**:386–97.
100. Hatakka A. Lignin-modifying enzymes from selected white-rot fungi: production and role from in lignin degradation. *FEMS Microbiol Rev* 1994;**13**(2-3):125–35.
101. Christy P, Merlin LR, Gopinath D. A review on anaerobic decomposition and enhancement of biogas production through enzymes and microorganisms. *Renew Sustain Energy Rev* 2014;**34**:167–73.

102. Rouches E, Herpoël-Gimbert I, Steyer JP, Carrere H. Improvement of anaerobic degradation by white-rot fungi pretreatment of lignocellulosic biomass: a review. *Renew Sustain Energy Rev* 2016;**59**:179–98.
103. Maintinguer SI, Fernandes BS, Duarte ICS, Saavedram NK, Adornom AT, Varesche B. Fermentative hydrogen production by microbial consortium. *Int J Hydrogen Energy* 2008;**33**(16):4309–17.
104. Gomes SD, Fuess LT, Penteado ED, Lucas SDM, Gotardo JT, Zaiat M. The application of an innovative continuous multiple tube reactor as a strategy to control the specific organic loading rate for biohydrogen production by dark fermentation. *Bioresour Technol* 2015;**197**:201–7.
105. Ekstrand EM, et al. High-rate anaerobic co-digestion of kraft mill fibre sludge and activated sludge by CSTRS with sludge recirculation. *Waste Manage.* 2016;**56**:166–72.
106. Luo G, Xie L, Zou Z, Wang W, Zhou Q, Shim H. Anaerobic treatment of cassava stillage for hydrogen and methane production in continuously stirred tank reactor (CSTR) under high organic loading rate (OLR). *Int J Hydrogen Energy* 2010;**35**:11733–7.
107. Seghezzo L, et al. A review: the anaerobic treatment of sewage in UASB and EGSB reactors. *Bioresour Technol* 1998;**65**(3):175–90.
108. Cabral CBG, Chernicharo CAL, Platzer CG, Barjenbruch M, Filho PB. Evaluation of biogas production and energy recovery potential in 5 full-scale WWTPS with UASB reactors. *Chem Eng Process Tech* 2017;**3**(3):1043–51.
109. Ferraz Jr AND, Zaiat M, Guta M, Elbeshbjshy E, Hafez H, Nakhlab G. Impact of organic loading rate on biohydrogen production in an up-flow anaerobic packed bed reactor (UANPBR). *Bioresour Technol* 2014;**164**:371–9.
110. Wu SC, Wang FJ, Pan C-L. Growth and survival of lactic acid bacteria during the fermentation and storage of seaweed oligosaccharides solution. *J Marine Sci Technol* 2007;**15**:104–14.
111. Achinas S, Vasileios A, Gerrit JWE. A technological overview of biogas production from biowaste. *Engineering* 2017;**3**(3):299–307.
112. Bachmann A, Beard VL, Mccarty PL. Performance characteristics of the anaerobic baffled reactor. *Water Res* 1985;**19**(1):99–106.
113. Guiot SR, Safi B, Frigon JC, Mercier P, Mulligan C, Tremblay R, et al. Performances of a full-scale novel multiplate anaerobic reactor treating cheese whey effluent. *Biotechnol Bioeng* 1995;**45**(5):398–405.
114. Khan IU, Othman MHD, Hashima H, Matsuura T, Ismail AF, Rezaei-DashtArzhandi M, et al. Biogas as a renewable energy fuel—a review of biogas upgrading, utilization and storage. *Energy Convers Manage* 2017;**150**:277–94.
115. Abatzoglou N, Boivin S. A review of biogas purification processes. *Biofuels Bioprod Biorefin* 2009;**3**(1):42–71.
116. Ryckebosch E, Drouillon M, Vervaeren H. Techniques for transformation of biogas to biomethane. *Biomass Bioenergy* 2011;**35**(5):1633–45.
117. Budzianowski WM. A review of potential innovations for production, conditioning and utilization of biogas with multiple-criteria assessment. *Renew Sustain Energy Rev* 2016;**54**:1148–71.
118. Syed M, Soreanu G, Falletta P, Béland M. Removal of hydrogen sulfide from gas streams using biological processes: a review. *Can Biosys Eng* 2006;**48**:2.
119. Miltner M, Makaruk A, Harasek M. Review on available biogas upgrading technologies and innovations towards advanced solutions. *J Clean Prod* 2017;**161**:1329–37.

120. Niesner J, Jecha D, Stehlík P. Biogas upgrading technologies: state of art review in European region. *Chem Eng Trans* 2013;**35**:517–22.
121. Kougias PG, Treu L, Benavente DP, Boe K, Campanaro S, Angelidaki I. Ex-situ biogas upgrading and enhancement in different reactor systems. *Bioresour Technol* 2017;**225**:429–37.
122. Scarlat N, Dallemand JF, Fahl F. Biogas: developments and perspectives in Europe. *Renew Energy* 2018;**129**:457–72.
123. Boodhun BSF, Mudhoo A, Kumar G, Kim SH, Lin CY. Research perspectives on constraints, prospects and opportunities in biohydrogen production. *Int J Hydrogen Energy* 2017;**42**(45):27471–81.
124. Hoa LQ, Sugano Y, Yoshikawa H, Saito M, Tamiya E. A biohydrogen fuel cell using a conductive polymer nanocomposite based anode. *Biosen Bioelectron* 2010;**25**(11):2509–14.

Chapter 8

Current challenges and advances in butanol production

Ravichandra C. Patil[1], Pravin G. Suryawanshi[2], Rupam Kataki[3] and Vaibhav V. Goud[1,2]
[1]Centre for Energy, Indian Institute of Technology Guwahati, Guwahati, India, [2]Department of Chemical Engineering, Indian Institute of Technology Guwahati, Guwahati, India, [3]Department of Energy, Tezpur University, Tezpur, India

8.1 Introduction

Energy is one of the most vital elements for the overall development and growth of a nation. The challenges of climate change, geopolitical instability, and fluctuations in crude oil prices have accelerated research in the energy sector in a search to find clean and alternative sources of energy. Progress in this field would lead to the overall economical and sustainable development of a nation. It is expected that by 2030, biomass will contribute the largest share of energy production (mostly ethanol) and have the largest share of all renewable energy sources.[1] Synthetic biology has the potential to make the bioeconomy much broader. It can produce different biological compounds, including an array of various biofuels.

Among these biofuels, butanol is found to have advantages over other forms of fuels as an efficient biofuel because of its unique physicochemical properties. 1-Butanol (butyl alcohol or *n*-butanol) is a four-carbon straight-chain alcohol with the molecular formula, C_4H_9OH (MW 74.12), and a boiling point of 117.7°C. Due to its wide range of advantages as listed in Table 8.1, it has attracted great attention as a biofuel. Also, it can be obtained from agricultural residue and industrial waste as feedstocks with insignificant or null pretreatment.

Butanol production, to large extent through acetone–butanol–ethanol (ABE) fermentation, is yet to be commercially viable due to its low concentration ratio, low productivity, high feedstock cost, and product feedback inhibition on microbial growth. These drawbacks make butanol production

Sustainable Bioenergy. DOI: https://doi.org/10.1016/B978-0-12-817654-2.00008-3

TABLE 8.1 Physical and chemical properties of butanol.

Molecular formula: C_4H_9OH (MW 74.12)
Boiling point: 117.7°C.
Energy density (MJ/L): 29.2
Water solubility (mL 100 m/L): 9.1
Vapor pressure: 6.4
Oxygen content (% mass): 21.6
Octane rate: 96
It is less evaporative/explosive with a Reid vapor pressure 7.5-times lower than ethanol
It is safer than ethanol because of its higher flash point and lower vapor pressure
It is more miscible with gasoline and diesel fuel, but less miscible with water

less commercially viable than the production of alternative biofuels like ethanol. These hurdles can be addressed in several ways using modern advances in the field of biology and chemical engineering. Genetic engineering [clustered regularly interspaced short palindromic repeats (CRISPR)/Cas] is a promising tool for improving strains in order to increase butanol yields and butanol titer. This chapter also reviews perspectives of genetic engineering and fermentation technology development in the butanol production process. These techniques deal with the use of various methods like gene overexpression, knock out/down, or the insertion of genes encoding important enzymes in the metabolic pathway of butanol production. Other ways to improve fermentation technology include production using the multistage continuous fermentation process, immobilized continuous fermentation process, the recycling of cells, and metabolic flux regulation by additional organic acids or electron carriers.

8.2 Uses of butanol

Butanol is not used widely as a biofuel, but holds great promise as an alternative fuel. It consists of pure alcohol with a longer hydrocarbon chain, making it nonpolar, with an energy content similar to that of petrol, and hence can be directly used as fuel in existing internal combustion engines by either blending or in its pure form. It is reported that butanol at 85% strength can serve as an alternative fuel in cars designed for gasoline (petrol) without any alteration to the engine (unlike with 85% ethanol).[2] Butanol can also be blended with diesel to reduce soot emissions.[3] Other uses of butanol include: in cosmetics, hydraulic fluids, detergent formulations, drugs, antibiotics, hormones, and vitamins, as well as a chemical intermediate in the production of butyl acrylate and methacrylate, in the synthesis of 2-butoxyethanol, in hydrogen production, and additionally as a potential extractant in the food and pharmaceutical industries.[4–6] It is used as a chemical precursor in the production of paint thinners, polymers, and plastics, as well as lacquers and ambient-cured enamels. It is also used as a component in hydraulic and brake fluids.[7]

8.3 Processes for butanol production

Butanol production from the biological fermentation of starchy raw materials was commercially introduced during World War I (WWI).[8] The challenges associated with this process resulted in a shift toward the synthesis of butanol using chemical processes. Thus the production of butanol can be carried out using both biological and chemical methods, with each having its own advantages and disadvantages. These methods are reviewed and explained here:

8.3.1 Chemical processes

Chemical synthesis in butanol production mostly includes methods like crotonaldehyde hydrogenation, the Guerbet reaction, oxo synthesis, and Reppe synthesis. Because of the efficient conversion yields of chemical methods, butanol is mostly produced by chemical route. However, the potential carcinogenicity of butanol produced chemically limits its use and shift towards fermentative production.[9] The chemical reactions of these processes are shown in Fig. 8.1.

8.3.1.1 Crotonaldehyde hydrogenation process

Crotonaldehyde hydrogenation was the first chemical process used for butanol production. Acetaldehyde is used as a raw material in this process. The process, as shown in Fig. 8.1A, consists of the steps: aldol condensation, dehydration, and hydrogenation.[10] In this process, the starting material,

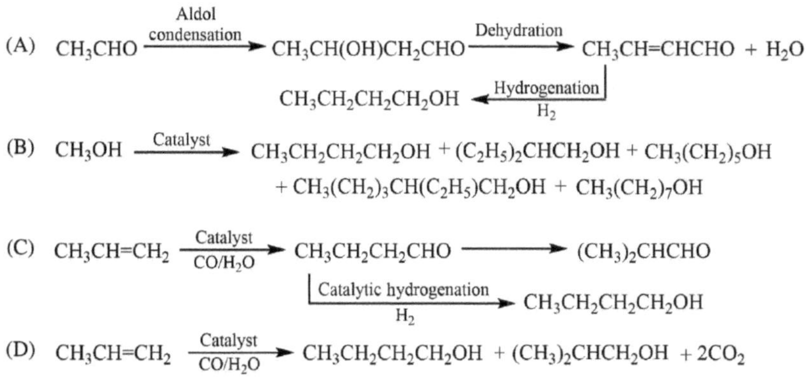

FIGURE 8.1 Chemical synthesis of butanol. (A) Crotonaldehyde hydrogenation, (B) Guerbet reaction, (C) oxo synthesis, (D) Reppe synthesis.

acetaldehyde, is produced from the dehydrogenation of ethanol. As ethanol can be produced from biomass resources, this process may be significant in the future.

8.3.1.2 Guerbet reaction

In this method, the direct conversion of ethanol into butanol takes place in the presence of a catalyst. *Saccharomyces cerevisiae* and other microbial species are used as hosts for this reaction. It is based on the catalytic conversion of ethanol into butanol. Iridium complexes,[11] manganese,[11,12] aryl diazenido, and aryl diazine complexes[13] are used as catalysts. Butanol has a higher energy content than ethanol (27 MJ/L compared to 19.6 MJ/L), that is a 30% higher energy content per gallon compared to that of ethanol as a biofuel, as well as being less corrosive and showing lower volatility.[14,15] The reaction is shown in Fig. 8.1B.

8.3.1.3 Oxo process

This method is also called hydroformylation or hydrocarbonylation.[16,17] 1-Butanol is produced by petrochemical route based on propylene oxo synthesis (Fig. 8.1C). In this process propylene undergoes hydroformylation to form aldehydes, these aldehydes are further hydrogenated to yield 1-butanol. The cost of propylene is extremely sensitive to the market price of crude oil and regulates the cost of synthetic butanol production. Other raw materials for butanol production by oxo synthesis are fossil oil derivatives, viz., ethylene and triethyl aluminum.

The first step of oxo synthesis involves the reaction of propylene with carbon monoxide and hydrogen in the presence of a cobalt or rhodium catalyst. In the second step, the resulting mixture of *n*- and isobutyraldehyde is hydrogenated to the corresponding *n*- and isobutyl alcohols and further distilled to recover the required quality of butanol.[18]

8.3.1.4 Reppe process

In the Reppe process, propylene is reacted with carbon monoxide and water under low pressure and temperature in the presence of a catalyst (Fig. 8.1D). Isobutyraldehyde and *n*-butyraldehyde are formed, from which *n*-butyraldehyde is converted into *n*-butanol. This process is a single step reaction and the costly processing technology required have limited its commercialization.[10,19]

8.3.2 Biological fermentation method

Reports on the isolation of butanol producing wild stains are reviewed in Table 8.2. These naturally available microorganisms are isolated and characterized by their butanol yield. Advances in this field include the use of

TABLE 8.2 Microbial strains used for butanol production.

Species	Substrate	Butanol concentration (g/L)	Reference
Clostridium acetobutylicum	Glucose	10.4	[20]
Clostridium a. KF 158795	Glycerol	13.57	[21]
Clostridium aurantibutyricum	Glucose	3.36[a]	[22]
Clostridium beijerinckii	Glucose	15.21[b]	[23]
Clostridium carboxidivorus	CO	0.37	[24]
Clostridium cadaveris	Glucose	0.829	[22]
Clostridium cellulolyticum[c*]	Crystalline cellulose	0.66	[25]
Clostridium cellulovorans[d*]	Crystalline cellulose	1.42	[26]
Clostridium pasteurianum	Glycerol	6.5[e]	[27]
C. pasteurianum GL11*	Birchwood xylan	1.48	[28]
Clostridium perfrigens	Glucose	0.019[f]	[29]
Clostridium saccharoperbutylacetonicum	Glucose	16.2	[30]
Clostridium saccharoperbutylicum	Glucose	9.7	[31]
Clostridium sporogenes BE01	Rice straw	5.52	[32]
*Clostridium tyrobutyricum**	Glucose	10.0	[33]
Clostridium thermocellum [g*]	Crystalline cellulose	5.4	[34]
Clostridium tetanomorphum	Glucose	3.5	[35]
*Bacillus subtilis**	Glycerol	0.024	[36]
*Escherichia coli**	Glucose	2.0[h]	[37]
*Lactobacillus brevis**	Glucose	0.303	[31]
*Lactobacillus buchneri**	Glucose	0.066	[38]
*Lactococcus lactis**	Glucose	0.028	[38]
*Pseudomonas putida**	Glycerol	0.12	[38]
*Saccharomyces cerevisiae**	Galactose	0.0025	[39]

[a]*Isopropanol as secondary by-product.*
[b]*Isopropanol as secondary by-product.*
[c]*Strain with introduction of* kivD, *2-ketoacid decarboxylase;* yqhD *alcohol dehydrogenase/aldehyde reductase;* alsS *acetolactate synthase;* ilvC *acetohydroxyacidisomeroreductase;* ilvD *dihydroxy-acid dehydrogenase.*
[d]*Strain with introduction of aldehyde/alcohol dehydrogenase* (adhE2).
[e]*1,3-Propanediolas secondary by-product.*
[f]*Ethanol as secondary by-product.*
[g]*Strain with introduction of* kivD, *2-ketoacid decarboxylase;* ilvBN *acetohydroxy acid synthase (AHAS);* ilvC *acetohydroxyacidisomeroreductase;* ilvD *dihydroxy-acid dehydrogenase.*
[h]*Propanol as secondary by-product.*
**Genetically modified.*

genetically modified strains to increase butanol yields. Mostly, the genus *Clostridium* dominates in the production of butanol through the fermentation process (Table 8.2). *Clostridium acetobutylicum* was first identified by Chaim Weizmann in 1916.[40] He discovered that it produced acetone, butanol, and ethanol from starch.[41] *Clostridium* is a spore forming, rod-shaped, strictly anaerobic bacteria. However, only a few species can produce significant amounts of butanol during fermentation under appropriate conditions. Genetically modified strains have also been established for butanol production using non-*Clostridium* genera as listed in Table 8.2.[20–39]

8.3.2.1 Acetone–butanol–ethanol pathway

The ABE fermentation process was first commercially started in the United Kingdom (1912) and quickly spread to other parts of the world during WWI (for acetone production) and WWII (for the production of butanol for paint lacquers).[42,43] The ABE fermentation pathway has been extensively studied and reported in the literature. The major products of this pathway are solvents (acetone, butanol, ethanol); gases (CO_2 and H_2); and acids [lactic acid, acetic acid (AA), and butyric acid].[44,45] For convenience, a pathway is reproduced here in Fig. 8.2. The ABE pathway produces the solvents: acetone, butanol, and ethanol, in a 3:6:1 ratio.[47] In 1861, the biological production of butanol was performed by Louis Pasteur for the first time.[48]

The metabolic pathway in *Clostridium* sp. mostly follows the ABE fermentation process. AA gets converted into ethanol or acetone, while butyric acid is converted into butanol in the metabolic pathway of ABE fermentation. Hexose (C6) and pentose (C5) sugars are utilized as substrates and converted into pyruvate by different pathways.[46] C6 sugars like glucose, fructose, sucrose, and galactose, are mostly utilized by the Embden–Meyerhof–Parnas (EMP) pathway, while C5 sugars like xylose, arabinose, and raffinose, are metabolized via the pentose phosphate pathway or the glycolic pathway. In these pathways, enzymes like transaldolase and ketolase play an important role in the conversion of fructose 6-phosphate and xylulose 5-phosphate substrates into glyceraldehyde 3-phosphate as a product, respectively.[49] These molecules then enter into the EMP pathway. During this process 5 mol of NADH, 5 mol of pyruvate, and 5 mol of ATP are formed.[50] Hydrogen ions released with exponential growth cause a decrease in pH and the formation of acetic and butyric acids. This are major products formed in this stage which is called the "acidogenesis phase." This stage is followed by the "solventogenesis phase" during which solvents like acetone, butanol, and ethanol are produced. At low external pH, solventogenesis is initiated by the enzymatic action as an adaptive response to pH which results in cell growth inhibition.[40] The important connecting node between acidogenesis and solventogenesis is acetyl-coenzyme A (A-CoA).[51] The presence of a small amount of mineral like zinc (0.001 g/L $ZnSO_4 \cdot 7H_2O$) in

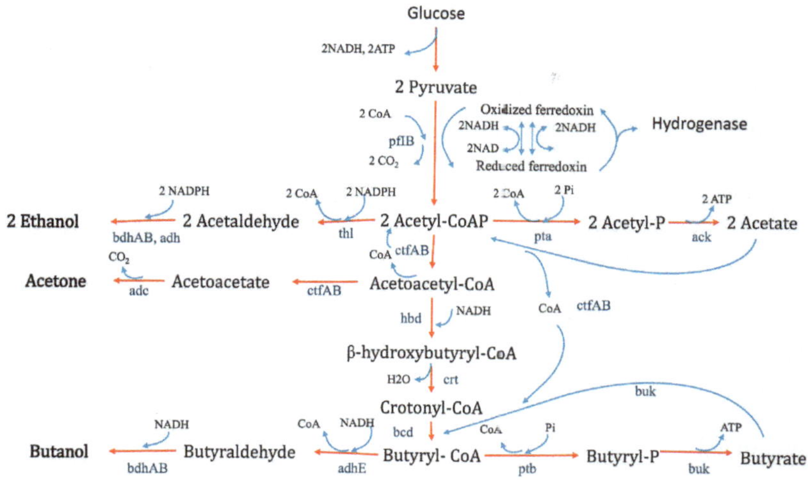

FIGURE 8.2 Metabolic pathways in *Clostridium acetobutylicum* ATCC 824 with glucose and xylose.Metabolic enzymes are: *tk*, transketolase; *ta*, transaldolase; *hydA*, hydrogenase; *pta*, phosphotransacetylase; *ack*, acetate kinase; *thl*, thiolase; *adc*, acetoacetate decarboxylase; *ctfAB*, acetoacetyl-CoA: acetate/butyrate:CoA-transferase; *hbd*, 3-hydroxybutyryl-CoA dehydrogenase; *crt*, crotonase; *bcd*, butyryl-CoA dehydrogenase; *buk*, butyrate kinase; *ptb*, phosphotransbutyrylase; (1) acetaldehyde dehydrogenase; (2) ethanol dehydrogenase; (3) butyraldehyde dehydrogenase; (4) butanol dehydrogenase. *Adapted from Shinto H, Tashiro Y, Kobayashi G, Sekiguchi T, Hanai T, Kuriya Y, et al. Kinetic study of substrate dependency for higher butanol production in acetone–butanol–ethanol fermentation. Process Biochem 2008;**43**(12):1452–61 with copyright permission form Elsevier.*

the fermentation broth also causes an earlier shift from acidogenesis.[52] Some strains like *Clostridium beijerinckii* can produce isopropanol instead of acetone, while *Clostridium aurantibutyricum* produces isopropanol along with the other two solvents.[35] The ratio of solvents produced is nearly the same irrespective of type.

Some of the enzymes involved in this process are acetate kinase (ack) and phosphor transacetylase (pta) which convert A-CoA to acetate. Butyrate kinase (buk) is responsible for butanol formation. Other important enzymes in the butanol synthesis pathway are butanol dehydrogenase (bdh), crotonase (crt), thiolase (thl) (plays a key role in both acid and solvent production), butyryl-CoA dehydrogenase (bcd), 3-hydroxybutyryl-CoA dehydrogenase (hbd), and butyraldehyde dehydrogenase. Acetoacetate decarboxylase (adc) converts acetoacetate into acetone and CO_2.[40] CoA-acylating aldehyde dehydrogenase (aad) and NADH dependent alcohol dehydrogenase (adhE) convert A-CoA into ethanol through catalysis. Control over the electron flow in the EMP pathway also plays an important role in regulating acetate, butyrate, and butanol formation.[5,53] In metabolic pathways, high NADPH levels have been proposed to induce the solventogenic phase, while ATP is generated during the acidogenic phase.[54]

8.4 Renewable feedstock resources as substrates

The scarcity of natural resources such as fossil fuels along with fluctuations in the market prices of these resources has raised the demand for the utilization of renewable and low-cost resources for biofuel production. The use of alternatively available resources would reduce dependency on fossil fuel consumption along with reducing CO_2 emissions. Progress in the development of scientific technologies over the years has resulted in the accessibility of divergent feedstocks as substrates for energy generation. The 21st century is looking for green resources and less pollutive processes for the production of valuable products and energy generation.[55] Developments in the utilization of feedstock for energy generation processes to overcome the limitations of prior stages have been addressed. The utilization of different generation feedstocks for energy generation can be summarized as:

8.4.1 First-generation feedstock

For first-generation ethanol, edible biomass is used as a substrate for ethanol production. The biomasses used for first-generation bioethanol production include sugarcane, corn, whey, sugar beets, barley, potato, and waste biomass.[56] However, these sources are largely food crops and thus are not sustainable for biofuel production.[55,57] The use of these resources in energy generation have several limitations, including the fact that they have a direct impact on food production in terms of food price and quality as well as soil usage for crop growth while providing only limited greenhouse gas emission reduction benefits. Due to these serious drawbacks, the search for alternate biomass sources was intensive which gave rise to the evolution of second-generation feedstocks.

8.4.2 Second-generation feedstock

Second-generation feedstocks are mostly related to the use of lignocellulosic biomass and industrial and municipal solid wastes for ethanol production. The use of these feedstocks not only results in greater greenhouse gas reductions, but also does not compete for agricultural land with food crops. The selected types of plants are specifically grown for bioenergy production (bioenergy crops) on marginal lands that are not suitable for use in food production. The nonedible parts of ordinary crops and forest trees could also be used to produce ethanol. The major barriers in the use of second-generation feedstocks are related to the cost of the treatment processes, biomass collection, and barriers in the utilization of multistep complex conversion technology.[58]

8.4.3 Third generation feedstock

The high cost of traditional substrates like glucose, corn, sugarcane molasses and whey permeate, adversely affects the commercial production of butanol. To reduce the cost of production, lignocellulosic biomass such as switchgrass, sugarcane bagasse, corn stover, wheat, and rice straw, corn fiber, miscanthus, and waste paper can be used as raw material for butanol production. A major part of lignocellulosic biomass contains cellulose, hemicellulose, and lignin along with other trace elements. Sugars like galactose, glucose, fructose, xylose, arabinose, rhamnose (found in agricultural biomass and hardwoods), and mannose (found in softwoods) are present in the form of cellulose and hemicelluloses.[59–63] Reports are available on the possibility of the conversion of hemicelluloses and lignin into transportation fuels like jet fuels.[64] It is important to note that, although the cost of these residual feedstocks is low, the additional processing steps of pretreatment and hydrolysis add up to an overall fermentation cost similar to that of butanol. Moreover, inhibitors are also formed during the process of pretreatment[65] which inhibit ABE fermentation.[66] ABE producing microorganisms are more sensitive to inhibitors produced from lignocellulosic biomass such as furfural, lignin derivatives, and hydroxymethyl furfural in comparison to the yeast used for ethanol production.[67]

8.4.4 Fourth generation feedstock

The production of biofuels from agricultural crop-based feedstock has always been limited by the insufficient photosynthetic efficiency of terrestrial plants as it converts less than 0.5% of the solar energy received at midlatitude.[68] In comparison to this, microalgae are able to convert more than 10% solar energy.[69] The basis of fourth generation biofuels is to exploit this photosynthetic efficiency and utilize it for the direct conversion of CO_2 into biofuels.[70,71] Unique microalgal properties like high photosynthetic efficiency, high biomass productivity, and rapid growth have made it a leading contender for biofuel production.[72] The principal objective of the fourth generation technique is to reduce the intermediate steps of converting solar energy and CO_2 into the desired product. This approach would reduce the overall initial capital cost along with the operational cost of biofuel production. Advances in the metabolic engineering of algae to increase the lipid content without compromising growth is regarded as an important milestone toward sustainable biodiesel production.[73] Engineered Cyanobacteria phototrophic organisms have been used as a "unit cell" for the production of ethanol, alkanes, and butanediol; and can be experimented for butanol production.[74–76] Engineered algae can bypass anabolic growth pathways and it is feasible to use engineered algae for the conversion of CO_2 into 1-butanol.[75] A designer pathway is expressed heterogeneously in the host algae which act upon one of the CO_2 fixing pathways to produce products like butanol. Atsumi et al.[77] demonstrated the production of isobutanol (450 mg/L), isobutyraldehyde (1100 mg/L) using cyanobacteria.

Another concept of "photanol," alternately known as photofermentation, has been reported. It is based on the construction of a metabolic chimera between the phototrophic and fermentative metabolisms in a single organism.[78] The feasibility of such conversion has already been proven in *Synechocystis* sp., where the fermentative enzyme, keto acid decarboxylase, from *L. lactis* has been expressed heterogeneously. The major drawback reported was of the oxygen sensitivity of the fermentative enzymes.[79] Still, the large gap between such technological development and its commercial viability limits its utilization for fourth generation biofuel.

Photosynthetic water splitting (i.e., water oxidation) using solar energy can become a major contributor to fuel production by artificial photosynthesis.[80] More scientific development in the field of fundamental research for creating "designer bacteria" with new properties of producing gaseous or liquid fuels is required.[81] The integrated production of algal biofuels with high-value chemicals using wastewater or sea water as culture media and the development of cost-effective bioreactors are technologies currently under heavy development. Progress in these fields in the near future will make algal biofuel production more profitable.

8.5 Fermentation and downstream regulation in butanol production

Mostly due to its ease of handling and bioprocess control, batch fermentation is favored over other types of fermentation.[82] The simultaneous hydrolysis of biomass in order to release sugars along with butanol production is directed to reduce overall batch fermentation time.[83] Various types of reactors are reported in the literature for ABE fermentation processes; these reactors are listed in Table 8.3.[84–92] A comparison of different fermentation processes used for ABE production from glucose substrate is represented in Table 8.4.[5,53,84,93]

Optimized process parameters and energy efficient solvent extraction can enhance the process efficiency and aid in the reduction of the cost of butanol production. Generally, distillation as a unit operation is widely used for solvent recovery.[93] The extraction of 1 t butanol solvent requires approximately 12 t of steam.[94] The problem of azeotrope formation in the separation of aqueous solutions from fermentation is a major disadvantage of distillation.[95] Mathematical models for continuous product recovery with integrated fermentation would lead to developing cost-effective downstream processes.[96]

Other methods like gas stripping, liquid–liquid extraction, adsorption, reverse osmosis (RO), pervaporation, and aqueous two-phase separation are reported in the literature for butanol extraction, out of which few have been

TABLE 8.3 Summary of acetone–butanol–ethanol (ABE) production in different physiological conditions.

Reactor type	Reactor life (h)	Productivity (g/L/h)	ABE produced (g/L)	Sugar used (g/L)	Feed concentration (g/L)	Reference
Batch reactor	68	0.29	17.6	45.4	59.9	[84]
Two stage immobilized batch in situ product removal	48	0.66	19.8	100	–	[85]
Batch/fed batch in situ butanol removal using biodiesel extraction	96	0.295	31.4	100	–	[86]
Integrated product recovery by cyclic vacuum	55	0.37	44.2 (in condensate)	100	–	[87]
Fed-batch Reactor	201	0.98	232.8	500.1	500	[84]
Continuous reactor	504	0.92	461.3	1125.0	250–500	[84,88,89]
Immobilized cell reactor	597	15.8	7.9	20.4	61.8–62.5	[90]
Immobilized cell reactor[a]	–	16.2	8.1	55.3	55.3	[90]
Cell recycle reactor	193	4.5	16.0	45.5	45.5	[91]
Cell recycle reactor	–	5.4	8.4	–	60.0	[92]

[a]*Integrated system with simultaneous product removal.*

TABLE 8.4 A comparison of novel ABE production systems using *C. beijerinckii* BA101 and *C. acetobutylicum* 824 strains and glucose as a substrate

Fermentation process	Sugar used (g/L[a])	Total ABE produced (g/L)	Yield	ABE productivity (g/L/h)	Reference
Batch (control)[d]	<60	<33	0.38–0.40	0.35	[5]
Product recovery by gas stripping:					
Batch[d]	161	75.9	0.4[b]	0.61	[53]
Fed-batch[d]	500	233	0.4[b]	1.16	[84]
Continuous[d]	1163	460	0.4	0.91	[93]
Product recovery by pervaporation:					
Fed-batch[e]	470[c]	155	0.31–0.35	0.13–0.26	[53]

[a]*g/L culture volume.*
[b]*values are higher than expected.*
[c]*calculated value.*
[d]*C. beijerinckii BA101.*
[e]*C. acetobutylicum 824.*

commercialized.[48,94,95] Among these, RO is the most cost-effective technology, but due to the drawback of membrane clogging, it is rarely used for butanol separation.

Gas stripping is a simple and efficient way to recover butanol from the fermentation broth. In this process, gas is bubbled through the broth and then passed through a condenser for solvent recovery. Gas stripping technology is patented by DuPont for the extraction of isobutanol solvent using oleyl alcohol.[42] Decanol and oleyl alcohol are mostly used as extractants in the gas stripping process for solvent recovery from fermentation broth.[97] The stripped gas can be recycled back into the fermenter until all the sugars are utilized and converted into product. Gas stripping enables the use of a concentrated sugar solution in the fermentor,[98] a reduction in butanol inhibition, an increase in cell mass, and a high sugar utilization.[84,90,99] Ennis et al.[100] reported a removal of ABE of 0.45 g/L/h with a 41% increase in productivity using gas striping in a batch process. During this process, the sugar (lactose) consumption was reported to have increased by 50%.

Liquid–liquid extraction has a high selectivity and capacity along with a higher operating expense.[48] The problem of toxicity when extracting solvent and the formation of emulsion are the major disadvantages of this method. To overcome these issues, the membrane extraction method is used and is also termed "perstraction."[5,53] Using this technique (a polydimethylsiloxane ultrafiltration membrane is mostly used) the productivity is reported to be increased by up to 2.34 g/L/h.[101] The drawback of membrane fouling can be observed over a period of time due to blockages caused by the fermentation broth and cell biomass.

8.6 Challenges in butanol production

The economics of butanol production are largely dependent on the cost of the substrate used in the fermentation process. For biobutanol production to have a meaningful impact as an alternative fuel, the biomass feedstock must be widely available at low cost.[102] Effective conversions of the feedstock with little or no negative environmental impact is also a key issue. Other negative aspects that make butanol production from petrochemicals more competitive than via bioprocesses include:

- relatively high substrate cost arising from the use of edible biomass
- low final butanol titer concentration (<20 g/L)
- low butanol yield (10–13 g/L) through ABE fermentation
- undesirable butanol selectivity (lower fraction of butanol in the final product)
- low volumetric butanol productivity (<0.5 g/L/h)
- the high cost of butanol recovery (conventional distillation is energy-intensive)

Other contributing technical problem include sluggish fermentations, uneconomical product recovery caused by low butanol titer, poor purity of butanol in the fermentation broth, and problems with the inhibition of microorganisms producing butanol as well as phage infections. Research in this field would play a vital role in making the production of biofuels feasible. Improvements and developments in product recovery processes and bioprocess technologies in applied microbiology and genetic engineering could result in an improvement of the economic competitiveness of biological production processes. Areas under research and development for ABE fermentation include alternative substrates for fermentation,[103] the removal of inhibitors present in biomass hydrolysate, product recovery, improvement of analytical methods for process monitoring, and the metabolic engineering of *Clostridium* species and other potential producers.

8.6.1 Feedstock pretreatment

The use of a pretreatment method for biomass depends on its composition, source, and the type of product. The cost of pretreatment contributes to a major portion of the overall cost of butanol production. In addition to the cost of feedstock, the kinetics and yield of the fermentation process also play a major role in the overall efficiency and economy of the process.

Various pretreatment methods used to release sugars from lignocellulosic biomass can be broadly classified as:

1. Mechanical methods: coarse size reduction, chipping, shredding, grinding, milling, ultrasonication, lysis—centrifuge, liquid shear, collision plate, and others.
2. Physicochemical methods: liquid hot water pretreatment, steam explosion, and CO_2 explosion.
3. Physiological methods: microwave,[104] ultrasound.[105]
4. Biological methods: aerobic, microaerophilic,[106] or anaerobic, and enzymatic.[107]
5. Thermal methods: steam explosion, low-temperature pyrolysis (dry torrefaction), wet torrefaction.[108]
6. Chemical methods: ammonia fiber/freeze explosion, ammonia recycle percolation, aqueous ammonia pretreatment, and organosolv pretreatment, oxidation, alkali treatments (NaOH, $Ca(OH)_2$, KOH), acid treatment, organic solvents, and ionic liquids.[109–112]

On the basis of pretreatment method used the hydrolysate contains a mixture of sugars which may or may not have inhibitors.[59–61,113] Glucose could suppress other sugars utilization from the mixture formed which is termed as carbon catabolite repression.[114]

8.6.2 Effect of an acid crash during the fermentation process

Acid crash is mostly found in the ABE fermentation process. The problem of acid crash develops due to uncontrolled pH drops during batch fermentation. In the first stage of exponential growth, sugars are converted into acids like AAs and butyric acids which results in a drop of pH followed by solventogensis. As the rate of acid consumptions is lower than the rate of acid production, cell growth is negatively affected.[44,115] It has been observed that acid crash takes place mostly when the concentration of undissociated acids exceeds 57–60 mmol/L.[116] To reduce this effect, the rate of glucose supply is limited in order to control the rate of acidogenesis. In one pH study performed under stable acidogenic and solventogenic conditions by Grimmler et al.[117] in *C. acetobutylicum*, it was reported that the expression levels of butanol producing genes: *adhE1*, *ctfA-ctfB*, and *adc,* were induced during the solventogenic phase (pH 4.5) compared to the acidogenic phase (pH 5.8).

8.6.3 Butanol selective yield and toxicity

An average ABE fermentation of 100 kg of starch produces acetone (11 kg), butanol (22.5 kg), and ethanol (2.7 kg) along with CO_2 (36 m^3) and hydrogen gas (24 m^3) as some of the by-products.[118] If the concentration of butanol exceeds 20 g/L, it causes the inhibition of clostridia.[119] Moreira et al.[120] attempted to elucidate the mechanism of butanol toxicity to *C. acetobutylicum* cells which showed that 0.1–0.15 M butanol causes a 50% inhibition of ATPase activity by reducing the sugar uptake thus resulting in low cell growth. Woods et al.[119] reported that a butanol concentration of 13 g/L in the medium can cause a 50% inhibition of cells, followed by the addition of ethanol and acetone up to 40 g/L.

Microbial strains like *Pseudomonas putida*,[121] *Bacillus subtilis*,[36] *Escherichia coli*[45,122] and *S. cerevisiae*[39] have inherent tolerance to solvents and the ability to withstand oxygen, making them good candidates for butanol production. The butanol-producing ability of these genetically modified strains needs to be improved as the butanol productivity was reportedly lower than 1 g/L.[36] A study indicated that gram-positive lactic acid bacteria can grow in the presence of up to 3% butanol with several other lactobacilli strains having the ability to uptake pentose as well as hexose sugars from lignocellulosic biomass.[123] Hence lactic acid bacteria are among the most potent hosts for engineering strains to produce butanol at a commercial scale.

8.6.4 Economic feasibility of production

The first two large-scale industrial butanol producing plants were set up in Canada and the United States during WWI. The end of the patent rights for butanol to Weizman in the 1930s resulted in the expansion of production facilities to other parts of the United States, Japan, and South Africa.[124] As compared to ethanol fermentation, feedstock consumption in ABE is almost double which is one of the challenges yet to be overcome.[125] Advances in the fermentation process along with molecular level modifications are used to make the process cost-effective. Due to the high production cost of butanol compared to its synthetic equivalents, its commercial production was reduced in parts of the United States and Europe in the 1950s. Some production plants utilizing the fermentation process remained active in Russia, South Africa, and China till the 1980s.[124,126,127] The maximum total concentration of butanol through ABE fermentation is 2%–4%,[44,128,129] whereas the commercial ethanol production is typically between 5%–9% and can be increased up to 16%.[63,130,131] The commercial butanol yield is approximately 12 g/L from *C. acetobutylicum*, while this strain has the potential to produce as high as 20 g/L butanol after mutation. In a commercial scale-up analysis done by Baral and Shah[132] the estimated cost of butanol production is US $1.8/L and US$1.5/L (with the absence of by-products). This price is required to be reduced to US$0.6/L to make it competitive with current gasoline prices. To achieve this, the use of integrated approaches from diverse fields would help in the coming future.

8.7 Creating genetically modified stable strains

Clostridium species are strictly anaerobic and grow slowly compared to well-characterized aerobic microorganisms. The underdeveloped gene manipulation system and complex pathways of clostridia have limited its use for butanol production. This limits the economic viability and scale-up of butanol production through conventional ABE fermentation using clostridia. On the other hand, aerobic host strains like *E. coli* and *S. cerevisiae* have been extensively studied for their entire gene sequence and are widely used as genetically engineered strains for the commercial production of valuable products.[133] Studies on the introduction of the ABE pathway into aerobic strains has been reported but has obtained limited success.

Atsumi et al.[134] successfully transformed *E. coli* for butanol biosynthesis by introducing genes like *thiL*, *crt*, *bcd*, *etfAB*, *hbd*, and *adhE2* (which encode for thiolase, crotonase, butyryl-CoA dehydrogenase, electron transfer proteins, and 3-hydroxybutyryl-CoA dehydrogenase, respectively) from *C. acetobutylicum*.[134] These genes are responsible for the stepwise conversion of acetoacetyl-CoA to butyryl-CoA. In this work, they also identified alternatives for thiolase and butyryl-CoA dehydrogenase. In microaerobic condition, the gene expression of $atoB_{E.\ coli}$ reported increased butanol titer, compared to $thl_{C.\ acetobutylicum}$ gene. A further increase in titer of 0.55 g/L butanol using 20 g/L glucose as a substrate was noted on removing *E. coli* competing pathway.[134]

Inui et al.[135] performed an experiment without the deletion of any competing pathways in *E. coli*. A yield of 1.2 g/L butanol was reported from 40 g/L of glucose by anaerobic culturing.[135,136] The yield obtained in this report is approximately the same as that of the results obtained by Atsumi et al.[134] Using metabolically engineered *E. coli* for the keto−acid pathway conversion, a butanol concentration of 2 g/L with propanol in the ratio of 1:1 was obtained by Shen and Liao.[37] With the introduction of the thiolase gene into *Lactococcus lactis* and *Lactobacillus buchneri* strains, butanol concentrations of 28 mg/L and 66 mg/L were reported, respectively, when the maximum butanol tolerance was higher than 2%.[38] Even though the butanol production achieved using engineered *E. coli* strains was lower compared to that from clostridia, these reports could help in further research for the use of heterologous aerobic strains in butanol production.

8.7.1 General genetic tools for the modification of strains

Efficient genetic tools are available for inserting or overexpressing genes of interest (GOI) in host vectors. Gene silencing has also gained prominence to control the metabolic regulation of cell cycles.[137] Insertional plasmid vectors with suitable marker techniques are used to transfer GOIs into a host while taking the size of the gene into consideration. Generally, replicative and nonreplicative plasmid transfer techniques are used. A limitation in the use of plasmid transfer technology is the low frequency of integration of target genes and gene loss with host generation cycles. In another method, group II intron-based TargeTron is used as an effective tool for targeted gene inactivation in *C. acetobutylicum*.[138] Antisense RNA has advantages over the mentioned methods as it is less time consuming and regulatory of protein expression without alteration to the gene.[139]

8.7.2 Genetic engineering in *Clostridium*

Suppression of the gene expressions responsible for acidogenec phase and to enhance solvetogenesis is reported in *C. acetobutylicum* species. A comprehensive range of genetic tools and methodologies are used to perform modifications in clostridial species. A common obstacle for the gene modification of *C. acetobutylicum* is the presence of a type II restriction endonuclease (*Cac824I*) against plasmid DNA.[140] In 1996, Green et al.[137,141] used integrative plasmid technology for disrupting butyrate and acetate formation in *C. acetobutylicum* which resulted in advanced butanol production. However, genome manipulations by homologous recombination in *C. acetobutylicum* are vital. The use of mobile introns for targeted gene inactivation is widely used for redirecting metabolic pathways and enhancing butanol production and yield.[141] Use of integrative plasmid technique can be addressed for inactivation of the key genes responsible for by-products synthesis. This would increase the carbon flux towards enhancement of butanol formation.[94,142,143] An in depth understanding of molecular mechanisms through systems biology and accessibility to modern genome editing tools are essential for the effective engineering of these anaerobic microbes.

The gene important for solvent formation is present at sol locus (*aad-ctfA-ctfB*) with its repressor gene *SolR*. In their work, Nair et al.[144] repressed the *SolR* gene, and yield of butanol was reported to be 17.8 g/L. This yield is much closer to the maximum possible concentration of butanol produced by clostridial strains (i.e., 20 g/L).[145,146] In an experiment by Wang et al., *C. acetobutylicum* TSH1 strain (anaerobic) and *Bacillus cereus* TSH2 strain (facultative anaerobic) were grown symbiotically and a yield of 12.3 ± 0.9 g/L butanol was reported.[147] The philosophy of using this symbiotic association was: the first strain had the ability to produce butanol while the second played a role in reducing oxygen in the media. The presence of even a small amount of oxygen can inhibit clostridial cell growth.[148,149] This approach can reduce the cost of oxygen removal from the fermentation media during ABE fermentation.

The suppression of the *maf* gene (related to cell division) does not suppress the cell growth in *E. coli*, while the overexpression of this gene inhibits cell division.[150,151] The role of this gene is not clear in the case of *C. acetobutylicum,* in which effective cell division takes place during the acidogenesis phase and not during solventogenesis. Mi et al.[152] found that when the m*af* gene was inactivated and grown symbiotically with *B. cereus* TSH2 strain, the yield of butanol increased to 13.9 ± 1.0 g/L which is higher than that of other symbiotic microbial cultures for butanol yield.

8.7.3 Genetic engineering to enhance solvent production

The first work and the concept of butanol selectivity were reported in 1994.[153] Reported techniques like the overexpression of solvent producing genes or the replacement of regulator genes show enhancement in solvent production. Harris et al.[154] successfully cloned *C. acetobutylicum* ACT 824 using the *spo0A* gene which has an important role in sporulation. Two recombinant strains, one with *spo0A* inactivation (*SKO1*) and the other with *spo0A* overexpression were created. From this trial, it was reported that inactivation resulted in a reduction in solvent production, whereas compared to the control, *spo0A* overexpression produced a higher butanol concentration. This suggests that *Spo0A* is a transcriptional regulator positively controlling sporulation and solvent production.[154] The overexpression of the *groESL* (class I heat shock protein) gene, a molecular chaperone in *C. acetobutylicum*, resulted in a 40% and 33% increase in the final solvent concentration in wild-type and plasmid control strains, respectively.[155] The activity level of hydrogenase was manipulated in order to obtain the desired solvent concentration since the production of hydrogen and solvent is closely associated. The literature on use of antisense technology for the downregulation of the hydrogenase gene cluster "*hupCBA*" in *C. saccharoperbutylacetonicum* N1-4 showed correlation with the decrease in hydrogen uptake. In this study, the level of acetone was increase by 1.6-fold with decrease in the butanol productivity by 75.6% in comparison to the control strain. This finding indicates that a shortage of hydrogen uptake leads to a shortage of NADH which leads to lower butanol production.[156]

The role of the *ofadhE1* gene (catalyzes the formation of acetoacetate and ethanol/butanol with acid reassimilation in solventogenesis) under the control of the ptb promoter was studied in the M5 mutant of wild *C. acetobutylicum* ATCC 824 by Jang et al.[136,157] An increased butanol selectivity of 0.84 g butanol/g in the modified strain was reported compared to a sensitivity of 0.57 g butanol/g in the wild ATCC 824 strain. In another study the *adc* gene (encoding acetoacetate decarboxylase) was disrupted using TargeTron technology. The butanol ration was reported to be increased by 10% and was further enhanced by the addition of methyl viologen for electron flow regulation. It altered the carbon flux from AA production to butanol production, thereby increasing the overall butanol yield from 57% to 70.8%.[158]

8.7.4 Genetic engineering to improve butanol tolerance

A high butanol concentration is a potent inhibitor of microbial growth, as it tends to partition into the cytoplasmic membrane and changes the membrane structure causing interference with the normal functioning of the cell. To make the production process more viable, it is essential to reduce the sensitivity of butanol-producing strains to butanol toxicity. The overexpression of the *Spo0A* gene in *C. acetobutylicum* shows an increase in butanol tolerance and prolonged metabolism in response to butanol stress.[159] Similarly, the overexpression of the *groESL*$_{C.\ acetobutylicum}$ (class I heat shock protein) gene, lead to an improved butanol tolerance since this gene prevents aggregation and assists in protein folding under butanol stress.[155] Reyes et al.[160] used a genomic library enrichment strategy for screening genes involved in *E. coli* butanol tolerance. The overexpression of *entC* (encodes isochorismate synthase) and *feoA* (encodes a small protein that is involved in iron transport), lead to an increased butanol tolerance by 32.8% and 49.1%, respectively. In contrast, the Δ*astE* gene (encodes a protein that belongs to the succinylglutamate desuccinylase/aspartoacylase family) showed a reported 48.7% increase in butanol resistance.[160] Knoshaug and Zhang[123] found that *Lactobacillus* strains can grow in and are tolerant to a butanol concentration of up to 3%. Adapted *P. putida* strains were reported to grow in the presence of up to 6% butanol, the highest butanol concentration tolerated by microorganisms.[121] The mentioned genetically engineered butanol-tolerant strains were capable of producing butanol; however, the achieved concentrations and yields of butanol are relatively low.

8.7.5 Genetic engineering to increase the ratio of butanol

The concentration of butanol can be increased by eliminating the production of by-products such as acetone (the major by-product of ABE fermentation) and ethanol. Using TargeTron technology for the disruption of the *adc* (produces acetoacetate decarboxylase) gene in *C. acetobutylicum* EA 2018, the butanol ratio was increased from 70% to 80.85%, whereas acetone production was reduced from approximately 2.83 g/L to 0.21 g/L.[161] In another report, antisense RNA technology was used for the downstream regulation of the effectiveness of enzymes in the acetone-formation pathway.[162] It was found that coenzyme A-transferase (CoAT) is a rate-limiting enzyme in the acetone formation process.

The inactivation of the *buk* gene (produces butyrate kinase) using integrational plasmid technology in mutant PJC4BK significantly decreased the acetone production by 50% and increased the butanol production by 10% compared to that of the wild strain.[137] Knockout technology (KO) was used in the creation of mutant M5 strain to study the effect of acetate kinase (AK) and butyrate kinase (BK) genes. This strain showed a reduction of more than 60% acetate or butyrate formation.[163]

8.7.6 Genetic engineering to allow the strain to grow on complex cellulosic substrates

C. acetobutylicum,[164] *Clostridium thermupapyrulyticum*,[165] and some other clostridial species[166] have the ability to convert different sugars into butanol. They cannot effectively utilize cellulose as a substrate, although they contain a complete cellulolytic enzyme complex called a "cellulosome."[167] *Clostridia* have been metabolically engineered to form a functional and efficient cellulosome complex for the breakdown of a complex substrate molecule. A gene encoding a miniCipA scaffolding protein, when expressed in *C. acetobutylicum*, shows the in vivo formation of a mini cellulosome. Although the resultant strain was not able to degrade cellulose, it was the first demonstration of the in vivo assembly of a recombinant mini cellulosome, and was demonstrated by Sabathé and Soucaille.[168] To promote the utilization of crystalline cellulose, heterologous minicellulosomes containing two different cellulases bound to a miniscaffolding were expressed in *C. acetobutylicum*.[169] In another study carried out by Mingardon et al.,[170] heterologous minicellulosomes with components from *Clostridium thermocellum* were created instead of *Clostridium acellulolyticum*. When the *man5K* (encoding mannanase) gene derived from *Clostridium cellulolyticum* was coexpressed with *cipC1* [encoding a truncated scaffolding (miniCipC1)] in *C. acetobutylicum*, full-length mannanase was secreted which helped to increase enzyme stability. It can be observed that *C. acetobutylicum* is a suitable host for the assembly and secretion of extracellular heterologous minicellulosomes though the utilization of the broad range of substrates present like syngas, glycerol, sucrose, and others as alternative carbon sources for ABE fermentation.[136]

8.7.7 Clustered regularly interspaced short palindromic repeats–Cas toolkit for genome editing

In order to develop a superior strain with desirable traits and high potential for industrial application, large-scale and efficient tools are essential for genome editing. These include multiple gene deletions, gene integrations, and gene mutations on the chromosome of the bacteria. The introduction of a new, disruptive, and precious genome editing technology, the CRISPR–Cas system, has provided the opportunity for a new era in biotechnology invention.[171,172] The use of this technology in the *Clostridium* community has been reported since 2015.[114] CRISPR and CRISPR-associated proteins (Cas) consist of an adaptive immune system in bacteria which slice exogenous DNA entering into the cells. To date, 1302 bacterial and archaeal strains having 2762 genome sequences were analyzed and recognized for CRISPR arrays.[173] The online database, CRISPRdb, is available for access to these identified genes which indicates that CRISPR–Cas systems are diverse and ubiquitous.[143]

8.8 Summary and future outlooks

To make ABE fermentation more cost-effective, the simultaneous production of commercially valuable by-products like riboflavin, isopropanol, and others have been reported in the literature.[174,175] A summary and future outlooks are presented here:

- The production of renewable alcoholic biofuel like biobutanol is a good approach to overcome society's reliance on limited fossil fuels for transportation.
- Among other alcoholic biofuels, biobutanol is the best candidate as a future alternative fuel to replace gasoline. This is because biobutanol has similar characteristics to gasoline and can be used in present engine and pipeline systems without any modifications.
- Producing butanol through biological routes like ABE fermentation by *Clostridium* sp., faces many challenges. One of the complexities observed when producing butanol through biological routes is the production of multiple by-products (gasses, acids, and ABE) which limits the yield of biobutanol. However, through genetic manipulation and modifying the fermentation conditions, the biobutanol yield and concentration can be enhanced.
- As *Clostridium* sp. is able to consume both pentose (C_5) and hexose (C_6) as a wide range of sugars, ABE fermentation is not only limited to glucose-based media, but can also be carried out using fermentable sugars obtained from lignocellulosic biomass.
- This situation provides some advantages to ABE fermentation since lignocellulosic biomass is cheap. However, utilizing lignocellulosic biomass as a substrate has some limitations such as the complex crystalline structure of the biomolecules, the hindrance in the utilization of sugars because of the complex cellulosic and hemicellulosic nature, a release of free sugars in the presence of lignin in the biomass, and others.
- In addition, the pretreatment steps for converting lignocellulosic biomass into biobutanol are less economically viable, and thus biobutanol production has yet to be commercialized on large scale. Extensive and progressive research adopting methods such as the use of subcritical water pretreatment technology can reduce costs with less processing steps, creating the possibility of the utilization of the whole cellulosic biomass as a substrate for ABE fermentation.
- Foreseeing the advantages, an increase in the market demand for butanol production is expected to expand by 2020.

8.9 Conclusion

Developments in butanol yields have an long history, yet currently have a wider scope for improvement. The large number of hurdles discussed in this chapter make the commercialization of the production process of butanol challenging. The gap in the full success and feasibility of its commercialization keep research in butanol production alive. Over the past few years, progress has been made by utilizing modern techniques and advances in the field of science to address these problems. The use of modern techniques like molecular engineering, gene editing, CRISPR–Cas in combination with bioprocessing approaches like continuous fermentation with high cell densities, and the addition of electron carriers or organic acids in the fermentation to increase metabolic flux can be studied for further development in this field.

The development in metabolic flux regulation has provided success in the reduction of acetone and its conversion into more valuable by-products like isopropanol. The production of similarly valuable by-products along with ABE fermentation would be a promising strategy. However, because of the complexity of the metabolic regulation process of proteins or enzymes involved in butanol susceptibility, there is scope to improve butanol titer.

Butanol production (<19 g/L) is still less competitive than ethanol production (up to 100 g/L in a batch fermentation mode). In the downstream in situ recovery of catalyst and biological separation, development would help in the reduction of production cost. Economic substrate selection would lead to the modeling of an eco-friendly commercial scale-up approach for biobutanol production. These strategies can provide a novel and potential outlook for an economic conversion of biomass into liquid transportation fuels.

References

1. Gerbens-Leenes PW, Lienden ARV, Hoekstra AY, van der Meer TH. Biofuel scenarios in a water perspective: the global blue and green water footprint of road transport in 2030. *Global Environ Chang* 2012;**22**(3):764–75.
2. Balat M, Balat H, Öz C. Progress in bioethanol processing. *Progr Energy Combust Sci* 2008;**34**(5):551–73.
3. Antoni D, Zverlov VV, Schwarz WH. Biofuels from microbes. *Appl Microbiol Biotechnol* 2007;**77**(1):23–35.
4. US-EPA. US-EPA, Assessing and managing chemicals under TSCA: US-EPA; [updated 22.06.2018]. Available from: <https://www.epa.gov/assessing-and-managing-chemicals-under-tsca>.
5. Ezeji TC, Qureshi N, Blaschek HP. Production of acetone butanol (AB) from liquefied corn starch, a commercial substrate, using *Clostridium beijerinckii* coupled with product recovery by gas stripping. *J Ind Microbiol Biotechnol* 2007;**34**(12):771–7.
6. Dewoolkar KD, Vaidya PD. New hybrid materials for improved hydrogen production by the sorption-enhanced steam reforming of butanol. *Energy Technol* 2017;**5**(8):1300–10.
7. Eikmanns BJ, Blombach B. Isobutanol. In: Bisaria VS, Kondo A, editors. *Bioprocessing of renewable resources to commodity bioproducts*. John Wiley & Sons; 2014. p. 327–52.

8. Sauer M. Industrial production of acetone and butanol by fermentation—100 years later. *FEMS Microbiol Lett* 2016;**363**(13):fnw134.
9. Formanek J, Mackie R, Blaschek HP. Enhanced butanol production by *Clostridium beijerinckii* BA101 grown in semidefined P2 medium containing 6 percent maltodextrin or glucose. *Appl Environ Microbiol* 1997;**63**(6):2306–10.
10. Jain S, Yadav MK, Kumar A. Production of butanol: a biofuel. In: Babu V, Thapliyal A, Pate GK, editors. *Biofuels production.* Hoboken, NJ: John Wiley & Sons; 2013. p. 255–83.
11. Koda K, Matsu-ura T, Obora Y, Ishii Y. Guerbet reaction of ethanol to *n*-butanol catalyzed by iridium complexes. *Chem Lett* 2009;**38**(8):838–9.
12. Kulkarni NV, Brennessel WW, Jones WD. Catalytic upgrading of ethanol to *n*-butanol via manganese-mediated Guerbet reaction. *ACS Catal* 2018;**8**(2):997–1002.
13. Farrell N, Sutton D. Aryldiazenido- and aryldiazene complexes: reaction of ethanol with aryldiazenido-complexes of iridium(III) to form orthometallated derivatives. *J Chem Soc, Dalton Trans* 1977;**21**:2124–8.
14. Sarchami T, Rehmann L. Optimizing acid hydrolysis of Jerusalem artichoke-derived inulin for fermentative butanol production. *BioEnergy Res* 2015;**8**(3):1148–57.
15. Lee S-M, Cho MO, Park CH, Chung Y-C, Kim JH, Sang B-I, et al. Continuous butanol production using suspended and immobilized *Clostridium beijerinckii* NCIMB 8052 with supplementary butyrate. *Energy Fuels* 2008;**22**(5):3459–64.
16. Kohlpaintner CW, Fischer RW, Cornils B. Aqueous biphasic catalysis: Ruhrchemie/Rhône-Poulenc oxo process. *Appl Catal A Gen* 2001;**221**(1-2):219–25.
17. Liu H, Wang G, Zhang J. The promising fuel-biobutanol. In: Fang Z, editor. *Liquid, gaseous and solid biofuels-conversion techniques.* InTech; 2013. p. 175–98.
18. García V, Päkkilä J, Ojamo H, Muurinen E, Keiski RL. Challenges in biobutanol production: how to improve the efficiency? *Renew Sustain Energy Rev* 2011;**15**(2):964–80.
19. Lee SY, Park JH, Jang SH, Nielsen LK, Kim J, Jung KS. Fermentative butanol production by *Clostridia. Biotechnol Bioeng* 2008;**101**(2):209–28.
20. Huesemann MH, Kuo L-J, Urquhart L, Gill GA, Roesijadi G. Acetone-butanol fermentation of marine macroalgae. *Bioresour Technol* 2012;**108**:305–9.
21. Yadav S, Rawat G, Tripathi P, Saxena R. A novel approach for biobutanol production by *Clostridium acetobutylicum* using glycerol: a low cost substrate. *Renew Energ* 2014;**71**:37–42.
22. George H, Johnson J, Moore W, Holdeman L, Chen J. Acetone, isopropanol, and butanol production by *Clostridium beijerinckii* (syn. *Clostridium butylicum*) and *Clostridium aurantibutyricum. Appl Environ Microbiol* 1983;**45**(3):1160–3.
23. Qureshi N, Saha BC, Hector RE, Dien B, Hughes S, Liu S, et al. Production of butanol (a biofuel) from agricultural residues: part II—use of corn stover and switchgrass hydrolysates. *Biomass Bioenergy* 2010;**34**(4):566–71.
24. Datar RP, Shenkman RM, Cateni BG, Huhnke RL, Lewis RS. Fermentation of biomass-generated producer gas to ethanol. *Biotechnol Bioeng* 2004;**86**(5):587–94.
25. Higashide W, Li Y, Yang Y, Liao JC. Metabolic engineering of *Clostridium cellulolyticum* for isobutanol production from cellulose. *Appl Environ Microbiol* 2011;**77**(8):2727–33.
26. Yang X, Xu M, Yang S-T. Metabolic and process engineering of *Clostridium cellulovorans* for biofuel production from cellulose. *Metab Eng* 2015;**32**:39–48.
27. Ahn J-H, Sang B-I, Um Y. Butanol production from thin stillage using *Clostridium pasteurianum. Bioresour Technol* 2011;**102**(7):4934–7.

28. Xin F, Wang C, Dong W, Zhang W, Wu H, Ma J, et al. Comprehensive investigations of biobutanol production by a non-acetone and 1,3-propanediol generating *Clostridium* strain from glycerol and polysaccharides. *Biotechnol Biofuels* 2016;**9**(1):220.
29. Boumba VA, Economou V, Kourkoumelis N, Gousia P, Papadopoulou C, Vougiouklakis T. Microbial ethanol production: experimental study and multivariate evaluation. *Forensic Sci Int* 2012;**215**(1-3):189–98.
30. Thang VH, Kanda K, Kobayashi G. Production of acetone–butanol–ethanol (ABE) in direct fermentation of cassava by *Clostridium saccharoperbutylacetonicum* N1-4. *Appl Biochem Biotechnol* 2010;**161**(1-8):157–70.
31. Berezina OV, Brandt A, Yarotsky S, Schwarz WH, Zverlov VV. Isolation of a new butanol-producing *Clostridium* strain: high level of hemicellulosic activity and structure of solventogenesis genes of a new *Clostridium saccharobutylicum* isolate. *Syst Appl Microbiol* 2009;**32**(7):449–59.
32. Gottumukkala LD, Parameswaran B, Valappil SK, Mathiyazhakan K, Pandey A, Sukumaran RK. Biobutanol production from rice straw by a non acetone producing *Clostridium sporogenes* BE01. *Bioresour Technol* 2013;**145**:182–7.
33. Yu M, Zhang Y, Tang I-C, Yang S-T. Metabolic engineering of *Clostridium tyrobutyricum* for *n*-butanol production. *Metab Eng* 2011;**13**(4):373–82.
34. Lin PP, Mi L, Morioka AH, Yoshino KM, Konishi S, Xu SC, et al. Consolidated bioprocessing of cellulose to isobutanol using *Clostridium thermocellum*. *Metab Eng* 2015;**31**:44–52.
35. Gottwald M, Hippe H, Gottschalk G. Formation of *n*-butanol from D-glucose by strains of the "*Clostridium tetanomorphum*" group. *Appl Environ Microbiol* 1984;**48**(3):573–6.
36. Nielsen DR, Leonard E, Yoon S-H, Tseng H-C, Yuan C, Prather KLJ. Engineering alternative butanol production platforms in heterologous bacteria. *Metab Eng* 2009;**11**(4-5):262–73.
37. Shen CR, Liao JC. Metabolic engineering of *Escherichia coli* for 1-butanol and 1-propanol production via the keto-acid pathways. *Metab Eng* 2008;**10**(6):312–20.
38. Liu S, Bischoff KM, Qureshi N, Hughes SR, Rich JO. Functional expression of the thiolase gene thl from *Clostridium beijerinckii* P260 in *Lactococcus lactis* and *Lactobacillus buchneri*. *New Biotechnol* 2010;**27**(4):283–8.
39. Steen EJ, Chan R, Prasad N, Myers S, Petzold CJ, Redding A, et al. Metabolic engineering of *Saccharomyces cerevisiae* for the production of *n*-butanol. *Microb Cell Fact* 2008;**7**(1):36.
40. Zheng J, Tashiro Y, Wang Q, Sonomoto K. Recent advances to improve fermentative butanol production: genetic engineering and fermentation technology. *J Biosci Bioeng* 2015;**119**(1):1–9.
41. Maddox IS. The acetone-butanol-ethanol fermentation: recent progress in technology. *Biotechnol Genet Eng Rev* 1989;**7**(1):189–220.
42. Green EM. Fermentative production of butanol—the industrial perspective. *Curr Opin Biotechnol* 2011;**22**(3):337–43.
43. Ndaba B, Chiyanzu I, Marx S. *n*-Butanol derived from biochemical and chemical routes: a review. *Biotechnol Rep* 2015;**8**:1–9.
44. Xue C, Zhao X-Q, Liu C-G, Chen L-J, Bai F-W. Prospective and development of butanol as an advanced biofuel. *Biotechnol Adv* 2013;**31**(8):1575–84.
45. Zheng Y-N, Li L-Z, Xian M, Ma Y-J, Yang J-M, Xu X, et al. Problems with the microbial production of butanol. *J Ind Microbiol Biotechnol* 2009;**36**(9):1127–38.

46. Shinto H, Tashiro Y, Kobayashi G, Sekiguchi T, Hanai T, Kuriya Y, et al. Kinetic study of substrate dependency for higher butanol production in acetone–butanol–ethanol fermentation. *Process Biochem* 2008;**43**(12):1452–61.
47. Qureshi N, Li XL, Hughes S, Saha BC, Cotta MA. Butanol production from corn fiber xylan using *Clostridium acetobutylicum. Biotechnol Prog* 2006;**22**(3):673–80.
48. Dürre P. New insights and novel developments in clostridial acetone/butanol/isopropanol fermentation. *Appl Microbiol Biotechnol* 1998;**49**(6):639–48.
49. Kruger NJ, von Schaewen A. The oxidative pentose phosphate pathway: structure and organisation. *Curr Opin Plant Biol* 2003;**6**(3):236–46.
50. Decker K, Jungermann K, Thauer R. Energy production in anaerobic organisms. *Angew Chem Int Ed Engl* 1970;**9**(2):138–58.
51. Hartmanis MG, Klason T, Gatenbeck S. Uptake and activation of acetate and butyrate in *Clostridium acetobutylicum. Appl Microbiol Biotechnol* 1984;**20**(1):66–71.
52. Wu Y-D, Xue C, Chen L-J, Bai F-W. Effect of zinc supplementation on acetone–butanol–ethanol fermentation by *Clostridium acetobutylicum. J Biotechnol* 2013;**165**(1):18–21.
53. Ezeji TC, Qureshi N, Blaschek HP. Bioproduction of butanol from biomass: from genes to bioreactors. *Curr Opin Biotechnol* 2007;**18**(3):220–7.
54. Grupe H, Gottschalk G. Physiological events in *Clostridium acetobutylicum* during the shift from acidogenesis to solventogenesis in continuous culture and presentation of a model for shift induction. *Appl Environ Microbiol* 1992;**58**(12):3896–902.
55. Naik SN, Goud VV, Rout PK, Dalai AK. Production of first and second generation biofuels: a comprehensive review. *Renew Sustain Energy Rev* 2010;**14**(2):578–97.
56. Lee RA, Lavoie J-M. From first-to third-generation biofuels: challenges of producing a commodity from a biomass of increasing complexity. *Anim Front* 2013;**3**(2):6–11.
57. Sasmal S, Goud VV, Mohanty K. Ultrasound assisted lime pretreatment of lignocellulosic biomass toward bioethanol production. *Energy Fuels* 2012;**26**(6):3777–84.
58. Nigam PS, Singh A. Production of liquid biofuels from renewable resources. *Progr Energy Combust Sci* 2011;**37**(1):52–68.
59. Karimi K, Shafiei M, Kumar R. *Progress in physical and chemical pretreatment of lignocellulosic biomass. Biofuel technologies*. Springer; 2013. p. 53–96.
60. Taherzadeh MJ, Karimi K. Acid-based hydrolysis processes for ethanol from lignocellulosic materials: a review. *BioResources* 2007;**2**(3):472–99.
61. Taherzadeh MJ, Karimi K. Enzymatic-based hydrolysis processes for ethanol from lignocellulosic materials: a review. *BioResources* 2007;**2**(4):707–38.
62. Taherzadeh MJ, Karimi K. Pretreatment of lignocellulosic wastes to improve ethanol and biogas production: a review. *Int J Mol Sci* 2008;**9**(9):1621–51.
63. Taherzadeh MJ, Karimi K. *Fermentation inhibitors in ethanol processes and different strategies to reduce their effects. Biofuels*. Elsevier; 2011. p. 287–311.
64. Bond JQ, Upadhye AA, Olcay H, Tompsett GA, Jae J, Xing R, et al. Production of renewable jet fuel range alkanes and commodity chemicals from integrated catalytic processing of biomass. *Energy Environ Sci* 2014;**7**(4):1500–23.
65. Gu T. *Pretreatment of lignocellulosic biomass using supercritical carbon dioxide as a green solvent. Green biomass pretreatment for biofuels production*. Dordrecht, Netherlands: Springer; 2013. p. 107–25.
66. Jönsson LJ, Martín C. Pretreatment of lignocellulose: formation of inhibitory by-products and strategies for minimizing their effects. *Bioresour Technol* 2016;**199**:103–12.

67. Kudahettige-Nilsson RL, Helmerius J, Nilsson RT, Sjöblom M, Hodge DB, Rova U. Biobutanol production by *Clostridium acetobutylicum* using xylose recovered from birch Kraft black liquor. *Bioresour Technol* 2015;**176**:71–9.
68. Smith VH, Sturm BS, Billings SA. The ecology of algal biodiesel production. *Trends Ecol Evol* 2010;**25**(5):301–9.
69. Li Y, Horsman M, Wu N, Lan CQ, Dubois-Calero N. Biofuels from microalgae. *Biotechnol Prog* 2008;**24**(4):815–20.
70. Wang Y, Ho S-H, Cheng C-L, Nagarajan D, Guo W-Q, Lin C, et al. Nutrients and COD removal of swine wastewater with an isolated microalgal strain Neochloris aquatica CL-M1 accumulating high carbohydrate content used for biobutanol production. *Bioresour Technol* 2017;**242**:7–14.
71. Wang Y, Ho S-H, Yen H-W, Nagarajan D, Ren N-Q, Li S, et al. Current advances on fermentative biobutanol production using third generation feedstock. *Biotechnol Adv* 2017;**35**:1049–59.
72. Difusa A, Mohanty K, Goud V. *Advancement and challenges in harvesting techniques for recovery of microalgae biomass. Environmental sustainability*. New Delhi, India: Springer; 2015. p. 159–69.
73. Aro E-M. From first generation biofuels to advanced solar biofuels. *Ambio* 2016;**45**(1):24–31.
74. Klinthong W, Yang Y-H, Huang C-H, Tan C-S. A review: microalgae and their applications in CO_2 capture and renewable energy. *Aerosol Air Qual Res* 2015;**15**(2):712–42.
75. Lan EI, Liao JC. Metabolic engineering of cyanobacteria for 1-butanol production from carbon dioxide. *Metab Eng* 2011;**13**(4):353–63.
76. Berla BM, Saha R, Immethun CM, Maranas CD, Moon TS, Pakrasi H. Synthetic biology of cyanobacteria: unique challenges and opportunities. *Front Microbiol* 2013;**4**:246.
77. Atsumi S, Higashide W, Liao JC. Direct photosynthetic recycling of carbon dioxide to isobutyraldehyde. *Nat Biotechnol* 2009;**27**(12):1177.
78. Hellingwerf K, De Mattos MT. Alternative routes to biofuels: light-driven biofuel formation from CO_2 and water based on the 'photanol' approach. *J Biotechnol* 2009;**142**(1):87–90.
79. Daroch M, Geng S, Wang G. Recent advances in liquid biofuel production from algal feedstocks. *Appl Energy* 2013;**102**:1371–81.
80. Inganäs O, Sundström V. Solar energy for electricity and fuels. *Ambio* 2016;**45**(1):15–23.
81. Lü J, Sheahan C, Fu P. Metabolic engineering of algae for fourth generation biofuels production. *Energy Environ Sci* 2011;**4**(7):2451–66.
82. Lee J, Jang Y-S, Choi SJ, Im JA, Song H, Cho JH, et al. Metabolic engineering of *Clostridium acetobutylicum* ATCC 824 for isopropanol-butanol-ethanol fermentation. *Appl Environ Microbiol* 2012;**78**:1416–23 AEM. 06382–06311.
83. Tran HTM, Cheirsilp B, Hodgson B, Umsakul K. Potential use of *Bacillus subtilis* in a co-culture with *Clostridium butylicum* for acetone–butanol–ethanol production from cassava starch. *Biochem Eng J* 2010;**48**(2):260–7.
84. Ezeji T, Qureshi N, Blaschek H. Acetone butanol ethanol (ABE) production from concentrated substrate: reduction in substrate inhibition by fed-batch technique and product inhibition by gas stripping. *Appl Microbiol Biotechnol* 2004;**63**(6):653–8.
85. Xue C, Zhao J, Liu F, Lu C, Yang S-T, Bai F-W. Two-stage in situ gas stripping for enhanced butanol fermentation and energy-saving product recovery. *Bioresour Technol* 2013;**135**:396–402.

86. Yen H-W, Wang Y-C. The enhancement of butanol production by in situ butanol removal using biodiesel extraction in the fermentation of ABE (acetone–butanol–ethanol). *Bioresour Technol* 2013;**145**:224–8.
87. Mariano AP, Filho RM, Ezeji TC. Energy requirements during butanol production and in situ recovery by cyclic vacuum. *Renew Energy* 2012;**47**:183–7.
88. Ezeji T, Milne C, Price ND, Blaschek HP. Achievements and perspectives to overcome the poor solvent resistance in acetone and butanol-producing microorganisms. *Appl Microbiol Biotechnol* 2010;**85**(6):1697–712.
89. Ezeji TC, Qureshi N, Blaschek HP. Microbial production of a biofuel (acetone–butanol–ethanol) in a continuous bioreactor: impact of bleed and simultaneous product removal. *Bioprocess Biosyst Eng* 2013;**36**(1):109–16.
90. Lienhardt J, Schripsema J, Qureshi N, Blaschek HP. Butanol production by *Clostridium beijerinckii* BA101 in an immobilized cell biofilm reactor. *Appl Biochem Biotechnol* 2002;**98**(1-9):591–8.
91. Continuous production of butanol by *Clostridium acetobutylicum* immobilized in a fibrous bed bioreactor. In: Huang W-C, Ramey DE, Yang S-T, editors. *Proceedings of the twenty-fifth symposium on biotechnology for fuels and chemicals held May 4–7, 2003, in Breckenridge, CO*. Totowa, NJ: Humana Press; 2004.
92. Afschar A, Biebl H, Schaller K, Schügerl K. Production of acetone and butanol by *Clostridium acetobutylicum* in continuous culture with cell recycle. *Appl Microbiol Biotechnol* 1985;**22**(6):394–8.
93. Mariano AP, Filho RM, Ezeji TC. Energy requirements during butanol production and in situ recovery by cyclic vacuum. *Renew Energy* 2012;**47**:183–7.
94. Ni Y, Sun Z. Recent progress on industrial fermentative production of acetone–butanol–ethanol by *Clostridium acetobutylicum* in China. *Appl Microbiol Biotechnol* 2009;**83**(3):415.
95. Vane LM. Separation technologies for the recovery and dehydration of alcohols from fermentation broths. *Biofuels Bioprod Biorefining* 2008;**2**(6):553–88.
96. Mariano AP, Costa CBB, Maciel MRW, Maugeri Filho F, Atala DIP, de Angelis DdF, et al. Dynamics and control strategies for a butanol fermentation process. *Appl Biochem Biotechnol* 2010;**160**(8):2424–48.
97. Evans PJ, Wang HY. Enhancement of butanol formation by *Clostridium acetobutylicum* in the presence of decanol-oleyl alcohol mixed extractants. *Appl Environ Microbiol* 1988;**54**(7):1662–7.
98. Qureshi N, Blaschek H. Recovery of butanol from fermentation broth by gas stripping. *Renew Energy* 2001;**22**(4):557–64.
99. Maddox I, Qureshi N, Roberts-Thomson K. Production of acetone-butanol-ethanol from concentrated substrate using *clostridium acetobutylicum* in an integrated fermentation-product removal process. *Process Biochem* 1995;**30**(3):209–15.
100. Ennis B, Marshall C, Maddox IS, Paterson A. Continuous product recovery by in-situ gas stripping/condensation during solvent production from whey permeate using *Clostridium acetobutylicum*. *Biotechnol Lett* 1986;**8**(10):725–30.
101. Izák P, Schwarz K, Ruth W, Bahl H, Kragl U. Increased productivity of *Clostridium acetobutylicum* fermentation of acetone, butanol, and ethanol by pervaporation through supported ionic liquid membrane. *Appl Microbiol Biotechnol* 2008;**78**(4):597–602.
102. Lynel L, Wyman C, Gerngross T. Biocommodity engineering. *Biotechnol Prog* 1999;**15**(5):777–93.

103. Pfromm PH, Amanor-Boadu V, Nelson R, Vadlani P, Madl R. Bio-butanol vs. bio-ethanol: a technical and economic assessment for corn and switchgrass fermented by yeast or *Clostridium acetobutylicum. Biomass Bioenergy* 2010;**34**(4):515–24.
104. Reshad AS, Tiwari P, Goud VV. Production of biodiesel from renewable resources. *Biofuels Product Future Perspect* 2016;217.
105. Achar SK, Borugadda VB, Goud VV. Ultrasonic-assisted transesterification of high free fatty acid Karanja oil. In: Acquaye A, editor. *Feedstocks: production practices, technologies and environmental impacts.* Nova Science Publishers; 2013. p. 47–72.
106. Keller FA, Hamilton JE, Nguyen QA. *Microbial pretreatment of biomass. Biotechnology for fuels and chemicals.* Totowa, NJ: Humana Press; 2003. p. 27–41.
107. Chang VS, Nagwani M, Kim C-H, Holtzapple MT. Oxidative lime pretreatment of high-lignin biomass. *Appl Biochem Biotechnol* 2001;**94**(1):1–28.
108. Yan W, Acharjee TC, Coronella CJ, Vasquez VR. Thermal pretreatment of lignocellulosic biomass. *Environ Prog Sustain Energy* 2009;**28**(3):435–40.
109. Liu Q-P, Hou X-D, Li N, Zong M-H. Ionic liquids from renewable biomaterials: synthesis, characterization and application in the pretreatment of biomass. *Green Chem* 2012;**14**(2):304–7.
110. Bugg TD, Ahmad M, Hardiman EM, Singh R. The emerging role for bacteria in lignin degradation and bio-product formation. *Curr Opin Biotechnol* 2011;**22**(3):394–400.
111. Palmowski L, Müller J. Influence of the size reduction of organic waste on their anaerobic digestion. *Water Sci Technol* 2000;**41**(3):155–62.
112. Zhu S, Wu Y, Chen Q, Yu Z, Wang C, Jin S, et al. Dissolution of cellulose with ionic liquids and its application: a mini-review. *Green Chem* 2006;**8**(4):325–7.
113. Karimi K, Brandberg T, Edebo L, Taherzadeh MJ. Fed-batch cultivation of Mucor indicus in dilute-acid lignocellulosic hydrolyzate for ethanol production. *Biotechnol Lett* 2005;**27**(18):1395–400.
114. Xue C, Zhao J, Chen L, Yang S-T, Bai F. Recent advances and state-of-the-art strategies in strain and process engineering for biobutanol production by *Clostridium acetobutylicum. Biotechnol Adv* 2017;**35**(2):310–22.
115. Kumar M, Gayen K. Developments in biobutanol production: new insights. *Appl Energy* 2011;**88**(6):1999–2012.
116. Maddox I, Steiner E, Hirsch S, Wessner S, Gutierrez N, Gapes J, et al. The cause of "Acid Crash" and "Acidogenic Fermentations" during the batch acetone-butanol-ethanol (ABE-) fermentation process. *J Mol Microbiol Biotechnol* 2000;**2**(1):95–100.
117. Grimmler C, Janssen H, Krauβe D, Fischer R-J, Bahl H, Dürre P, et al. Genome-wide gene expression analysis of the switch between acidogenesis and solventogenesis in continuous cultures of *Clostridium acetobutylicum. J Mol Microbiol Biotechnol* 2011;**20**(1):1–15.
118. Sun Z, Shi Z. *The acetone-butanol (ABE) fermentation industries in China 2006.* Available from: <http://dc.engconfintl.org/cgi/viewcontent.cgi?article = 1006&context = bioenergy_i>.
119. Woods DR. The genetic engineering of microbial solvent production. *Trends Biotechnol* 1995;**13**(7):259–64.
120. Moreira A, Ulmer D, Linden J, editors. Butanol toxicity in the butylic fermentation. *Biotechnol. Bioeng. Symp.* 1981; 11.
121. Rühl J, Schmid A, Blank LM. Selected *Pseudomonas putida* strains able to grow in the presence of high butanol concentrations. *Appl Environ Microbiol* 2009;**75**(13):4653–6.

122. Huffer S, Roche CM, Blanch HW, Clark DS. *Escherichia coli* for biofuel production: bridging the gap from promise to practice. *Trends Biotechnol* 2012;**30**(10):538–45.
123. Knoshaug EP, Zhang M. Butanol tolerance in a selection of microorganisms. *Appl Biochem Biotechnol* 2009;**153**(1-3):13–20.
124. Zverlov V, Berezina O, Velikodvorskaya G, Schwarz W. Bacterial acetone and butanol production by industrial fermentation in the Soviet Union: use of hydrolyzed agricultural waste for biorefinery. *Appl Microbiol Biotechnol* 2006;**71**(5):587–97.
125. Karimi K, Tabatabaei M, Sárvári Horváth I, Kumar R. Recent trends in acetone, butanol, and ethanol (ABE) production. *Biofuel Res J* 2015;**2**(4):301–8.
126. Chiao J-s, Sun Z-h. History of the acetone-butanol-ethanol fermentation industry in China: development of continuous production technology. *J Mol Microbiol Biotechnol* 2007;**13**(1-3):12–14.
127. Jones DT, Keis S. Origins and relationships of industrial solvent-producing clostridial strains. *FEMS Microbiol Rev* 1995;**17**(3):223–32.
128. Huang H-J, Ramaswamy S, Liu Y. Separation and purification of biobutanol during bioconversion of biomass. *Sep Purif Technol* 2014;**132**:513–40.
129. Ye C, Takigawa T, Burtovvy O, Langsdorf L, Jablonski D, Bell A, et al. Impact of nanostructure on mechanical properties of norbornene-based block copolymers under simulated operating conditions for biobutanol membranes. *ACS Appl Mater Interfaces* 2015;**7**(22):11765–74.
130. Breisha GZ. Production of 16% ethanol from 35% sucrose. *Biomass Bioenergy* 2010;**34**(8):1243–9.
131. Taherzadeh MJ, Karimi K. Bioethanol: market and production processes. *Biofuels Refining Perform* 2008;69–106.
132. Baral NR, Shah A. Techno-economic analysis of cellulosic butanol production from corn stover through acetone–butanol–ethanol fermentation. *Energy Fuels* 2016;**30**(7):5779–90.
133. Jarboe L, Grabar T, Yomano L, Shanmugan K, Ingram L. *Development of ethanologenic bacteria. Biofuels*. Berlin, Heidelberg: Springer; 2007. p. 237–61.
134. Atsumi S, Cann AF, Connor MR, Shen CR, Smith KM, Brynildsen MP, et al. Metabolic engineering of *Escherichia coli* for 1-butanol production. *Metab Eng* 2008;**10**(6):305–11.
135. Inui M, Suda M, Kimura S, Yasuda K, Suzuki H, Toda H, et al. Expression of *Clostridium acetobutylicum* butanol synthetic genes in *Escherichia coli*. *Appl Microbiol Biotechnol* 2008;**77**(6):1305–16.
136. Jang YS, Lee J, Malaviya A, Cho JH, Lee SY. Butanol production from renewable biomass: rediscovery of metabolic pathways and metabolic engineering. *Biotechnol J* 2012;**7**(2):186–98.
137. Green EM, Boynton ZL, Harris LM, Rudolph FB, Papoutsakis ET, Bennett GN. Genetic manipulation of acid formation pathways by gene inactivation in *Clostridium acetobutylicum* ATCC 824. *Microbiology* 1996;**142**(8):2079–86.
138. Shao L, Hu S, Yang Y, Gu Y, Chen J, Yang Y, et al. Targeted gene disruption by use of a group II intron (targetron) vector in *Clostridium acetobutylicum*. *Cell Res* 2007;**17**(11):963.
139. Desai RP, Harris LM, Welker NE, Papoutsakis ET. Metabolic flux analysis elucidates the importance of the acid-formation pathways in regulating solvent production by *Clostridium acetobutylicum*. *Metab Eng* 1999;**1**(3):206–13.

140. Dong H, Zhang Y, Dai Z, Li Y. Engineering *Clostridium* strain to accept unmethylated DNA. *PLoS One* 2010;**5**(2):1–8.
141. Green EM, Bennett GN. Inactivation of an aldehyde/alcohol dehydrogenase gene from *Clostridium acetobutylicum* ATCC 824. *Appl Biochem Biotechnol* 1996;**57**(1):213.
142. Bharathiraja B, Sowmya V, Sridharan S, Yuvaraj D, Jayamuthunagai J, Praveenkumar R. Biodiesel production from microbial oil derived from wood isolate *Trichoderma reesei. Bioresour Technol* 2017;**239**:538–41.
143. Luo J. Elucidates the CRISPR gene editing technology, and how it may affect cancer therapy in the future. *Oncology (Williston Park, NY)* 2016;**30**(10):879.
144. Nair RV, Green EM, Watson DE, Bennett GN, Papoutsakis ET. Regulation of the sol locus genes for butanol and acetone formation in *Clostridium acetobutylicumatcc* 824 by a putative transcriptional repressor. *J Bacteriol* 1999;**181**(1):319–30.
145. Xue C, Zhao J, Lu C, Yang ST, Bai F, Tang IC. High-titer *n*-butanol production by *clostridium acetobutylicum* JB200 in fed-batch fermentation with intermittent gas stripping. *Biotechnol Bioeng* 2012;**109**(11):2746–56.
146. Jiang W, Zhao J, Wang Z, Yang S-T. Stable high-titer *n*-butanol production from sucrose and sugarcane juice by *Clostridium acetobutylicum* JB200 in repeated batch fermentations. *Bioresour Technol* 2014;**163**:172–9.
147. Wang G, Wu P, Liu Y, Mi S, Mai S, Gu C, et al. Isolation and characterisation of non-anaerobic butanol-producing symbiotic system TSH06. *Appl Microbiol Biotechnol* 2015;**99**(20):8803–13.
148. Choi SJ, Lee J, Jang Y-S, Park JH, Lee SY, Kim IH. Effects of nutritional enrichment on the production of acetone-butanol-ethanol (ABE) by *Clostridium acetobutylicum. J Microbiol* 2012;**50**(6):1063–6.
149. Han B, Ujor V, Lai LB, Gopalan V, Ezeji TC. Use of proteomic analysis to elucidate the role of calcium on acetone-butanol-ethanol (ABE) fermentation in *Clostridium beijerinckii* NCIMB 8052. *Appl Environ Microbiol* 2013;**79**(1):282–93.
150. Butler YX, Abhayawardhane Y, Stewart GC. Amplification of the *Bacillus subtilis* maf gene results in arrested septum formation. *J Bacteriol* 1993;**175**(10):3139–45.
151. Jin J, Wu R, Zhu J, Yang S, Lei Z, Wang N, et al. Insights into the cellular function of YhdE, a nucleotide pyrophosphatase from *Escherichia coli. PLoS One* 2015;**10**(2):1–15.
152. Mi S, Gu C, Wu P, Liu H, Yan X, Li D, et al. Improvement of butanol production by the development and co-culture of *C. acetobutylicum* TSH1 and B. cereus TSH2. *Appl Microbiol Biotechnol* 2018;1–11.
153. Nair RV, Papoutsakis ET. Expression of plasmid-encoded aad in *Clostridium acetobutylicum* M5 restores vigorous butanol production. *J Bacteriol* 1994;**176**(18):5843–6.
154. Harris LM, Welker NE, Papoutsakis ET. Northern, morphological, and fermentation analysis of spo0A inactivation and overexpression in *Clostridium acetobutylicum* ATCC 824. *J Bacteriol* 2002;**184**(13):3586–97.
155. Tomas CA, Welker NE, Papoutsakis ET. Overexpression of groESL in *Clostridium acetobutylicum* results in increased solvent production and tolerance, prolonged metabolism, and changes in the cell's transcriptional program. *Appl Environ Microbiol* 2003;**69**(8):4951–65.
156. Nakayama S-i, Kosaka T, Hirakawa H, Matsuura K, Yoshino S, Furukawa K. Metabolic engineering for solvent productivity by downregulation of the hydrogenase gene cluster hupCBA in *Clostridium saccharoperbutylacetonicum* strain N1-4. *Appl Microbiol Biotechnol* 2008;**78**(3):483–93.

157. Jang YS, Malaviya A, Lee J, Im JA, Lee SY, Lee J, et al. Metabolic engineering of *Clostridium acetobutylicum* for the enhanced production of isopropanol-butanol-ethanol fuel mixture. *Biotechnol Prog* 2013;**29**(4):1083–8.
158. Jiang Y, Yang Y, Yang S, Jiang W. Disruption of acetoacetate decarboxylase gene in solvent-producing *Clostridium acetobutylicum* increases butanol ratio. *J Biotechnol* 2008;**136**:284–91.
159. Alsaker KV, Spitzer TR, Papoutsakis ET. Transcriptional analysis of spo0A overexpression in *Clostridium acetobutylicum* and its effect on the cell's response to butanol stress. *J Bacteriol* 2004;**186**(7):1959–71.
160. Reyes LH, Almario MP, Kao KC. Genomic library screens for genes involved in *n*-butanol tolerance in *Escherichia coli*. *PLoS One* 2011;**6**(3):1–9.
161. Jiang Y, Xu C, Dong F, Yang Y, Jiang W, Yang S. Disruption of the acetoacetate decarboxylase gene in solvent-producing *Clostridium acetobutylicum* increases the butanol ratio. *Metab Eng* 2009;**11**(4-5):284–91.
162. Tummala SB, Welker NE, Papoutsakis ET. Design of antisense RNA constructs for downregulation of the acetone formation pathway of *Clostridium acetobutylicum*. *J Bacteriol* 2003;**185**(6):1923–34.
163. Sillers R, Chow A, Tracy B, Papoutsakis ET. Metabolic engineering of the non-sporulating, non-solventogenic *Clostridium acetobutylicum* strain M5 to produce butanol without acetone demonstrate the robustness of the acid-formation pathways and the importance of the electron balance. *Metab Eng* 2008;**10**(6):321–32.
164. Nölling J, Breton G, Omelchenko MV, Makarova KS, Zeng Q, Gibson R, et al. Genome sequence and comparative analysis of the solvent-producing bacterium *Clostridium acetobutylicum*. *J Bacteriol* 2001;**183**(16):4823–38.
165. Mendez BS, Pettinari MJ, Ivanier SE, Ramos CA, Siñeriz F. *Clostridium thermopapyrolyticum* sp. nov., a cellulolytic thermophile. *Int J Syst Evol Microbiol* 1991;**41**(2):281–3.
166. Virunanon C, Chantaroopamai S, Denduangbaripant J, Chulalaksananukul W. Solventogenic-cellulolytic clostridia from 4-step-screening process in agricultural waste and cow intestinal tract. *Anaerobe* 2008;**14**(2):109–17.
167. Koukiekolo R, Cho H-Y, Kosugi A, Inui M, Yukawa H, Doi RH. Degradation of corn fiber by *Clostridium cellulovorans* cellulases and hemicellulases and contribution of scaffolding protein CbpA. *Appl Environ Microbiol* 2005;**71**(7):3504–11.
168. Sabathé F, Soucaille P. Characterization of the CipA scaffolding protein and in vivo production of a minicellulosome in *Clostridium acetobutylicum*. *J Bacteriol* 2003;**185**(3):1092–6.
169. Perret S, Casalot L, Fierobe H-P, Tardif C, Sabathe F, Belaich J-P, et al. Production of heterologous and chimeric scaffoldins by *Clostridium acetobutylicum* ATCC 824. *J Bacteriol* 2004;**186**(1):253–7.
170. Mingardon F, Perret S, Bélaïch A, Tardif C, Bélaïch J-P, Fierobe H-P. Heterologous production, assembly, and secretion of a minicellulosome by *Clostridium acetobutylicum* ATCC 824. *Appl Environ Microbiol* 2005;**71**(3):1215–22.
171. Al-Shorgani NKN, Isa MHM, Yusoff WMW, Kalil MS, Hamid AA. Isolation of a *Clostridium acetobutylicum* strain and characterization of its fermentation performance on agricultural wastes. *Renew Energy* 2016;**86**:459–65.
172. Shinwari ZK, Tanveer F, Khalil AT. Ethical issues regarding CRISPR mediated genome editing. *Curr Issues Mol Biol* 2017;**26**:103–10.

173. Luo ML, Leenay RT, Beisel CL. Current and future prospects for CRISPR-based tools in bacteria. *Biotechnol Bioeng* 2016;**113**(5):930–43.
174. Dürre P. Fermentative production of butanol—the academic perspective. *Curr Opin Biotechnol* 2011;**22**(3):331–6.
175. Inokuma K, Liao JC, Okamoto M, Hanai T. Improvement of isopropanol production by metabolically engineered *Escherichia coli* using gas stripping. *J Biosci Bioeng* 2010;**110**(6):696–701.

Chapter 9

Mesoporous and other types of catalysts for conversion of non-edible oil to biogasoline via deoxygenation

Lee Eng Oi[1], Min-Yee Choo[1], Hwei Voon Lee[1], Noorsaadah Abdul Rahman[2] and Joon Ching Juan[1,3]

[1]*Nanotechnology and Catalysis Research Center (NANOCAT), University of Malaya, Kuala Lumpur, Malaysia,* [2]*Department of Chemistry, University of Malaya, Kuala Lumpur, Malaysia,* [3]*Monash University, Sunway Campus, Subang Jaya, Malaysia*

9.1 Introduction

Modern material culture is highly dependent on fossil fuels, primarily oil, natural gas, and coal. Fossil fuel reserves are finite which means they are going to be exhausted over the next two centuries. As the Earth's population and standard of living increases, the demands for energy progressively grow year by year (Fig. 9.1). Currently about 90% of the world's energy demand is supplied by fossil fuel.[3] In the meantime, there is strong scientific evidence that global warming has accelerated due to the ever-increasing emissions of carbon dioxide from the combustion of fossil fuels. Based on the report from the United States Environmental Protection Agency the energy production and consumption of the transportation sector (40% of primary energy) in 2010 released about 71% of greenhouse gases, and this value has increased by 35% over the past two decades.[4] Apart from the petroleum fuel price fluctuation, the health risk and environmental deterioration associated with the utilization of fossil fuels are much more worrying.[5]

The scientific society is sourcing for alternatives to these problems by employing a holistic approach with the complementary use of renewable energy (e.g., hydropower, solar energy, and wind energy) in response to the energy and climate challenge. Current transportation technology has been

Sustainable Bioenergy. DOI: https://doi.org/10.1016/B978-0-12-817654-2.00009-5

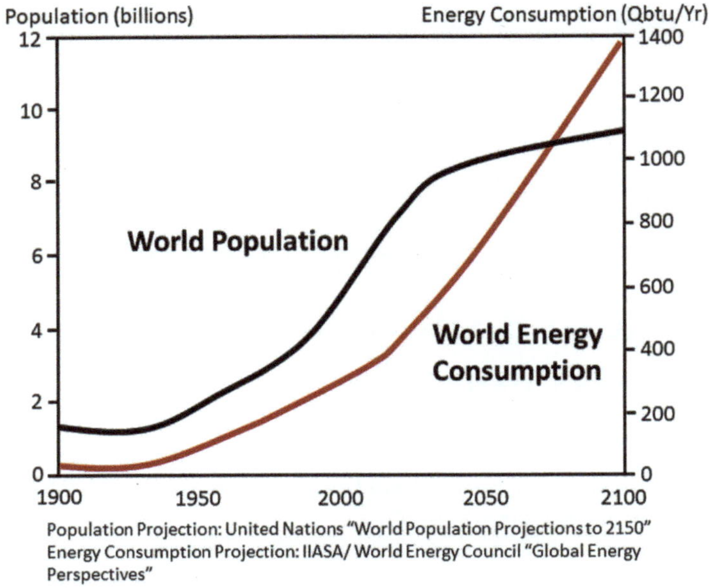

FIGURE 9.1 World population and energy demand growth. *Adapted and modified from population division of the population division of the United Nations, S. United Nations World Population Projections to 2150.* Popul Dev Rev *1998;**24**(1):183–9.*[1] *and Grubler A, Jefferson M, Nakicenovic N. Global energy perspectives: a summary of the joint study by the International Institute for Applied Systems Analysis and World Energy Council. Technol Forecast Soc Change 1996;**51**(3):237–64*[2]*; open access articles.*

established based on liquid fuels (e.g., gasoline, diesel, and kerosene). Thus there is a restricted storage potential of electrical energy which hinders the complete usage of electrical energy in the transportation sector. Renewable liquid fuels can be derived from biomass (e.g., starch, triglycerides, and lignocellulose) from which bio-oil can be produced.[6] Bio-oil, an oxygen-rich liquid with a very complex composition, needs substantial deoxygenation (DO) before it can be transformed into green fuels.

The viable conversion of biomass to liquid fuels via different technologies is shown in Fig. 9.2. There are three important routes to convert biomass into liquid fuels: (1) gasification of biomass to produce syngas followed by converting syngas into liquid fuels; (2) pyrolysis/liquefaction to produce bio-oil followed by the upgrading of bio-oil through DO; and (3) acid hydrolysis of biomass to monomer units followed by conversion into liquid.

The conversion of biomass involves the removal of oxygen to form CO_2 or H_2O. This process involves the removal of functionality from a molecule with low-thermal stability. Biomass typically consists of 40%–45% of oxygen.[7] Thus bio-oil produced from biomass contains highly oxygenated compounds.[8] The oxygen content in bio-oil leads to undesirable fuel properties such as low-heating value, high viscosity, and it is corrosive.

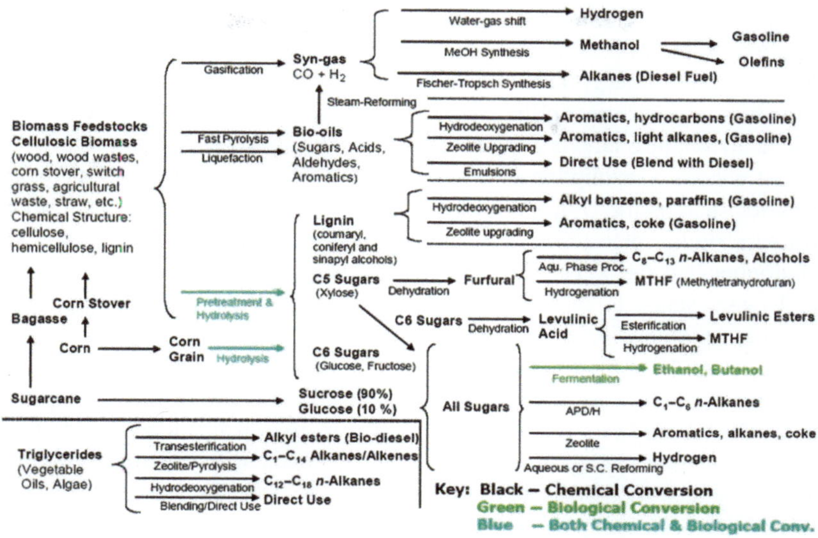

FIGURE 9.2 Roadmap for the production of hydrocarbon biofuels from biomass and triglycerides. *Adapted from Huber GW, Iborra S, Corma A. Synthesis of transportation fuels from biomass: chemistry, catalysts, and engineering. Chem Rev 2006;**106**(9):4044–98 with copyright permission from American Chemical Society.*

These properties hinder the direct application of biofuels as a petroleum fuel substitute. The lack of storage stability is also caused by the highly reactive organic compounds such as the oxygenated compounds.[9] Therefore it is important to stabilize these oxygenate compounds by removing the oxygen and increasing the hydrocarbon content via DO to obtain high-quality biofuels from low-cost and environment-friendly biofuels.

In this chapter we focus on the conversion of triglycerides or their derivatives as well as the upgrading of bio-oil to produce hydrocarbon-like green fuel via DO.

9.2 The evolution of biofuels

The necessity for energy security has motivated many countries to reduce dependence on fossil fuels and adopt renewable energy strategies. Renewable energy has a wide variety of sources. However, the main source of energy for transport is biofuel. The European Union (EU) has established the Directive 2003/30/EC to encourage the transportation sector to adopt biofuels and renewable fuels by enacting the national laws which are consistent with the Directive.[10] The EU has set the ambitious target of a 20% increase in the consumption of renewable energy by 2020, including a 10% share in the transportation sector. In addition, the Directive aims to reduce at least 60% of greenhouse gas emission in 2018 for biofuel produced in installations as compared to the EU fossil fuel comparator for heat (313.2 kg CO_2/MWh). The blending of biofuels contribute to CO_2 reductions for fossil fuel suppliers. The Renewable Energy Directive 2009/28/EC certainly triggers rapid growth of the biofuel industries in the EU. The majority of EU countries are keeping their 2020 binding targets for renewable energy on track, as around 17% of the EU energy consumption in 2016 was from renewable sources. Developed countries such as the United States, Canada, Australia, and Japan have implemented a similar plan to promote biofuels over the past few years. It is worth mentioning that the EU is the world's leading producer of biodiesel whereas the United States is the world's largest ethanol producer. Primarily, first-generation biofuels are derived from edible vegetable oil crops such as corn, sunflower, soybean, and rapeseed. These first-generation of biofuels (e.g., ethanol and biodiesel) are commercially obtainable and the supply is approximately 50 billion liters per annum.[11] However, first-generation biofuels have sparked the food versus fuel debate in recent years with the main concerns being the global food crisis, ecosystem imbalance, land use, and other environmental issues caused by deforestation for plantation purposes.[11–13]

In the year 2002, the Renewable Energy Directive limited the use of food-based biofuels to 5% which stimulated the development of alternative, second-generation biofuels from non-edible feedstock such as ceiba oil, jatropha oil, waste cooking oil, agricultural waste, and cellulosic energy crops.[12,14–16] Although second-generation biofuels have been successfully produced, the outcome is far from meeting the global energy demands.[17] Therefore there is a need for continuous research in the exploitation of non-edible oil as a feedstock for a transportation fuel substitute. Third-generation biofuels are known as advanced generation non-edible oils which are derived from microalgae to produced jet fuel or diesel. This is a promising alternative due to the high energy yield with simple growth requirement, photosynthetic ability to reduce the carbon footprint, and, most importantly, production does not require arable land, so avoiding direct competition with food crops.[18,19] The unique feature of the engineered microalgae makes it a promising source to produce green hydrocarbons.

9.3 Biofuels production via deoxygenation

Catalytic DO is an important method to convert oxygenated compounds into green hydrocarbons and involves all the reactions that remove oxygen from a molecule.[20] Table 9.1 summarizes the possible reaction and thermodynamic data for DO. The primary routes of DO in the liquid phase include hydrodeoxygenation (HDO), decarboxylation (DCO_2), and decarbonylation (DCO). In the vapor phase the main reactions are water gas shift (WGS) and methanation. HDO is similar to hydrotreating technology which is commonly used in petroleum refineries and means that HDO can be carried out in the existing petroleum refinery infrastructure.[22] The HDO, DCO_2, and DCO

TABLE 9.1 Thermodynamic data of plausible deoxygenation reactions.[21]

Liquid phase	Reaction	ΔG°_{533K} (kJ/mol)	ΔH°_{533K} (kJ/mol)
Hydrodeoxygenation	$R-COOH+3H_2 \rightarrow R-CH_3+2H_2O$	−88.0	−112.6
Decarbonylation	$R-COOH+H_2 \rightarrow R'-H+CO+H_2O$	−59.5	49.7
Decarboxylation	$R-COOH \rightarrow R-H+CO_2$	−78.6	10.1
Gas phase	**Reaction**	**ΔG°_{533K} (kJ/mol)**	**ΔH°_{533K} (kJ/mol)**
Water gas shift	$CO+H_2O \rightarrow CO_2+H_2$	−19.1	−39.6
Methanation	$CO+3H_2 \rightarrow CH_4+H_2O$	−88.4	−215.3
	$CO_2+4H_2 \rightarrow CH_4+2H_2O$	−69.2	−175.7

R, Saturated carbon chain; *R'*, unsaturated carbon chain.

pathways can be differentiated by determining the carbon chain length of the products. In the presence of hydrogen, HDO occurred by eliminating the oxygen to form n-alkanes and H_2O. The hydrocarbon produced in the HDO pathway retains the same carbon chain length as the corresponding feedstock. On the contrary, the DCO_2 and DCO pathways can occur with minimum or without hydrogen to form hydrocarbon with one carbon atom less than the parent fatty acid while oxygen is released in the form of CO_2 and CO, respectively. However, DCO_2 and DCO have a limitation; the CO_2 and CO that are attached to the catalyst surface tend to poison the catalyst when hydrogen is insufficient, and this leads to the catalyst deactivation.[21,23]

The DO reaction is promising as the reaction is performed under mild reaction temperature (250°C–380°C) with minimal or no hydrogen consumption. This reaction is also able to improve the selectivity of the targeted hydrocarbons that are compatible with vehicle engines and have similar properties as transport fuels.[24] HDO requires a high volume of hydrogen while DCO_2 consumes about 0–1 mol of hydrogen per mol of fatty acids.[25] The hydrogen demand for the DO process is in the order of $DCO_2 < DCO < HDO$.[26] The theoretical yield of hydrocarbon for HDO and DCO_2 routes are 85% and 80%, respectively.[27] Indeed, the DO technology with low hydrogen requirement is reliable to be implemented on a small scale. It is more economical and an environmentally desirable approach to produce green hydrocarbons.

9.3.1 The triglycerides feedstock

Triglycerides and fatty acids with a hydrocarbon chain, in particular, C_{16} and C_{18} ranging from C_4 to C_{24} are the main components in natural oils and fats.[28] Triglycerides are derived from glycerol bound to three fatty acids (Fig. 9.3). Notably, all the fatty acids have an even carbon number chain length. The general DO of triglycerides and fatty acids from edible and non-edible feedstocks are widely discussed in the literature.[29–32] Researchers have extensively studied the reaction and mechanism by evaluating various model compounds, oils, and fats such as vegetable oils, methyl and ethyl esters, and fatty acids due to their structural similarity. The fatty acid composition of various feedstocks is shown in Table 9.2.[33–40] Upon performing the DO of the triglycerides or derivatives, the key objective is to produce saturated or unsaturated hydrocarbon-liked biofuels.

9.3.2 Reaction mechanism

The understanding of the initial breakdown mechanism of the triglyceride and its dependence on the reaction condition is essential for a DO study. This is particularly so when oils and fats are used as feedstocks. The β-elimination is the predominant pathway for the cleavage of triglyceride and it is widely accepted by researchers. The reaction starts with metal-catalyzed hydrogenation to remove the double bonds on the triglyceride and then the hydrogenolysis of the unsaturated triglyceride through β-elimination.[38] At this point, one fatty acid is released from the triglyceride and becomes an unsaturated glycol difatty acid ester (UGDE). Thus the triglyceride is able to produce fatty acid intermediates in an inert condition. Subsequent β-elimination of the UGDE takes place to produce fatty acids and propane in the presence of hydrogen.[41] The UGDE will further crack or cleave into shorter chain fatty acids and alkanes at high temperature. On the other hand, researchers had proposed that triglyceride can be broken down through hydrolysis, γ-hydrogen transfer, and direct DO (Fig. 9.4). In an H_2-modest condition when the γ-hydrogen transfer mechanism looks favorable, the hydrocarbon produced is two carbon numbers lower than the original fatty acid at a temperature of 450°C.

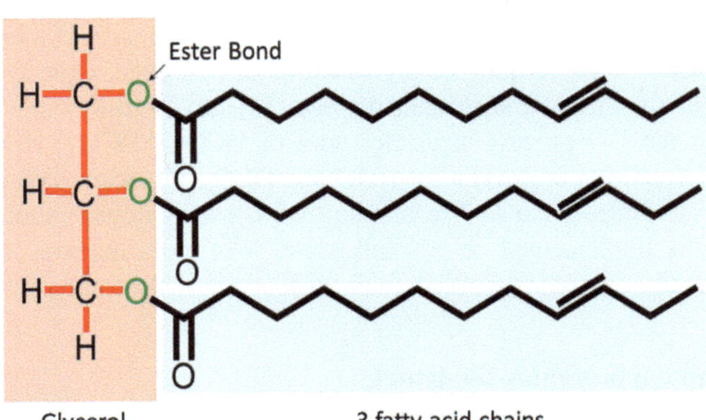

FIGURE 9.3 Structure of triglycerides and fatty acids.

TABLE 9.2 Fatty acid profile of common edible and nonedible oils/fats.

Fatty acid composition (%)	SFA			MUFA			PUFA	
	Myristic	Palmitic	Stearic	Palmitoleic	Oleic	Eicosenic	Linoleic	Linolenic
	$C_{14:0}$	$C_{16:0}$	$C_{18:0}$	$C_{16:1}$	$C_{18:1}$	$C_{20:1}$	$C_{18:2}$	$C_{18:3}$
Edible oil								
Palm oil[a]	0.5	40	5	0.3	36	0	9	0.2
Sunflower oil[a]	0	6	3	0	17	0	74	0
Canola oil[a]	0.1	5.5	2.2	1.1	55	1.4	24	8.8
Rice bran oil[b]	0	15	2	0	43	0	39	1
Rapeseed oil[b]	0	4	1	0	65	0	22	8
Nonedible oil								
Waste cooking oil[c]	1.0	39.0	4.2	0	45.1	0	11.8	0
Jatropha oil[d]	0	20.2	7.2	0	39.8	0	31.5	0
Sterculla oil[e]	0.2	18.4	7.6	0.3	4.0	2.3	11.1	0.7
Microalgae oil[f]	0.04	4.4	4.4	0	32.2	0.4	56.2	0
Animal fats								
Chicken fat (poultry fat)[b]	1	23	6	8	42	2	17	1
Fish oil (menhaden)[b]	11	20	3	4	15	0	2	1

(Continued)

TABLE 9.2 (Continued)

Fatty acid composition (%)	SFA			MUFA			PUFA	
	Myristic	Palmitic	Stearic	Palmitoleic	Oleic	Eicosenic	Linoleic	Linolenic
	$C_{14:0}$	$C_{16:0}$	$C_{18:0}$	$C_{16:1}$	$C_{18:1}$	$C_{20:1}$	$C_{18:2}$	$C_{18:3}$
Model compounds								
Triolein[g]	1.5	3.8	1.9	4.3	83.3	0.9	0.4	0.2
Tristreain[h]	0.9	25.2	71	0	0	0	0	0

MUFA, Monounsaturated fatty acids; *PUFA*, polyunsaturated fatty acids; *SFA*, saturated fatty acids.
[a]*Ref. [33].*
[b]*Ref. [34].*
[c]*Ref. [35].*
[d]*Ref. [36].*
[e]*Ref. [37].*
[f]*Ref. [38].*
[g]*Ref. [39].*
[h]*Ref. [40].*

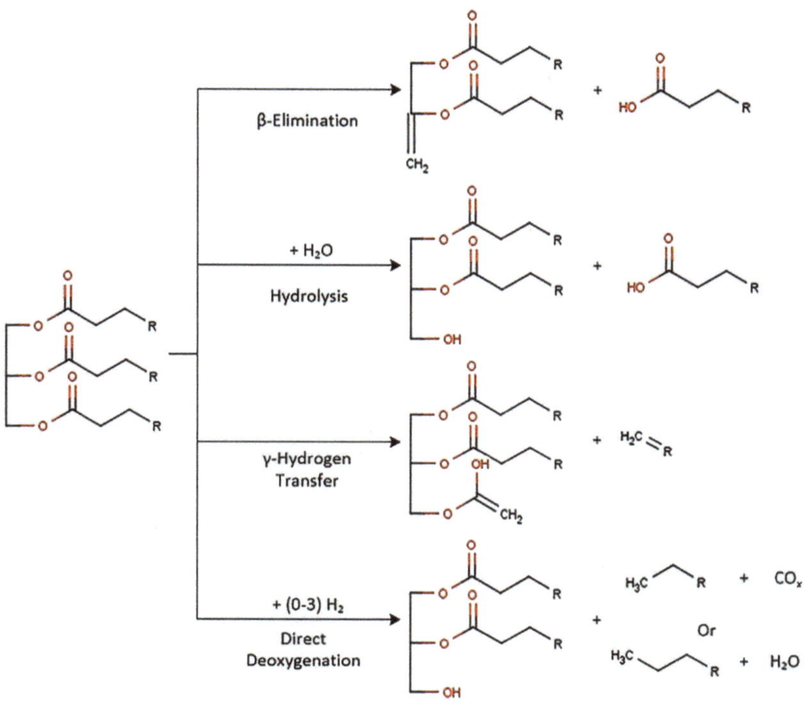

FIGURE 9.4 Breakdown of triglycerides. *Adapted from Rogers KA, Zheng Y. Selective deoxygenation of biomass-derived bio-oils within hydrogen-modest environments: a review and new insights. ChemSusChem 2016;**9**(14):1750–72 with copyright permission from John Wiley and Sons.*

The overall reaction pathway for the conversion of fatty acid is shown in Fig. 9.5. There are three reaction pathways that lead to oxygen removal from a fatty acid. The first pathway is the HDO route and involves hydrogenolysis leading to the removal of the hydroxyl group as water. In an H_2 atmosphere, the hydrogenation of the carboxylic group of fatty acids leads to the formation of corresponding aldehydes. The formed aldehyde will undergo hydrogenation to form alcohol and subsequently produce paraffin as a final product.[38] In an H_2-modest atmosphere, the olefin produced from the dehydration of alcohol will go through cracking and cyclization to form cycloalkanes and lighter hydrocarbons. As the C–C bond is more stable, it is more susceptible to cracking. It has been shown that unsaturated compounds are precursors for coke formation, giving rise to the side products (e.g., aromatic compounds).[42] This drawback is less profound in the presence of hydrogen as these compounds are hydrogenated to form saturated compounds which are less prone to cleaving and cracking. Meanwhile, the isomerization forms isomerized hydrocarbons which is advantageous for aviation fuel production.

The second reaction pathway is DCO_2 which starts with the dehydrogenation of an α-carbon atom of the fatty acid which subsequently leads to the

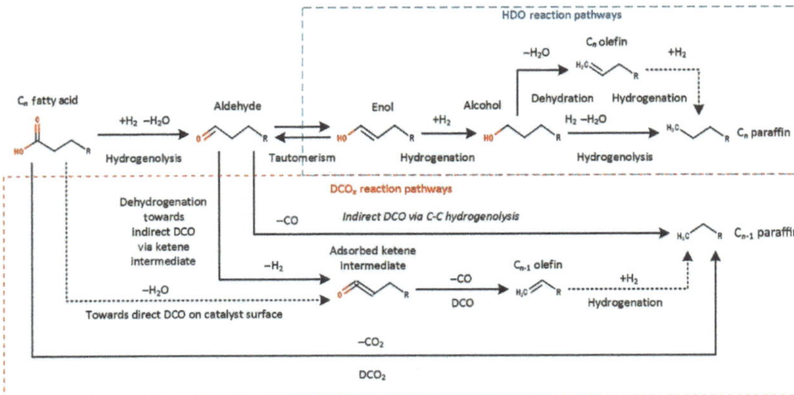

FIGURE 9.5 The plausible reaction for DO of fatty acids to produce biofuels. *DO*, Deoxygenation. *Adapted from Rogers KA, Zheng Y. Selective deoxygenation of biomass-derived bio-oils within hydrogen-modest environments: a review and new insights.* ChemSusChem *2016;**9**(14):1750–72 with copyright permission from John Wiley and Sons.*

strongly absorbed organic compounds. After the scission of the C–C bond, hydrogenation takes place to desorb the product as paraffin and release CO_2. The DCO reaction pathway is similar to DCO_2 and begins with the dehydrogenation of the α-carbon atom of the fatty acid. However, in the H_2-modest condition (0.01 bar) the fatty acid will undergo dehydrogenation of the β-carbon followed by the removal of a hydroxyl group.[43] On the contrary, in the H_2-rich condition, the fatty acids undergo the dehydrogenation of the hydroxyl group, prior to the scission of the C–C bond, the β-carbon will undergo dehydrogenation. Lu et al.[43] proposed indirect DCO that absorbs the alkanoyl group which will then form a ketene intermediately on the catalyst surface; the researchers also suggested that the ketonization reaction leads to DCO_2. This finding was supported by Peng et al.[21]

Another plausible side reaction includes cracking to produce short chain hydrocarbon at high temperatures >300°C.[44–46] The CO released from the DCO may react with water in the system to form hydrogen and CO_2 via WGS. In addition, methanation may occur when CO and CO_2 react with hydrogen.

9.4 Catalysts for deoxygenation

The key challenge for the success of high-quality biofuels production is the selection of a catalyst. The efficiency of removing oxygen from biomass via HDO or DCO is highly dependent on the type of catalyst and catalyst support. The selection of a catalyst is also important to prevent undesirable catalyst deactivation[47] and decomposition[48] which can occur during DO due to the complex composition of the bio-oil. Table 9.3 shows

TABLE 9.3 Overview of catalysts investigated for catalytic deoxygenation.

No	[a]Catalyst	Feed	Reactor	T (°C)	t (h)	Solvent	Cond (bar)	Conv (%)	Catalytic performance	Ref.
1	3.3%Ni15%MoS/TiO_2	Rapeseed oil	Batch	300	4	NA	$P_{H2} = 35$	78	56% C_{18}, 5% C_{17}	[20]
2	3.3%Ni15%MoS/SiO_2	Rapeseed oil	Batch	300	4	NA	$P_{H2} = 35$	95	20% C_{18}, 55% C_{17}	[20]
3	3.3%Ni15%MoS/Al_2O_3	Rapeseed oil	Batch	300	4	NA	$P_{H2} = 35$	80	55% C_{18}, 25% C_{17}	[20]
4	10%Ni/HBEA (Si/Al = 180)	Microalgae oil	Batch	260	8	NA	$P_{H2} = 60$	78	60% C_{18}, 7.1% C_{17}	[41]
5	0.75%Pt/SAPO-11	Microalgae oil	Batch	375	1	NA	$P_{H2} = 10$	83	32.8% C_{9-15}, 4.2 *i/n*	[49]
6	5%Ru/TiO_2	Pyrolysis oil	Batch	350	4	NA	$P_{H2} = 200$	65	77% O_2 removal	[50]
7	5% Ru/C	Pyrolysis oil	Batch	350	4	NA	$P_{H2} = 200$	53	86% O_2 removal	[50]
8	5%Ru/Al_2O_3	Pyrolysis oil	Batch	350	4	NA	$P_{H2} = 200$	35	79% O_2 removal	[50]
9	20%Co–Cao	Triolein	Semi	350	1	NA	N_2	73.5	76% C_{8-20}, 41% $C_{17} + C_{15}$	[51]
10	20%Ni–Cao	Triolein	Semi	350	1	NA	N_2	53.7	73% C_{8-20}, 40% $C_{17} + C_{15}$	[51]
11	20%Zn–Cao	Triolein	Semi	350	1	NA	N_2	58.5	70% C_{8-20}, 41% $C_{17} + C_{15}$	[51]
12	20%Fe–Zao	Triolein	Semi	350	1	NA	N_2	61.1	60% C_{8-20}, 41% $C_{17} + C_{15}$	[51]
13	5%Co10%Ca/ SiO_2–Al_2O_3	Triolein	Semi	350	1	NA	N_2	NA	73% C_{8-20}, 48% C_{17}	[37]
14	CaO–La_2O_3/C	Triolein	Semi	330	3	NA	N_2	NA	72% C_{8-20}	[52]
15	10% Ni/HMS	Triolein	Semi	380	2	NA	Vacuum	92.5	95.2% C_{11-20}	[53]
16	1% Pt/C	Tristearin	Batch	350	4	NA	N_2	42	83% HC	[40]

(*Continued*)

TABLE 9.3 (Continued)

No	[a]Catalyst	Feed	Reactor	T (°C)	t (h)	Solvent	Cond (bar)	Conv (%)	Catalytic performance	Ref.
17	5% Pd/C	Tristearin	Batch	350	4	NA	N_2	29	93% HC	[40]
18	20% Ni/C	Tristearin	Batch	350	4	NA	N_2	85	56% HC	[40]
19	5% Pd/C	Stearic acid	Semi	300	6	Dodecane	$P_{He} = 60$	100	95% $C_{17\text{-sat}}$, 3% $C_{17\text{-unsat}}$	[54]
20	5% Pt/C	Stearic acid	Semi	300	6	Dodecane	$P_{He} = 60$	86	87% $C_{17\text{-sat}}$, 8% $C_{17\text{-unsat}}$	[54]
21	5% Ru/C	Stearic acid	Semi	300	6	Dodecane	$P_{He} = 60$	13.2	24% $C_{17\text{-sat}}$, 41% $C_{17\text{-unsat}}$	[54]
22	5% Os/C	Stearic acid	Semi	300	6	Dodecane	$P_{He} = 60$	6.9	29% $C_{17\text{-sat}}$, 24% $C_{17\text{-unsat}}$	[54]
23	1% Rh/C	Stearic acid	Semi	300	6	Dodecane	$P_{He} = 60$	19.9	18%$C_{17\text{-sat}}$, 66% $C_{17\text{-unsat}}$	[54]
24	16% Ni/Al_2O_3	Stearic acid	Semi	300	6	Dodecane	$P_{He} = 60$	17.8	29% $C_{17\text{-sat}}$, 16% $C_{17\text{-unsat}}$	[54]
25	60% Ni/SiO_2	Stearic acid	Semi	300	6	Dodecane	$P_{He} = 60$	18.1	19% $C_{17\text{-sat}}$, 38% $C_{17\text{-unsat}}$	[54]
26	5% Ru/SiO_2	Stearic acid	Semi	300	6	Dodecane	$P_{He} = 60$	7.2	6% $C_{17\text{-sat}}$, 18% $C_{17\text{-unsat}}$	[54]
27	1% Ir/SiO_2	Stearic acid	Semi	300	6	Dodecane	$P_{He} = 60$	4.6	14% $C_{17\text{-sat}}$, 55% $C_{17\text{-unsat}}$	[54]
28	5%Pt/MWCNT	Stearic acid	Batch	330	0.5	Water	NA	52.4	97% C_{17}	[55]
29	5% Pt/C	Stearic acid	Batch	330	0.5	Water	NA	52.4	57% C_{17}	[55]
30	5% Ru/C	Stearic acid	Batch	330	0.5	Water	NA	42.3	42.9% C_{17}	[55]
31	5% Pd/C	Stearic acid	Semi	300	5	Dodecane	He	100	98% $C_{17\text{-sat}}$, 2% $C_{17\text{-unsat}}$	[56]
32	0.6% Pd/SBA-15	Stearic acid	Semi	300	5	Dodecane	5%H_2/Ar	96	98% C_{17}	[57]
33	$Ni(OAc)_2$	Stearic acid	Batch	350	2	NA	N_2	72	57.4% HC yield, 41.3% C_{17}	[58]
34	30% MO/Al_2O_3-TiO_2	Palmatic acid	Batch	280	4	NA	$P_{H2} = 40$	100	93.2% $C_{17\text{-sat}}$, 6.8% $C_{17\text{-unsat}}$	[59]

35	4% Ni13%MoS/ γ-Al_2O_3	Palmatic acid	Batch	360	2	NA	$P_{H2} = 34$	98.7	56.2% HC, HDO:DCO = 6	[27]
36	4% Co/H-ZSM22	Palmitic acid	Batch	260	4	Decane	$P_{H2} = 10$	100	66.9% C_{16}, 31.9% C_{15}	[60]
37	5% Ni/HBEA (Si/Al = 180)	Palmitic acid	Batch	260	6	Dodecane	$P_{H2} = 12$	100	49% C_{16}, 24% C_{15}	[21]
38	5% Ni/Al_2O_3	Palmitic acid	Batch	260	6	Dodecane	$P_{H2} = 12$	51	74% C_{15}	[21]
39	5% Ni/SiO_2	Palmitic acid	Batch	260	6	Dodecane	$P_{H2} = 12$	41	59% C_{15}	[21]
40	5% Ni/ZrO_2	Palmitic acid	Batch	260	6	Dodecane	$P_{H2} = 12$	100	90% C_{15}	[21]
41	5% Pt/ZrO_2	Palmitic acid	Batch	260	6	Dodecane	$P_{H2} = 12$	99	61% C_{15}	[21]
42	5% Pd/ZrO_2	Palmitic acid	Batch	260	6	Dodecane	$P_{H2} = 12$	98	98% C_{15}	[21]
43	10% Ni/ZrO_2	Palmitic acid	Batch	260	6	Dodecane	$P_{H2} = 12$	100	85% C_{15}	[21]
44	4 Ru/TiO_2	Phenol	Batch	300	1	NA	$P_{H2} = 45$	12	85% DDO (Benzene)	[61]
45	1% Pt/TiO_2	m-cresol	Batch	300	1	NA	$P_{H2} = 10$	17	88% toluene	[62]

C17-sat, heptadecane; *C17-unsat*, heptadecene; *Cond*, condition; *Conv.*, conversion; *DDO*, direct deoxygenation, *i/n*, ratio of isomers over straight-chain hydrocarbon; *HC*, hydrocarbon; *NA*, Not available in the reference; *PT*, pretreatment; *T*, temperature; *t*, time.
[a]*Percentage for metal loading on catalyst refer to wt.%.*

a comprehensive summary of the relevant literature on DO by providing the reaction condition which is related to the feedstock, catalyst used, and performance.[20,21,27,37,40,41,49–62]

9.4.1 Hydrotreating catalyst

The conventional hydrotreating catalyst (NiMo and CoMo sulfided supported on Al_2O_3) is an important catalyst in industries and removes oxygen, sulfur, and nitrogen from petroleum. Triglycerides can be hydrotreated with conventional NiMo and CoMo sulfided catalysts which are commonly employed in petroleum refineries. It can improve the selectivity of the diesel-range hydrocarbon ranging from C_{12} to C_{18}.[63–65] These catalysts are more active in a sulfided form compared to the oxide form.[66] Based on the literature, almost complete conversion (98.7%) of palmitic acid is achieved using NiMoS/γ-Al_2O_3. The final HDO/DCO ratio was about 6.[27] This result was in good agreement with other findings where the sulfided catalyst proceeded via the HDO route.[67] However, the sulfided catalysts suffered from product contamination due to sulfur leaching, and catalyst deactivation due to insufficient sulfur supplied for the reaction occurred.[27,68] Therefore nonsulfided and highly efficient catalysts are preferred to replace the conventional sulfide catalysts.

It is interesting to note that the activity of the hydrotreating catalyst can be affected by the modification and selection of the support. Kubička et al.[20] proved that the selectivity could be tuned even with the hydrotreating catalyst (NiMoS). The hydrotreating of rapeseed oil via NiMoS/Al_2O_3 and NiMoS/TiO_2 can produce straight-chain C_{18} as the major product and minor share of unsaturated C_{17}. The ratios of C_{18}/C_{17} are 2 and 10, respectively, which demonstrates that the selectivity of the HDO is more prominent. On the other hand, the NiMoS/SiO_2 showed increased selectivity of DCO_2. This study clearly shows that the selectivity could be tuned by the nature of support such as acidity, surface area, and pore size.

9.4.2 Precious metal-based catalysts

Precious metals such as ruthenium (Ru), rhodium (Rh), iridium (Ir), palladium (Pd), and platinum (Pt) are frequently employed as an active phase for the DO of fatty acids.[56,69–73] Precious metal catalysts show high activity and selectivity for fatty acid conversion.[74] Snåre et al.[54] studied various noble metals (Ru, Pd, Pt, Ir, Os, and Rh) and supports (carbon, alumina, and silica) on the DO with stearic acid in order to determine their performance. It was found that a metal-supported carbon catalyst was very selective of the DCO_2 product. 5 wt.% Pd/C was the best catalyst for the reaction; it achieved 100% conversion of the stearic acid with DCO_2 selectivity. According to Morgan et al.,[40] Immer et al.,[56] Lestari et al.,[57] Peng et al.,[21] and Snåre et al.,[54] the Pd and Pt catalysts exhibited superior DCO_2 selectivity, which produced a high amount of C_{17} hydrocarbons. Wildschut et al.[50] performed a similar study on the conversion of pyrolysis oil via Ru supported on TiO_2, Al_2O_3, and C. They observed that the DO yield was in the order of $TiO_2 > C > Al_2O_3$, while the ability to remove oxygen was in the order of $C > TiO_2 > Al_2O_3$.

Yang et al.[55] studied the conversion of stearic acid and the selectivity of C_{17} with Pt supported on MWCNT (multiwall carbon nanotubes). They also demonstrated the effect of active metal and catalyst support by comparing its performance with Pt/C and Ru/C. It was surprising that both Pt/MWCNT and Pt/C showed a similar conversion (52.4%). However, the selectivity to C_{17} for Pt/MWCNT (97.0%) was 1.7-times higher than Pt/C. Pt/C had lower selectivity due to the side reactions catalyzed by the organic groups present in the activated carbon.[75] However, they exhibited much lower conversion (<42%) when converting triglycerides under a nitrogen atmosphere.[40] In addition, they observed that Pt exhibited higher activity compared to Ru in the DCO_2 reaction.

Although precious metals (especially Pt and Pd) have been proven to be highly active and selective in the DO process, their high cost are a major drawback from an economic standpoint and limit their wide application in the industrial scale. Hence, an alternative which is low-cost and has high activity[76] is desirable to satisfy the requirements of these catalytic chemical transformations.

9.4.3 Transition metal-based catalysts

The performances of precious and nonprecious metals are still far from being understood. Thus Peng et al.[41] attempted to compare the performance of nickel (Ni) with Pt and Pd. Interestingly, the DO performance of Ni supported on ZrO_2 had proven to be more active than Pd and Pt. Among the catalysts with the same Ni content, zeolite exhibited the highest catalytic activity. The DO activity of the supported catalyst exhibited the following trend: Ni/HBEA (Si/Al = 180) > $Ni/ZrO_2 > Pd/ZrO_2 > Pt/ZrO_2 > Ni/Al_2O_3 > Ni/SiO_2$. Another study by Snåre et al.[54] revealed the effect of metals on the DO of stearic acid shown in the descending order: Pd > Pt > Ni > Rh > Ir > Ru > Os. In addition, Li et al.[58] reported that the outstanding performance of $Ni(OAc)_2$ alone could achieve 72% of the conversion of stearic acid and 57.4% selectivity of hydrocarbon, with 41.3% DCO_2 product. It is in accordance with other findings that group 10 metals (Pt, Pd, and Ni) tend to select the DCO_2 pathway.[38] Based on the literature, it is obvious that the performance of Ni supported catalysts are comparable with the precious metal-based catalysts. Furthermore, the price of Ni is approximately 2500- and 1000-times cheaper compared to Pt and Pd, respectively.[77] Hence, Ni is a potentially good and cheaper alternative to replace the use of precious metals.

Asikin-Mijan et al.[51] studied the transition metals (Ni, Co, Zn, and Fe) supported on a CaO basic catalyst for the DO of triolein under nitrogen flow. The results showed that all the catalysts rendered the desired reaction with 60%–76% of the saturated and unsaturated hydrocarbon ranging from C_8 to C_{20}. Among the catalysts, Co–CaO achieved the highest conversion (73.5%) with the hydrocarbon selectivity of 76% and 41% DCO_2 of the product due to the strong acid–base properties. Bimetallic Co–Ca supported on SiO_2–Al_2O_3 exhibited an excellent performance of 73% hydrocarbon selectivity and 48% DCO_2 product. The researchers suggested that a large surface area of CoCa/SiO_2–Al_2O_3 enhanced the catalytic activity. The transition metal-supported catalysts have the promising prospect for catalytic DO.

As mentioned above, the Ni-based catalyst is prominent to the DCO_2 selectivity. However, there is an exceptional case owing to the acid properties of the catalyst support, for example, Ni supported on zeolite which exhibited a selectivity of the HDO products. The conversion of palmitic acid via Ni/HBEA achieved 100% with the HDO/DCO_2 ratio of 2. This was contributed by the strong acid strength and existence of the Brönsted acid site on zeolite which promoted dehydration compared to Lewis acid site.[21] This is in agreement with the result of Cao et al.[60] who reported 100% conversion of palmitic acid over Co/HZSM22 which gave a C_{16}/C_{15} ratio of 2, illustrating that zeolite favor HDO over DCO_2. Again, this provided clear evidence that the selectivity can be tuned by catalyst support.

9.4.4 Mesoporous support catalysts

Basically, there are four categories of catalyst support:

1. Solid acids support likes zeolites[45]
2. Reducible oxides support such as titania,[78–80] ceria,[81] and zirconia[82]
3. Refractory oxides support likes alumina[83] and silica[81]
4. Carbon-based support likes activated carbon,[84] graphene, and carbon nanotubes.[85]

Every group of support revealed different features which affected the activity and selectivity. Solid acid supports have strong Brönsted acid sites which promoted a fast dehydration step of alcohol and tends to follow the HDO route.[22] For reducible oxides support, due to the inherent oxophilicity which allows oxygenated compounds to adsorb onto oxygen vacancies, it plays a vital role in the reductive conversion of carboxylic acids to aldehydes. With the refractory oxides support, the reaction proceeds to both the HDO and DCO routes which have no preference for HDO or DCO. Lastly, with the carbon support, which is neutral with the large surface area, the DO reaction is affected by the metals used.

Porous materials have attracted significant interest because of the enhanced accessibility and high surface area which allows interaction with atoms, ions, or molecules throughout the surface and internal pore channels.[86,87] Hence, the existence of a porous framework in nanomaterials significantly enhances the physicochemical properties of the catalyst. According to IUPAC, the porosity can be categorized into three primary groups: (1) microporous (pore size up to 2 nm); (2) mesoporous (pore size in between 2 and 50 nm); and (3) macroporous (pore size larger than 50 nm)[88–90] (Fig. 9.6).

The DO and upgrading of bio-oil are challenging due to the complex chemical composition of the reactant that normally involves multiple reactions. The existence of the bulky oxygenates intermediates make mesoporous materials crucial as the catalyst supports in the process. Theoretically, the larger pore diameter can open the door to process bulky molecules and provide better accessibility for the reaction. Fig. 9.7 demonstrates the effect of the pore diameter on mesoporous carbon supported Au catalyst to the selectivity.[91] Based on the study, mesoporous carbon with a smaller pore diameter tended to form highly oxidized products, particularly glyceric acid. When the pore diameter increased from 5 to 20 nm, the selectivity was progressively shirted to partially-oxidized dihydroxyacetone. This provided clear evidence that the selectivity of the reaction was greatly affected by the diffusional ability which can be controlled by pore diameter. In this case, the mesoporous catalyst was preferable in order to avoid the undesired overreaction.

Considering the diffusional limit involving bulky molecules, hexagonal mesoporous silica (HMS) with a high surface area (512 m^2/g) and large pore diameter of 5.39 nm could have an advantage in the production of green diesel because the diffusional limit and mass transfer resistance will dissolve into nothingness.[53] Zulkepli et al.[53] reported the outstanding performance of

FIGURE 9.6 Example of TiO_2 with various porosity: (A) microporous TiO_2 with 1.5 nm pore; (B) mesoporous TiO_2 with 30 nm pore size; and (C) macroporous TiO_2 with 130 µm pore size. *(A) Adapted from Chandra D, Bhaumik A. Super-microporous TiO synthesized by using new designed chelating structure directing agents. Micropor Mesopor Mater 2008;**112**:533–41 with copyright permission from Elsevier, (B) adapted from Guldin S, Hüttner S, Tiwana P, Orilall MC, Ülgüt B, Stefik M, et al. Improved conductivity in dye-sensitised solar cells through block-copolymer confined TiO2 crystallisation. Energy Environ Sci 2011;**4**:225 with copyright permission from Royal Society of Chemistry, (C) adapted from Chen L, Huang C, Xu G, Hutton SL, Miao L. Macroporous TiO2 foam with mesoporous walls. Mater Character 2013;**5**:8–12 with copyright permission from Elsevier.*

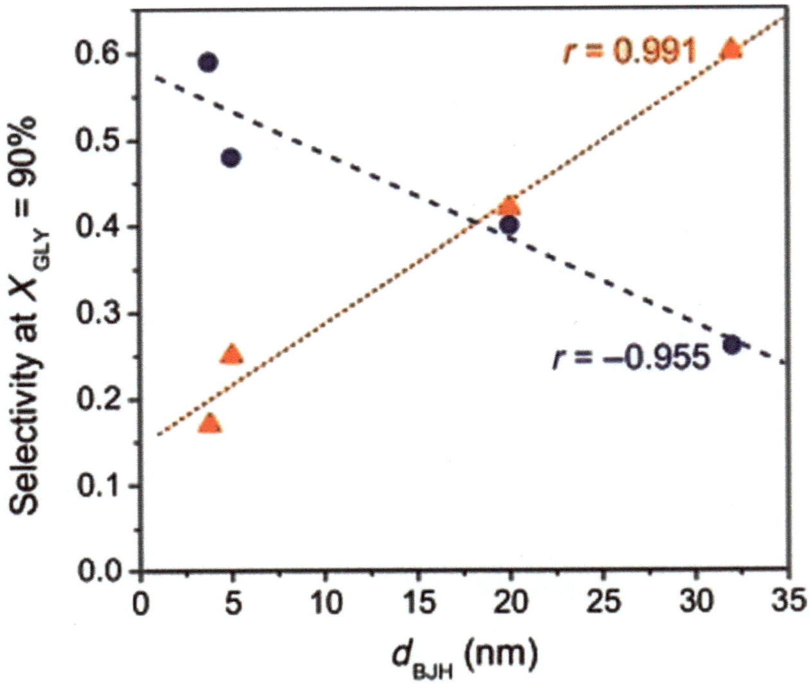

FIGURE 9.7 Correlations between selectivity to dihydroxyacetone (▲) and glyceric acid (●) and the main pore size for Au nanoparticles supported on mesoporous carbon (calculated by the BJH (Barrett, Joyner and Halenda) method). *Adapted from Rodrigues EG, Pereira MFR, Órfão JJM. Glycerol oxidation with gold supported on carbon xerogels: tuning selectivities by varying mesopore sizes. Appl Catal B: Environ 2012;**115–116**:1–6 with copyright permission from Elsevier.*

mesoporous Ni/HMS which achieved 92.5% conversion and 95.2% selectivity of hydrocarbon ranging from C_{11} to C_{20}. The mesoporous structure of HMS allows facile diffusion of the bulky triglycerides molecule to active sites in the internal pore channels. With reference to Fig. 9.8, the molecular dimensions of triglycerides were estimated from 6.252 to 43.706 Å. With this in mind, it can be suggested that the pore diameter must be bigger than 43.706 Å to allow the triolein molecules to enter the pores of the catalyst and react with the active site along the interpores wall.[92]

Besides the DO reaction, Verma et al.[49] suggested that the mesoporosity on the catalyst promotes isomerization of hydrocarbon which leads to green jet-fuel with a carbon chain length of C_9 to C_{15}. Diesel-range hydrocarbon could be obtained through the hydrotreating of microalgae oil via NiMoS/Al_2O_3 and is further upgraded through the DO via Pt/SAPO-11 Notably, the Pt/SAPO-11 with both micropore and mesopore furnished 83% conversion and yielded 32.8% jet-fuel hydrocarbon. The catalyst exhibited high isomerization selectivity of 4.2 over straight-chain hydrocarbon.

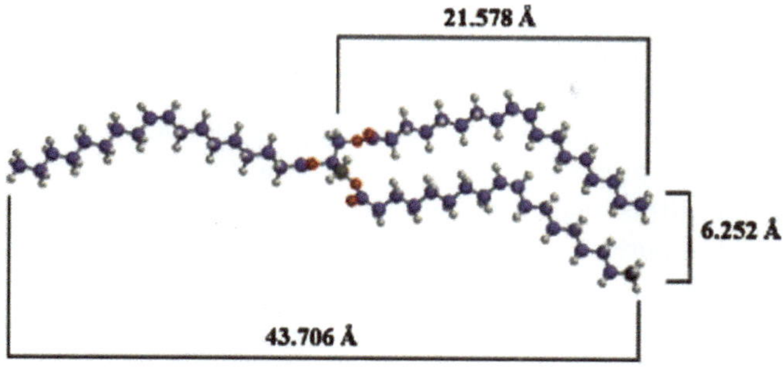

FIGURE 9.8 Molecular geometry of triglycerides determined via semiempirical AM-1 calculations using Spartan software. *Adapted from Benson TJ, Hernandez R, French WT, Alley EG, Holmes WE. Elucidation of the catalytic cracking pathway for unsaturated mono-, di-, and triacylglycerides on solid acid catalysts.* J Mole Catal A: Chem *2009;**303**(1):117–23 with copyright permission from Elsevier.*

9.5 Conclusion and future prospective

Currently, biofuels produced from non-edible oils contain high oxygenated compounds (10–40%). The presence of these oxygenated compounds in biofuels leads to undesirable fuel properties and hinders its direct application as a petroleum fuel substitute. Therefore it is important to stabilize the oxygenate compounds, remove the oxygen, and increase the hydrocarbon content in order to obtain high-quality biofuels from this low-production cost and environment-friendly approach. DO has developed to become a key process in biofuels production due to its practical potential in converting triglycerides feedstock into biofuels by removing the oxygenated compounds. The main component found in first-generation or second-generation feedstocks were triglycerides and fatty acids. The primary DO reaction includes HDO, DCO_2, and DCO accompanied by several reactions such as cracking and isomerization to produce high-quality green fuel.

The key challenge to produce high-quality biofuels is the selection of the catalyst. The selection of catalyst is also essential to prevent undesirable catalyst deactivation and decomposition which can occur during DO caused by the complex composition of bio-oil. Various catalysts and supports have been explored to produce biofuels. Conventional hydrotreating catalyst (NiMo, CoMo) is commercially available and favors the HDO pathway. However, this catalyst rapidly deactivates without an auxiliary sulfur supply and leads to sulfur contamination. Precious metal-based catalysts, especially Pt and Pd, are proven candidates with outstanding performance and high selectivity to DCO_2 product. Nonetheless, the expensive price of precious metals reduces the cost-effectiveness of the catalyst. Transition metal-based catalysts with lower cost has attracted more attention in developing an efficient catalyst for DO. Ni exhibits enormous potential due to the excellent performance comparable to precious metals, and it has the tendency to DCO_2 selectivity. Despite the active metals, the catalytic selectivity is tunable by the nature of the support. Metal supported on zeolite with the strong Bronsted acid site is prominent to HDO and cracking reaction. In addition, the selectivity of the DO is controllable by the pore diameter, preferable mesopores with a diameter from 2 to 50 nm to enhance the diffusional limit and accessibility of bulky molecules. The presence of mesopores on the catalyst is able to catalyze isomerization of n-alkanes to produce higher amount of isomers. This proposes new advances for mesoporous catalysts in DO.

References

1. The population division of the United Nations, S. United Nations World Population Projections to 2150. *Popul Dev Rev* 1998; 24(1): 183-189.
2. Grubler A, Jefferson M, Nakicenovic N. Global energy perspectives: a summary of the joint study by the International Institute for Applied Systems Analysis and World Energy Council. *Technol Forecast Soc Change* 1996;**51**(3):237–64.
3. Yen HW, Hu IC, Chen CY, Ho SH, Lee DJ, Chang JS. Microalgae-based biorefinery – from biofuels to natural products. *Bioresour Technol* 2013;**135**(0):166–74.
4. Mohan D, Pittman CU, Steele PH. Pyrolysis of wood/biomass for bio-oil: a critical review. *Energy Fuels* 2006;**20**(3):848–89.
5. Ho SH, Ye X, Hasunuma T, Chang JS, Kondo A. Perspectives on engineering strategies for improving biofuel production from microalgae – a critical review. *Biotechnol Adv* 2014;**32**(8):1448–59.
6. Climent MJ, Corma A, Iborra S. Conversion of biomass platform molecules into fuel additives and liquid hydrocarbon fuels. *Green Chem* 2014;**16**(2):516–47.
7. Huber GW, Iborra S, Corma A. Synthesis of transportation fuels from biomass: chemistry, catalysts, and engineering. *Chem Rev* 2006;**106**(9):4044–98.
8. Capunitan JA, Capareda SC. Hydrotreatment of corn stover bio-oil using noble metal catalysts. *Fuel Proc Technol* 2014;**125**:190–9.
9. Lu Q, Li WZ, Zhu XF. Overview of fuel properties of biomass fast pyrolysis oils. *Energy Convers Manage* 2009;**50**(5):1376–83.
10. Romero MJA, Pizzi A, Toscano G, Busca G, Bosio B, Arato E. Deoxygenation of waste cooking oil and non-edible oil for the production of liquid hydrocarbon biofuels. *Waste Manage* 2016;**47**(Part A):62–8.
11. Naik SN, Goud VV, Rout PK, Dalai AK. Production of first and second generation biofuels: a comprehensive review. *Renew Sustain Energy Rev* 2010;**14**(2):578–97.
12. Atabani AE, Silitonga AS, Ong HC, Mahlia TMI, Masjuki HH, Badruddin IA, et al. Non-edible vegetable oils: a critical evaluation of oil extraction, fatty acid compositions, biodiesel production, characteristics, engine performance and emissions production. *Renew Sustain Energy Rev* 2013;**18**:211–45.
13. Parmar A, Singh NK, Pandey A. Cyanobacteria and microalgae: a positive prospect for biofuels. *Bioresour Technol* 2011;**102**(22):10163–72.
14. Galadima A, Muraza O. Biodiesel production from algae by using heterogeneous catalysts: a critical review. *Energy* 2014;**78**:72–83.
15. Kumar A, Sharma S. Potential non-edible oil resources as biodiesel feedstock: an Indian perspective. *Renew Sustain Energy Rev* 2011;**15**(4):1791–800.

16. Fischer G, Prieler S, van Velthuizen H, Lensink SM, Londo M, de Wit M. Biofuel production potentials in Europe: sustainable use of cultivated land and pastures. Part I: Land productivity potentials. *Biomass Bioenergy* 2010;**34**(2):159–72.
17. Farooq M, Ramli A, Naeem A. Biodiesel production from low FFA waste cooking oil using heterogeneous catalyst derived from chicken bones. *Renew Energy* 2015;**76**:362–8.
18. Chen CY, Zhao XQ, Yen HW, Ho SH, Cheng CL, Lee DJ, et al. Microalgae-based carbohydrates for biofuel production. *Biochem Eng J* 2013;**78**:1–10.
19. Tan CH, Show PL, Chang JS, Ling TC, Lan JCW. Novel approaches of producing bioenergies from microalgae: a recent review. *Biotechnol Adv* 2015;**33**(6, Part 2):1219–27.
20. Kubička D, Horáček J, Setnička M, Bulánek R, Zukal A, Kubičková I. Effect of support-active phase interactions on the catalyst activity and selectivity in deoxygenation of triglycerides. *Appl Catal B: Environ* 2014;**145**:101–7.
21. Peng B, Zhao C, Kasakov S, Foraita S, Lercher JA. Manipulating catalytic pathways: deoxygenation of palmitic acid on multifunctional catalysts. *Chem: A Eur J* 2013;**19**(15):4732–41.
22. Kumar P, Yenumala SR, Maity SK, Shee D. Kinetics of hydrodeoxygenation of stearic acid using supported nickel catalysts: effects of supports. *Appl Catal A: Gen* 2014;**471**:28–38.
23. Mäki-Arvela P, Snåre M, Eränen K, Myllyoja J, Murzin DY. Continuous decarboxylation of lauric acid over Pd/C catalyst. *Fuel* 2008;**87**(17–18):3543–9.
24. Zhao C, Brück T, Lercher JA. Catalytic deoxygenation of microalgae oil to green hydrocarbons. *Green Chem* 2013;**15**(7):1720.
25. Rogers KA, Zheng Y. Selective deoxygenation of biomass-derived bio-oils within hydrogen-modest environments: a review and new insights. *ChemSusChem* 2016;**9**(14):1750–72.
26. Srifa A, Faungnawakij K, Itthibenchapong V, Assabumrungrat S. Roles of monometallic catalysts in hydrodeoxygenation of palm oil to green diesel. *Chem Eng J* 2014;**278**:249–58.
27. Zhou L, Lawal A. Evaluation of presulfided NiMo/γ-Al2O3 for hydrodeoxygenation of microalgae oil to produce green diesel. *Energy Fuels* 2015;**29**(1):262–72.
28. Snåre M, Kubičková I, Mäki-Arvela P, Chichova D, Eränen K, Murzin DY. Catalytic deoxygenation of unsaturated renewable feedstocks for production of diesel fuel hydrocarbons. *Fuel* 2008;**87**(6):933–45.
29. He Z, Hu M, Wang X. Highly effective hydrodeoxygenation of guaiacol on Pt/TiO_2: promoter effects. *Catal Today* 2018;**302**:136–45.
30. Barrios AM, Teles CA, de Souza PM, Rabelo-Neto RC, Jacobs G, Davis BH, et al. Hydrodeoxygenation of phenol over niobia supported Pd catalyst. *Catal Today* 2018;**302**:115–24.
31. Di L, Yao S, Song S, Wu G, Dai W, Guan N, et al. Robust ruthenium catalysts for the selective conversion of stearic acid to diesel-range alkanes. *Appl Catal B: Environ* 2017;**201**(Suppl. C):137–49.
32. Kordulis C, Bourikas K, Gousi M, Kordouli E, Lycourghiotis A. Development of nickel based catalysts for the transformation of natural triglycerides and related compounds into green diesel: a critical review. *Appl Catal B: Environ* 2016;**181**:156–96.
33. Taufiqurrahmi N, Bhatia S. Catalytic cracking of edible and non-edible oils for the production of biofuels. *Energy Environ Sci* 2011;**4**(4):1087–112.
34. Satyarthi JK, Chiranjeevi T, Gokak DT, Viswanathan PS. An overview of catalytic conversion of vegetable oils/fats into middle distillates. *Catal Sci Technol* 2013;**3**(1):70–80.

35. Chuah LF, Jiří-Jaromír K, Suzana Y, Awais B, Majid MA. Influence of fatty acids in waste cooking oil for cleaner biodiesel. *Clean Technol Environ Policy* 2017;**19**(3):859–68.
36. Asikin-Mijan N, Lee HV, Abdulkareem-Alsultan G, Afandi A, Taufiq-Yap YH. Production of green diesel via cleaner catalytic deoxygenation of *Jatropha curcas* oil. *J Cleaner Prod* 2016;**167**:1048–59.
37. Asikin-Mijan N, Lee HV, Juan JC, Noorsaadah AR, Ong HC, Razali SM, et al. Promoting deoxygenation of triglycerides via Co-Ca loaded SiO_2-Al_2O_3 catalyst. *Appl Catal A: Gen* 2018;**552**:38–48.
38. Choo MY, Oi LE, Show PL, Chang JS, Ling TC, Ng EP, et al. Recent progress in catalytic conversion of microalgae oil to green hydrocarbon: a review. *J Taiwan Inst Chem Eng* 2017;**79**(Suppl. C):116–24.
39. Asikin-Mijan N, Lee HV, Taufiq-Yap YH, Abdulkrem-Alsultan G, Mastuli MS, Ong HC. Optimization study of SiO_2-Al_2O_3 supported bifunctional acid–base NiO-CaO for renewable fuel production using response surface methodology. *Energy Convers Manage* 2017;**141**:325–38.
40. Morgan T, Grubb D, Santillan-Jimenez E, Crocker M. Conversion of triglycerides to hydrocarbons over supported metal catalysts. *Topics Catal* 2010;**53**(11-12):820–9.
41. Peng B, Yao Y, Zhao C, Lercher JA. Towards quantitative conversion of microalgae oil to diesel-range alkanes with bifunctional catalysts. *Angew Chem* 2012;**124**(9):2114–17.
42. Gosselink RW, Hollak SA, Chang SW, van Haveren J, de Jong KP, Bitter JH, et al. Reaction pathways for the deoxygenation of vegetable oils and related model compounds. *ChemSusChem* 2013;**6**(9):1576–94.
43. Lu J, Behtash S, FaheemM, Heyden A. Microkinetic modeling of the decarboxylation and decarbonylation of propanoic acid over Pd(111) model surfaces based on parameters obtained from first principles. *J Catal* 2013;**305**:56–66.
44. Bernas H, Eränen K, Simakova I, Leino AR, Kordás K, Myllyoja J, et al. Deoxygenation of dodecanoic acid under inert atmosphere. *Fuel* 2010;**89**(8):2033–9.
45. Hengst K, Arend M, Pfützenreuter R, Hoelderich WF. Deoxygenation and cracking of free fatty acids over acidic catalysts by single step conversion for the production of diesel fuel and fuel blends. *Appl Catal B: Environ* 2015;**174–175**(0):383–94.
46. Sotelo-Boyás R, Liu Y, Minowa T. Renewable diesel production from the hydrotreating of rapeseed oil with Pt/Zeolite and NiMo/Al_2O_3 Catalysts. *Ind Eng Chem Res* 2011;**50**(5):2791–9.
47. Yang X, Chatterjee S, Zhang Z, Zhu X, Pittman CU. Reactions of phenol, water, acetic acid, methanol, and 2-hydroxymethylfuran with olefins as models for bio-oil upgrading. *Ind Eng Chem Res* 2010;**49**(5):2003–13.
48. Zhang Z, Wang Q, Tripathi P, Pittman Jr CU. Catalytic upgrading of bio-oil using 1-octene and 1-butanol over sulfonic acid resin catalysts. *Green Chem* 2011;**13**(4):940–9.
49. Verma D, Kumar R, Rana BS, Sinha AK. Aviation fuel production from lipids by a single-step route using hierarchical mesoporous zeolites. *Energy Environ Sci* 2011;**4**(5):1667–71.
50. Wildschut J, Mahfud FH, Venderbosch RH, Heeres HJ. Hydrotreatment of fast pyrolysis oil using heterogeneous noble-metal catalysts. *Ind Eng Chem Res* 2009;**48**(23):10324–34.
51. Asikin-Mijan N, Lee HV, Juan JC, Noorsaadah AR, Taufiq-Yap YH. Catalytic deoxygenation of triglycerides to green diesel over modified CaO-based catalysts. *RSC Adv* 2017;**7**(73):46445–60.
52. Alsultan GA, Asikin-Mijan N, Lee HV, Albazzaz AS, Taufiq-Yap YH. Deoxygenation of waste cooking to renewable diesel over walnut shell-derived nanorode activated carbon supported CaO-La_2O_3 catalyst. *Energy Convers Manage* 2017;**151**(Suppl. C):311–23.

53. Zulkepli S, Juan JC, Lee HV, Rahman NSA, Show PL, Ng EP. Modified mesoporous HMS supported Ni for deoxygenation of triolein into hydrocarbon-biofuel production. *Energy Convers Manage* 2018;**165**:495–508.
54. Snåre M, Kubičková I, Mäki-Arvela P, Eränen K, Murzin DY. Heterogeneous catalytic deoxygenation of stearic acid for production of biodiesel. *Ind Eng Chem Res* 2006;**45** (16):5708–15.
55. Yang C, Nie R, Fu J, Hou Z, Lu X. Production of aviation fuel via catalytic hydrothermal decarboxylation of fatty acids in microalgae oil. *Bioresour Technol* 2013;**146**:569–73.
56. Immer JG, Kelly MJ, Lamb HH. Catalytic reaction pathways in liquid-phase deoxygenation of C18 free fatty acids. *Appl Catal A: Gen* 2010;**375**(1):134–9.
57. Lestari S, Mäki-Arvela P, Eränen K, Beltramini J, Max Lu GQ, Murzin DY. Diesel-like hydrocarbons from catalytic deoxygenation of stearic acid over supported Pd nanoparticles on SBA-15 catalysts. *Catal Lett* 2010;**134**(3):250–7.
58. Li W, Gao Y, Yao S, Ma D, Yan N. Effective deoxygenation of fatty acids over $Ni(OAc)_2$ in the absence of H_2 and solvent. *Green Chem* 2015;**17**(8):4198–205.
59. Vo TK, Kim WS, Kim SS, Yoo KS, Kim J. Facile synthesis of $Mo/Al_2O_3-TiO_2$ catalysts using spray pyrolysis and their catalytic activity for hydrodeoxygenation. *Energy Convers Manage* 2018;**158**:92–102.
60. Cao Y, Shi Y, Bi Y, Wu K, Hu S, Wu Y, et al. Hydrodeoxygenation and hydroisomerization of palmitic acid over bi-functional Co/H-ZSM-22 catalysts. *Fuel Process Technol* 2018;**172**:29–35.
61. Newman C, Zhou X, Goundie B, Ghampson IT, Pollock RA, Ross Z, et al. Effects of support identity and metal dispersion in supported ruthenium hydrodeoxygenation catalysts. *Appl Catal A: Gen* 2014;**477**(0):64–74.
62. Nie L, Resasco DE. Kinetics and mechanism of m-cresol hydrodeoxygenation on a Pt/SiO_2 catalyst. *J Catal* 2014;**317**(0):22–9.
63. Šimáček P, Kubička D, Šebor G, Pospíšil M. Fuel properties of hydroprocessed rapeseed oil. *Fuel* 2010;**89**(3):611–15.
64. Han J, Duan J, Chen P, Lou H, Zheng X. Molybdenum carbide-catalyzed conversion of renewable oils into diesel-like hydrocarbons. *Adv Synt Catal* 2011;**353**(14-15):2577–83.
65. Huber GW, O'Connor P, Corma A. Processing biomass in conventional oil refineries: production of high quality diesel by hydrotreating vegetable oils in heavy vacuum oil mixtures. *Appl Catal A: Gen* 2007;**329**(0):120–9.
66. Şenol Oİ, Ryymin EM, Viljava TR, Krause AOI. Effect of hydrogen sulphide on the hydrodeoxygenation of aromatic and aliphatic oxygenates on sulphided catalysts. *J Mole Catal A: Chem* 2007;**277**(1-2):107–12.
67. Şenol Oİ, Viljava TR, Krause AOI. Hydrodeoxygenation of methyl esters on sulphided $NiMo/\gamma\text{-}Al_2O_3$ and $CoMo/\gamma\text{-}Al_2O_3$ catalysts. *Catal Today* 2005;**100**(3-4):331–5.
68. Gutiérrez OY, Kaufmann C, Lercher JA. Influence of potassium on the synthesis of methanethiol from carbonyl sulfide on sulfided Mo/Al_2O_3 catalyst. *ChemCatChem* 2011;**3** (9):1480–90.
69. Fu J, Lu X, Savage PE. Hydrothermal decarboxylation and hydrogenation of fatty acids over Pt/C. *ChemSusChem* 2011;**4**(4):481–6.
70. Kim SK, Lee HS, Hong MH, Lim JS, Kim J. Low-temperature, selective catalytic deoxygenation of vegetable oil in supercritical fluid media. *ChemSusChem* 2014;**7**(2):492–500.
71. Shimada I, Kato S, Hirazawa N, Nakamura Y, Ohta H, Suzuki K, et al. Deoxygenation of triglycerides by catalytic cracking with enhanced hydrogen transfer activity. *Ind Eng Chem Res* 2017;**56**(1):75–86.

72. Emori EY, Hirashima FH, Zandonai CH, Ortiz-Bravo CA, Fernandes-Machado NRC, Olsen-Scaliante MHN. Catalytic cracking of soybean oil using ZSM5 zeolite. *Catal Today* 2017;**279**:168–76.

73. Xia Q, Zhuang X, Li MMJ, Peng YK, Liu G, Wu TS, et al. Cooperative catalysis for the direct hydrodeoxygenation of vegetable oils into diesel-range alkanes over Pd/$NbOPO_4$. *Chem Commun* 2016;**52**(29):5160–3.

74. Simakova IL, Simakova OA, Romanenko AV, Murzin DY. Hydrogenation of vegetable oils over Pd on nanocomposite carbon catalysts. *Ind Eng Chem Res* 2008;**47**(19):7219–25.

75. Fu J, Shi F, Thompson LT, Lu X, Savage PE. Activated carbons for hydrothermal decarboxylation of fatty acids. *ACS Catal* 2011;**1**(3):227–31.

76. Jin S, Xiao Z, Li C, Chen X, Wang L, Xing J, et al. Catalytic hydrodeoxygenation of anisole as lignin model compound over supported nickel catalysts. *Catal Today* 2014;**234**:125–32.

77. Santillan-Jimenez E, Crocker M. Catalytic deoxygenation of fatty acids and their derivatives to hydrocarbon fuels via decarboxylation/decarbonylation. *J Chem Technol Biotechnol* 2012;**87**(8):1041–50.

78. Zhang J, Li Y, Zhang Y, Chen M, Wang L, Zhang C, et al. Effect of support on the activity of Ag-based catalysts for formaldehyde oxidation. *Sci Rep* 2015;**5**:12950.

79. Luo W, Sankar M, Beale AM, He Q, Kiely CJ, Bruijnincx PCA, et al. High performing and stable supported nano-alloys for the catalytic hydrogenation of levulinic acid to γ-valerolactone. *Nat Commun* 2015;**6**:6540.

80. Oi LE, Choo MY, Lee HV, Ong HC, Hamid SBA, Juan JC. Recent advances of titanium dioxide (TiO_2) for green organic synthesis. *RSC Adv* 2016;**6**(110):108741–54.

81. Saadi A, Merabti R, Rassoul Z, Bettahar MM. Benzaldehyde hydrogenation over supported nickel catalysts. *J Mole Catal A: Chem* 2006;**253**(1-2):79–85.

82. Ardiyanti AR, Gutierrez A, Honkela ML, Krause AOI, Heeres HJ. Hydrotreatment of wood-based pyrolysis oil using zirconia-supported mono- and bimetallic (Pt, Pd, Rh)catalysts. *Appl Catal A: Gen* 2011;**407**(1–2):56–66.

83. Kent PD, Mondloch JE, Finke RG. A four-step mechanism for the formation of supported-nanoparticle heterogenous catalysts in contact with solution: the conversion of Ir(1,5-COD) Cl/γ-Al_2O_3 to Ir(0)$\sim$170/γ-Al_2O_3. *J Am Chem Soc* 2014;**136**(5):1930–41.

84. Mäki-Arvela PI, Rozmysłowicz B, Lestari S, Simakova O, Eränen K, Salmi T, et al. Catalytic deoxygenation of tall oil fatty acid over palladium supported on mesoporous carbon. *Energy Fuels* 2011;**25**(7):2815–25.

85. Yan Y, Miao J, Yang Z, Xiao FX, Yang HB, Liu B, et al. Carbon nanotube catalysts: recent advances in synthesis, characterization and applications. *Chem Soc Rev* 2015;**44**(10):3295–346.

86. Pavel CC, Palkovits R, Schüth F, Schmidt W. The benefit of mesopores in ETS-10 on the vapor-phase Beckmann rearrangement of cyclohexanone oxime. *J Catal* 2008;**254**(1):84–90.

87. Linares N, Silvestre-Albero AM, Serrano E, Silvestre-Albero J, García-Martínez J. Mesoporous materials for clean energy technologies. *Chem Soc Rev* 2014;**43**(22):7681–717.

88. Chandra D, Bhaumik A. Super-microporous TiO synthesized by using new designed chelating structure directing agents. *Micropor Mesopor Mater* 2008;**112**:533–41.

89. Guldin S, Hüttner S, Tiwana P, Orilall MC, Ülgüt B, Stefik M, et al. Improved conductivity in dye-sensitised solar cells through block-copolymer confined TiO_2 crystallisation. *Energy Environ Sci* 2011;**4**:225.

90. Chen L, Huang C, Xu G, Hutton SL, Miao L. Macroporous TiO_2 foam with mesoporous walls. *Mater Character* 2013;**5**:8–12.
91. Rodrigues EG, Pereira MFR, Órfão JJM. Glycerol oxidation with gold supported on carbon xerogels: tuning selectivities by varying mesopore sizes. *Appl Catal B: Environ* 2012;**115–116**:1–6.
92. Benson TJ, Hernandez R, French WT, Alley EG, Holmes WE. Elucidation of the catalytic cracking pathway for unsaturated mono-, di-, and triacylglycerides on solid acid catalysts. *J Mole Catal A: Chem* 2009;**303**(1):117–23.

Further reading

Lin YC, Huber GW. The critical role of heterogeneous catalysis in lignocellulosic biomass conversion. *Energy Environ Sci* 2009;**2**(1):68–80.

Chapter 10

Third generation biofuels: an overview

João Moreira Neto[1], Andrea Komesu[2], Luiza Helena da Silva Martins[3], Vinicius O.O. Gonçalves[4], Johnatt Allan Rocha de Oliveira[5] and Mahendra Rai[6]

[1]Department of Engineering, Federal University of Lavras (UFLA), Lavras, Brazil, [2]Department of Marine Sciences, Federal University of São Paulo (UNIFESP), Santos, Brazil, [3]Department of Natural Sciences and Technology, State University of Pará (UEPA), Belém, Brazil, [4]Institute of Chemistry, Department of Physical Chemistry, Federal University of Rio de Janeiro, Rio de Janeiro, Brazil, [5]Institute of Health Sciences, Nutrition College, Federal University of Pará (UFPA), Belém, Brazil, [6]Nanobiotechnology Lab, Department of Biotechnology, Sant Gadge Baba Amravati University, Amravati, India

10.1 Introduction

Rapid industrialization and high population growth are two major factors contributing to the global energy crisis.[1] Today, approximately 88% of worldwide energy consumption is a consequence of the combustion of fossil fuels.[2] Unfortunately, the massive use of fossil fuels has led to problems such as the depletion of reserves, price fluctuations, negative environmental impacts, and climatic change.[1] Therefore alternative energy sources that are readily available, facilely accessible, and environment-friendly are very much needed.[3]

Biofuels have emerged as a promising alternative to fossil fuel resources. The term "biofuels" refers to energy enriched chemicals generated through biological processes or derived from the biomass of living organisms such as microalgae, plants, and bacteria.[4] The energy in biofuel is stabilized through the process of biological carbon fixation, in which carbon dioxide (CO_2) is converted into sugar that is only found in living organisms such as plants. In contrast to fossil fuels, biofuel is produced by, or derived from, living organisms in a relatively short period of time rather than being derived by the decomposition of organic matter over several million years.[5]

Currently, there are three generations of biofuels that have flourished and that are based on different feedstock. "First generation" biofuels such as

Sustainable Bioenergy. DOI: https://doi.org/10.1016/B978-0-12-817654-2.00010-1

bioethanol and biodiesels are produced directly from starch, sugar, and oil present in food crops like corn, wheat, and soybeans.[5] Important characteristics of "first generation" biofuels include their ability to be blended with petroleum-based fuels and their efficiency in internal combustion engines, as well as their compatibility with flexible fuel vehicles.[2] The most controversial issue with first generation biofuels is that it is necessary to choose between either fuel or food alternatives.[5] Additionally, food crop feedstocks require large agricultural areas to produce sufficient quantities of biomass. In terms of environmental issues, such increased agricultural yield and subsequent harvesting may lead to enhanced land clearing, loss in biodiversity due to habitat destruction, water depletion, and air pollution.[2]

"Second generation" biofuels are generally produced from nonfood crops mainly lignocellulosic feedstocks such as grass, wood, and other organic wastes.[5] The advantages of these feedstocks are their ease of availability, they do not compete with the food industry, and eventually have a lesser impact on the environment.[1] Nevertheless, second generation biofuels also face steep challenges in the form of technical difficulties during pretreatment processes and inefficient conversion of lignocellulosic materials due to their complex structures.[2]

In the search for viable and cost-effective alternatives to fossil fuels, studies have reported superior capabilities of algae-derived biomass for the production of an improved version of biofuels: "third generation" biofuels.[2] Algae (Alien to Light Green Antenna Entity) are highly diverse micro- and macro-plants that are found almost everywhere on planet Earth. They play an important role in many ecosystems; providing food, feed, and vital nutrients as well as supplying up to 60% of the required oxygen to all living beings vital for their survival.[6]

Lately, algal cultures have been used to produce vitamins, feed, and other products on an industrial scale. However, the rapid growth in world energy demand has led to the development and analysis of efficient sources capable of producing fuels and chemicals. Thus microalgae are reported to be a promising alternative for the production of fuels and chemicals due to their high photosynthetic conversion efficiency. These organisms are unicellular, contain carbon, hydrogen, oxygen, and nitrogen, and are categorized as aquatic biomass. The growth of microalgae depends heavily on the availability and intensity of light, the availability of nutrients like phosphorus and nitrogen, CO_2 and O_2 levels, temperature, and the type of culture system used.[7]

Algae represent a promising, everlasting source of fuel and other valuable products.[6] Algae have the ability to produce crude oil which may, then, be easily processed into diesel and gasoline. Particular algae species which may be categorized as either microalgae or macroalgae may be genetically modified so that the carbon metabolic pathway facilitates the production of important end-products like ethanol.[2] In addition, algae have the ability to produce more energy per acre of land compared to other conventional feedstock crops such as sugarcane and corn.[8] For these reasons, microalgae are touted to have the maximum potential to act as a substitute for petroleum-derived transport diesel without adversely affecting food supply and other crop products.[2]

The main goal of this chapter is to explore the potential of algae for the production of different biofuels. Moreover, the physicochemical composition of microalgae and types of treatments applied for the deconstruction of microalgae are discussed.

10.2 Technologies for biomass conversion into biofuel

The most widely used processes for transforming microalgal biomass into different energy sources can be classified as biochemical conversion, thermochemical conversion, chemical reactions, and direct combustion, as shown in Fig. 10.1.[9]

Biochemical conversion is a process of converting biomass into others fuels through anaerobic digestion and alcoholic fermentation. The example of processes belonging to biochemical conversion include fermentation for ethanol production[10] and anaerobic digestion for methane[11] and hydrogen production.[12] Thermochemical conversion is a process of thermal decomposition of biomass into fuels products. Such thermochemical processes include gasification,[13] pyrolysis,[14] and liquefaction.[14] An example of chemical conversion process involves the conversion of the extracted lipid from microalgae cells to biodiesel via a transesterification reaction.[15] In the direct

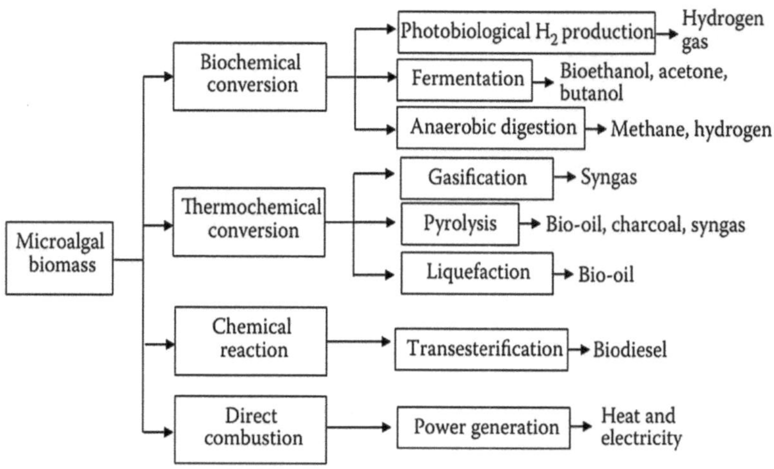

FIGURE 10.1 Biofuels production via the microalgal biomass conversion processes. *Adapted from Wan B, Li Y, Wu N, Lan CQ. CO_2 bio-mitigation using microalgae.* Appl Microbiol Biotechnol *2008;**79**:707–18 with copyright permission from Springer Nature.*

combustion, the algal biomass is burned in the presence of air, converting the chemical energy of the algal biomass into heat or electricity.[14]

10.3 Potential of microalgae as raw material for biofuels production

Biofuels could be an alternative source of energy to meet the energy needs generated by the ever increasing world population.[4] This scenario has led to the development and analysis of efficient sources to produce fuels and chemicals.[16] The emergence of third generation bioethanol provides more benefits compared to first and second generation bioethanol. Third generation bioethanol is focused on the use of marine organisms such as algae.[1]

Several studies performed in the past have demonstrated that among different sources, microalgae have significant potential for the production of bioenergy and chemicals.[16,17] This is due to their high lipid and carbohydrate contents, high proton conversion, easy cultivation in a wide variety of water environments, relatively low land usage and high carbon dioxide (CO_2) absorption.[18] With new approaches involving the production of microbial biofuels, the cultivation of microalgae has been recommended as a possible strategy to obtain higher amounts of biomass for the production of fuel. This biomass could be obtained by culturing biofilms of microalgae or cyanobacteria.[19,20]

The production of biofuels can be achieved by several species of algae, but species rich in carbohydrates, proteins, and lipids, are preferably used in the production of biofuels.[21] Microalgae are considered as potential alternative feedstocks for biofuel production as they are a sustainable energy source and do not compete with other edible feedstocks. The cultivation of microalgae also creates a carbon sink for greenhouse gasses, thus, making it environment-friendly.[3]

Sugar is the basic molecular substrate for the production of bioethanol and biomethanol.[22] The use of photosynthetic organisms as a source of biofuel is cheap and feasible, and is based on the main component found in this type of material; carbohydrate.[21] Compared to other traditional biofuel crops, it is a better source for oil extraction and may be utilized to synthesize a multitude of diverse biofuels such as biomethane (via anaerobic digestion) and biodiesel (using microalgal oil), as well as biohydrogen (via photobiological synthesis).[8] Their photosynthetic efficiency is higher than that of other plants, and some species are considered to be among the fastest growing plants in the world.[23] It has been reported that the photosynthetic efficiencies of algae range from 3% to 8%, compared with the 0.5% of many terrestrial crops.[24]

Algae are simple plants ranging from microscopic (microalgae) to large seaweeds (macroalgae), from small green dots to larger cyanobacteria and the tinsel diatoms. There are nine major groups of algae which include cyanobacteria (*Cyanophyceae*), green algae (*Chlorophyceae*), diatoms (*Bacillariophyceae*), yellow-green algae (*Xanthophyceae*), golden algae (*Chrysophyceae*), red algae (*Rhodophyceae*), brown algae (*Phaeophyceae*), dinoflagellates (*Dinophyceae*), and "picoplankton" (*Prasinophyceae* and *Eustigmatophyceae*).[25]

Microalgae like *Spirulina*, *Chlorella*, *Dunaliella*, and *Haematococcus* are currently cultivated commercially to produce photosynthetically grown biomass from a few tons to several hundred tons annually. They grow at an exceptionally fast rate; 100 times faster than terrestrial plants, and they can double their biomass in less than 1 day.[26] Harvesting microalgal biomass is a major challenge because of their small size and their low concentration in culture media. In the overall production process, microalgae are initially grown to reach a maximum biomass concentration of 0.02%–0.5%. From the culture medium, the biomass is concentrated to 15%–20%, either in a single step or in a series of concentration steps, before they can be processed further via drying, extraction, or other downstream processing steps.[8,27,28]

10.4 Physicochemical composition of microalgae

The existence of nearly 300,000 species of algae, with varied oil contents and growth rates, was reported by Scott et al.[29] Microalgal biomass is mainly constituted of lipids, carbohydrates, and proteins. A microscope is required in order to observe microalgae because of their size, being either unicellular or multicellular, but always smaller than 0.4 mm in diameter. The category includes groups such as diatoms and cyanobacteria that have high growth rates.[5] The three most abundant classes of microalgae are *Bacillariophyceae*, *Chlorophyceae*, and *Chrysophyceae*.[5]

Different microalgal species have significantly different compositions, where the average lipid oil content typically varies between 8% and 31%.[2] The carbohydrate content was also found to be relatively high which can reach up to 50% of the dry weight for some species such as *Scenedesmus*, *Chlorella*, and *Chlamydomonas*.[1] Table 10.1 summarizes the dry weight composition of protein, carbohydrate, and lipid of some species of microalgae.

Factors such as light, temperature, nutrient content, pH, O_2, and CO_2 level, salinity, and toxic chemicals influence the lipid and carbohydrate contents of microalgae.[1] The lipid content of microalgae varies in accordance with the culture conditions. It is possible to increase the lipid concentration by almost 80% above natural levels, by optimizing the growth determining factors.[5] Some of the common components in the cell wall of microalgae are cellulose, protein, lignin, pectin, hemicelluloses, and other carbohydrates which can be converted into monomers through acid or enzymatic hydrolysis to produce bioethanol.[1]

TABLE 10.1 Composition of different species of microalgae on a dry matter basis.

Microalgal species	Protein	Carbohydrate	Lipid
Anabaena cylindrica	43–56	25–30	4–7
Chlamydomonas reinhardtii	48	17	21
Chlorella pyrenoidosa	57	26	2
Chlorella vulgaris	51–58	12–17	14–22
Dunaliella bioculata	49	4	8
Porphyridium cruentum	28–39	40–57	9–14
Prymnesium parvum	28–45	25–33	22–39
Scenedesmus dimorphus	8–18	21–52	16–40
Scenedesmus obliquus	50–56	10–17	12–14
Spirogyra sp.	6–20	33–64	11–21
Spirulina maxima	60–71	13–16	6–7
Spirulina platensis	46–63	8–14	4–9
Synechococcus sp.	63	15	11
Tetraselmis maculata	52	15	3

Source: Adapted from Um BH, Kim YS. Review: a chance for Korea to advance algal-biodiesel technology. *J Ind Eng Chem* 2009;**15**:1–7 with copyright permission from Elsevier.

10.5 General use of bioethanol

Bioethanol is mostly used in motor vehicles as fuel and as a gasoline additive. More than any other country, Brazil relies on ethanol as a motor fuel for several motor vehicles. Gasoline sold in Brazil contains at least 25% anhydrous ethanol. Hydrated ethanol (95% ethanol) can be used as fuel in more than 90% of new cars sold in the country. The United States uses mixtures of ethanol/gasoline, known as Gasohol (with a maximum of 10% ethanol), and 85% ethanol. Ethanol can also be used as rocket fuel in lightweight rocket-powered aircrafts.[30] Australian law limits the use of pure ethanol from sugarcane residues to 10% in automobiles. It has been recommended that older cars should have their valves updated or replaced.[30]

Direct hydrated ethanol as an automotive fuel has been widely used in Brazil for pure ethanol vehicles and more recently for flexible vehicles. Ethanol fuel is distilled near to the azeotropic mixture of 95.63 wt% ethanol and 4.73 wt% water. By the end of 2012, there were approximately 20 million flexible vehicles running on Brazilian highways. Hydrated ethanol imposes a limitation on the normal operation of these vehicles because the lower evaporative pressure of ethanol (compared to gasoline) causes problems when the engine starts cold at temperatures below 288K. For this reason, both pure and flex-fuel vehicles are built with an additional small gasoline reservoir inside the engine compartment to aid in starting the engine when cold, initially by injecting gasoline.[30]

10.6 Bioethanol production

In general, bioethanol production from biomass involves pretreatment, enzymatic hydrolysis, fermentation, and distillation.[10] The pretreatment of lignocellulosic feedstocks is a necessary step to unwind the cellulose from the hemicellulose and lignin in which cellulose is embedded, to make the cellulose more susceptible in order to be easily accessed by enzymes during enzymatic hydrolysis, and to enhance the rate and yield of reducing sugars.[31,32]

Based on the type of treatment process involved, lignocellulosic biomass pretreatment methods are broadly classified into different categories: physical (e.g., milling, microwave, ultrasound, etc.,), chemical (e.g., dilute acid, organosolv, mild alkali, ionic liquid, etc.,), physicochemical (e.g., steam explosion, liquid hot water, ammonia based, oxidative pretreatment, etc.,), and biological (e.g., bacterial, brown, white, and soft-rot fungi, etc.,) methods.[32] Such pretreatments, depending on the process conditions, can generate a range of inhibitory byproducts (e.g., hydroxymethylfurfural, furfural, phenolic compounds, or organic acids) which can impact on the viability of yeast in subsequent fermentation processes.[33,34] Due to the absence of lignin and low hemicellulose content, algae generally require relatively mild conditions for processing as compared to lignocellulosic biomass.[5]

The composition of carbohydrates in algal biomass may differ significantly from species to species.[1] It is of great importance to select algae with high carbohydrate productivity as well as suitable sugar composition for bioethanol production. Although macroalgae are a potential source of sugar, the majority of carbohydrates in macroalgae biomass (e.g., brown macroalgae) are present in the form of polysaccharides such as laminarin, mannitol, carrageenan, and alginate.[34] The degradation of these polysaccharides is slow and requires specific enzymes.[35] The development of conversion methods to release monosaccharides from these polysaccharides and the choice of industrial microorganisms for bioethanol production from macroalgae are bottlenecks to the feasibility of the utilization of seaweed as a feedstock.[34]

Wargacki et al.[36] were capable of co-fermenting alginate, mannitol, and glucan from a brown algal species (*Saccharina japonica*) using *Escherichia coli*, yielding 0.41 g of ethanol/g of carbohydrate sugars. Sunwoo et al.[37] performed a fermentation using mixed yeast, i.e., *Saccharomyces cerevisiae* KCTC 1126 and *Pichia angophorae* KCTC 17574, to galactose and mannitol from waste seaweed obtained from Gwangalli beach, Busan, Korea. The maximum ethanol concentration and ethanol yield obtained were 13.54 g/L and 0.45%, respectively.

Carbohydrates (cellulose) in microalgae biomass are mainly present in the cell wall and starch in the plastids which can be readily converted into fermentable sugars that more suitable for ethanol production.[38] The harvesting of microalgae is more expensive than macroalgae. Macroalgae can be harvested using nets which requires less energy, while microalgae are harvested by conventional processes which include filtration, flocculation, centrifugation, sedimentation, froth floatation, and ultrasonic separation.[39] Several studies have been conducted on the utilization of different micro- and macroalgae for the production of bioethanol, such as *Chlorella vulgaris*,[40] *Gracilaria verrucosa*,[41] *Gracilaria salicornia*,[42] *Gelidium amansii*,[43] *Sargassum* spp.,[10] *Laminaria digitate*,[34] *Chondrus crispus*,[34] *Palmaria palmata*,[34] and *Kappaphycus alvarezii*.[44]

Saccharification is usually a rate-limiting step in biofuel production using lignocellulosic materials or microalgal biomass that contain a cellulose source.[45] Saccharification or hydrolysis consist of the depolymerization of the polysaccharide contents (such as starch, cellulose, alginates, fucans, laminarin, and carrageenans) in algal cell walls into free monomer molecules which can be fermented into bioethanol.[1] The absence or near absence of lignin and hemicellulose contents in algae, in comparison to lignocellulosic biomass, makes the enzymatic hydrolysis of algal cellulose easier.[46]

The main methods that are applied to produce sugars from microalgae can be categorized into two groups: enzymatic saccharification and acid saccharification.[45] In acid hydrolysis, a wide range of acids have been used, of which sulfuric acid (H_2SO_4) is most preferred. It has been found that the polysaccharides of the three classes of macroalgae (brown, red, and green) can be effectively hydrolyzed to monosaccharides by treatment with diluted H_2SO_4 at high temperatures. The acid is responsible for breaking the bonds of the long chains of the polysaccharides. In the initial stage, the destruction of hydrogen bonds occurs in order to break up the chains of polysaccharides, thus transforming them into molecules in the amorphous state. Polysaccharides are extremely susceptible to acid hydrolysis at this point. The acid will then serve as a catalyst where the cleavage of the polysaccharide will occur by the hydrolysis of the glycosidic bonds. At the end of this process, any addition or dilution with water at a moderate temperature will provide complete hydrolysis and the hydrolysate will be rich in monosaccharides.[1]

In enzymatic hydrolysis, cellulases are employed as enzymes to degrade polysaccharides and can be categorized into three major types, namely endoglucanases, exoglucanases, and β-glycosidase. Among these, endoglucanases have the ability to hydrolyze the complex sugars of the raw material, attacking the inner parts of the amorphous region of the cellulose. Exoglucanases degrade cellulose by the cleavage of cellobiose units (union of two glucose molecules) from the nonreducing end of a cellulose fiber which allows for enzymatic attack. With the combined efforts of β-glycosidase, the residues of cellobiose finally split into two units of glucose.[1]

Although chemical hydrolysis usually gives a higher sugar production rate, this method requires harsh reaction conditions and the hydrolysates may contain inhibitory byproducts (such as hydroxymethylfurfural and furfural) to the subsequent bioethanol fermentation. On the other hand, enzymatic hydrolysis produces high amounts of fermentable sugars under mild conditions without producing sugar-degradation products or toxic byproducts.[47] In order to improve the production of high concentrations of fermentable sugars, acid and enzymatic saccharification can be combined. As red seaweeds contain galactan and glucans and galactan cannot be hydrolyzed by the same enzyme used for the hydrolysis of glucans, Yanagisawa et al.[48] combined the acid hydrolysis of galactan to produce galactose with the subsequent enzymatic hydrolysis of glucan to obtain glucose.

There are different groups of microorganisms like yeast, bacteria, and fungi that can be used for the fermentation of saccharified algal biomass under anaerobic conditions for the production of bioethanol.[39] *S. cerevisiae* is the most commonly employed strain in the fermentation of bioethanol due to its high selectivity, low accumulation of byproducts, high ethanol yield, and high rate of fermentation.[1] In order to increase the production rate in the fermentation process, new process configurations were developed such as separated hydrolysis and fermentation (SHF) and simultaneous saccharification and fermentation (SSF). In SHF, the saccharification and fermentation of the biomass are performed in two distinct reactors. The saccharification is conducted first to degrade the biomass into monomer sugars using an enzyme. This process is followed by the fermentation reaction which utilizes the sugars formed in the hydrolysis stage. However, one major problem associated with the SHF process is the end-product inhibition by sugars that are formed during the hydrolysis stage.[1] In SSF, the saccharification and fermentation processes are conducted simultaneously in a single reactor. During the reaction, the biomass, enzyme, and yeast are put together in an orderly manner so that the sugars released are rapidly converted into bioethanol. SSF can limit end-product inhibition by removing the residual sugar.[1] Thus SSF processes can achieve better ethanol yields than methods that apply separate hydrolysis and fermentation processes.[31]

El-Dalatony et al.[49] studied both SHF and SSF processes for bioethanol production from microalgal (*Chlamydomonas mexicana*) biomass. SSF was selected as an efficient process to enhance bioethanol yields through repeated batches using immobilized *S. cerevisiae* YPH499 cells. The combined sonication and enzymatic hydrolysis of *C. mexicana* generated 10.5 and 8.48 g/L of ethanol in SSF and SHF, respectively. Similarly, Tan and Lee[50] found that the SSF of seaweed solid wastes with *S. cerevisiae* was more effective as compared to SHF for bioethanol production. They were able to obtain a bioethanol yield of 55.9% for the SHF process and a yield of 90.9% for the SSF process.

10.7 Production of fuel from microalgae through pyrolysis

Thermochemical processes depend on the degradation of biomass following chemical reactions occurring at moderate to high temperatures. There are three common thermochemical processes for converting biomass into biofuels and chemicals: pyrolysis, liquefaction, and gasification. The main thermochemical route for biomass conversion is pyrolysis which is conducted in the absence of oxygen and at temperatures between 300°C and 700°C. This process converts biomass into pyrolytic oil (or bio-oil), char (also called biochar), and light gases (H_2, CO, CO_2, and C1–C4 hydrocarbons). Biochar is similar to conventional coal but contains some oxygen. Pyrolytic bio-oil is a complex mixture containing several oxygenated compounds. Generally, pyrolysis bio-oils contain mainly water, aldehydes, carboxylic acids, carbohydrates, phenols, furfural, alcohols, and ketones. The water content in bio-oil is relatively high (up to 40 wt.%) which often creates an aqueous-rich phase and an organic-rich phase.[51]

The operational conditions for pyrolysis are varied and the process is mostly classified into fast, flash, and slow and catalytic pyrolysis. The operational conditions can be set in order to favor the production of biochar, bio-oil, or gas fractions. For example, fast pyrolysis and flash pyrolysis (low residence times) produce higher yields of bio-oil (up to 80%). In contrast, slow pyrolysis produces larger amounts of char due to the increased degradation time of the biomass components. Another possibility is to use microwave assisted pyrolysis which shows advantages over traditional processes such as rapid and uniform heating which reduces residence times and accelerates chemical reactions. Regarding the use of fuels and chemicals, bio-oil is preferable due to the variety of compounds in its composition and its flexibility in upgradation, thus being a platform for both renewable chemicals and fuels. In addition, the use of a catalyst is generally related to decreases in the reaction temperature, the removal of the oxygen content, and obtaining more interesting fractions related to certain fuels.

The dynamics of biomass production as well as its habitat are important factors for a sustainable chain of renewable fuel. Thus the production of biofuel from algae is of great interest given the high growth rate of some species, their interesting ability to sequester CO_2, their robustness in growing in places that do not compete with arable lands, and their noncompetition with human alimentation.[51] Promising aspects of the pyrolysis of many algal species such as *Chlorella protothecoides*, *C. vulgaris*, *Dunaliella*, *Emilianiahuxleyi*, *Nannochloropsis residue*, *Plocamium*, *Sargassum*, *Spirulina*, *Synechococcus*, *Scenedesmus*, *Spirulina platensis*, *Tertiolecta*, and *Tetraselmis* have been reported. Studies showed encouraging results that depict potential future industrial applications.[52]

Since lignocellulosic biomasses and algae are different in composition; algal bio-oils are quite different compared to lignocellulosic bio-oils. The pyrolysis of algae produces a bio-oil with a considerably higher number of linear hydrocarbons, fatty oxygenates, and nitrogenated compounds resulting from the cleavage of lipid and protein bonds. In addition, since algae have no lignin in their composition, the number of phenolic compounds is low. It is also reported that the pyrolysis of algae produces less biochar and a more stable bio-oil with improved properties which could be more feasibly upgraded.[53]

It is noteworthy that the use of micro- and macroalgae as raw materials for biofuel production is constantly increasing. However, large-scale aspects concerning environmental impacts need to be further studied. Moreover, the direct application of bio-oil (either from algae or lignocellulose) is of major concern due to its high oxygen content which leads to several undesired properties such as poor heating values, corrosiveness, and high viscosity. In this context, bio-oil production through catalytic pyrolysis and bio-oil hydrotreatment upgrading are major research topics in the production of liquid fuels or chemicals in the biorefinery context. Concerning the complexity of bio-oil, upgrading algae bio-oil shows advantages compared to lignocellulosic bio-oil.[54]

10.8 Protein biofuels

Microalgae offer great potential for the large-scale and cost-effective production of recombinant proteins. Huo et al.[55] studied the feasibility of the use of proteins as raw material for the bioreflex process through the application of a metabolic engineering strategy centralized in nitrogen. They integrated the merits of microorganisms, including rapid growth and ease of cultivation, with those from higher plants in posttranslational modification and photosynthesis.[6,56]

Large-scale protein production using microalgae through natural selection under frequent or continuous harvesting conditions may favor robust and fast-growing microorganisms which generally contain a high protein content and are fully adapted to the local environment. Biofuels are currently produced from the carbohydrates and lipids found in this raw material. Proteins, in contrast, have not been used to produce fuels due to the difficulties of deaminating protein hydrolysates. Thus recombinant strategies have proven to be viable for the large-scale production of enzymes that can be used to produce biofuels. Huo et al.,[55] as mentioned by Gajraj et al.,[6] applied metabolic engineering to generate *E. coli* that can deaminate protein hydrolysates, allowing cells to convert proteins into C4 and C5 alcohols which present a theoretical yield of 56%. They introduced three exogenous cycles of transamination and deamination which provided an irreversible metabolic force to drive the deamination reactions to conclusion.[6]

Huo et al.[55] demonstrated that *S. cerevisiae*, *E. coli*, *Bacillus subtilis*, and microalgae can be used as protein sources; producing up to 4035 mg/L of alcohols from biomass containing ~22 g/L amino acids. These results showed the feasibility of using proteins in biorefineries, in which high protein microalgae could be used as feedstocks, with the possibility of maximizing algal growth and total CO_2 fixation.

To test the viability of algal and bacterial proteins as raw materials, Huo et al.[55] cultivated green algae (*C. vulgaris*), red algae (*Porphyridium purpureum*), blue-green algae (*S. platensis*), and cyanobacterium (*Synechococcus elongatus*), as well as *E. coli* and *B. subtilis*, in 1 L of fluorine or a 30 L tank. The biomass was harvested, digested with protease, and used as a feedstock for the production of biofuel with the engineered *E. coli* strain YH83. In all experiments, the protein concentration was adjusted to 21.6 g/L which was equivalent to the amount of protein in 4% of yeast extract.[6]

10.9 Direct combustion

For this process, the algal biomass is burned in the presence of air in a furnace, boiler, or steam turbine, converting the chemical energy of the algal biomass into heat or electricity. The direct combustion of microalgae is only feasible with a moisture content of less than 50% of its dry weight. However, the process suffers from a limitation because it requires pretreatment of the algae (like drying) which can lead to a decrease in energy efficiency. Instead, the production of algae and coal-biomass is an effective way to generate electricity. During algal biomass aging, algae are cultivated and fed with a CO_2 recycling stream from a biorefinery plant, and the cultured algae are fed to the plant together with charcoal.[14]

10.10 Ultrasonication to aid biofuel yields

According to Chye et al.,[2] the ultrasound process requires the incorporation of acoustic energy with frequencies in the range of 10 kHz to 20 MHz which is applied to form microbubbles in constant expansion and shrinkage. The microbubbles destabilize as the acoustic energy exceeds a certain limit. When it resonates with the natural frequency of the microbubbles, it collapses, causing high-speed microdata of more than 100 m/s which carry more than 100 MPa of pressure toward the surface of the biomass. This causes cavitation and the process improves biochemical reactions and may help maximize lipid yields during the lipid extraction of microalgae or cause the rupture of the cell wall to aid in the hydrolysis step for bioethanol fermentation or anaerobic digestion. Suali and Sarbatly[57] demonstrated that ultrasound is able to reduce the lipid extraction time by one-tenth of the conventional methods and reach a 50% to more than 500% yield.

10.11 Conclusions

Microalgae and macroalgae are promising raw materials for the production of biofuels. Some technologies discussed in this chapter show the suitability and applicability of biofuel products which may be alternatives to fossil energy. The physicochemical composition of algal biomass based on their lipid, carbohydrate, and protein contents and types of treatments applied for the deconstruction of microalgae are crucial factors for obtaining products of interest in the chemical, food, and pharmaceutical industries. Although microalgae are reported to be a promising alternative for the production of fuels and chemicals due to their high photosynthetic conversion efficiency, further research and development are necessary to establish an economical industrial-scale production of algal biofuels.

References

1. Jambo SA, Abdulla R, Azhar SHM, Marbawi H, Gansau JA, Ravindra P. A review on third generation bioethanol feedstock. *Renew Sustain Energy Rev* 2016;**65**:756–69.
2. Chye JTT, Jun LY, Yon LS, Pan S, Danquah MK. Biofuel production from algal biomass. In: Konur O, editor. *Bioenergy and biofuels*. Boca Raton, FL: CRC Press, Taylor & Francis Group; 2018. p. 87–117.
3. Leong W, Lim J, Lam M, Uemura Y, Ho Y. Third generation biofuels: a nutritional perspective in enhancing microbial lipid production. *Renew Sustain Energy Rev* 2018;**91**:950–61.
4. Rodionova MV, Poudyal RS, Tiwari I, Voloshin RA, Zharmukhamedov SK, Nam HG, et al. Biofuel production: challenges and opportunities. *Int J Hydrogen Energy* 2017;**42**:8450–61.
5. Alaswad A, Dassisti M, Prescott T, Olabi AG. Technologies and developments of third generation biofuel production. *Renew Sustain Energy Rev* 2015;**51**:1446–60.
6. Gajraj RS, Singh GP, Kumar A. Third-generation biofuel: algal biofuels as a sustainable energy source. In: Kumar, et al., editors. *Biofuels: greenhouse gasmitigation and global warming, A*. Springer (India) Pvt. Ltd; 2018. p. 307–25.
7. Moncada J, Tamayo JA, Cardona CA. Integrating first, second, and third generation biorefineries: incorporating microalgae into the sugarcane biorefinery. *Chem Eng Sci* 2014;**118**:126–40.
8. Chisti Y. Biodiesel from microalgae. *Biotechnol Adv* 2007;**25**(3):294–306.
9. Wang B, Li Y, Wu N, Lan CQ. CO_2 bio-mitigation using microalgae. *Appl Microbiol Biotechnol* 2008;**79**:707–18.
10. Borines MG, Leon RL, Cuello JL. Bioethanol production from the macroalgae *Sargassum* spp. *Bioresour Technol* 2013;**138**:22–9.
11. Sangeetha P, Babu S, Rangasamy R. Potential of green alga *Chaetomorpha litorea* (Harvey) for biogas production. *Int J Curr Sci* 2011;**1**:24–9.
12. Park JH, Yoon JJ, Park HD, Kim YJ, Lim DJ, Kim SH. Feasibility of biohydrogen production from *Gelidium amansii*. *Int J Hydrogen Energy* 2011;**36**:13997–4003.
13. Onwudili JA, Lea-Langton AR, Ross AB, Williams PT. Catalytic hydrothermal gasification of algae for hydrogen production: composition of reaction products and potential for nutrient recycling. *Bioresour Technol* 2013;**127**:72–80.

14. Milano J, Ong HC, Masjuki HH, Chong WT, Lam MK, Loh PK, et al. Microalgae biofuels as an alternative to fossil fuel for power generation. *Renew Sustain Energy Rev* 2016;**58**:180–97.
15. Um BH, Kim Y-S. Review: a chance for Korea to advance algal-biodiesel technology. *J Ind Eng Chem* 2009;**15**:1–7.
16. Sánchez-Tuirán E, El-Halwagi MM, Kafarov V. Integrated utilization of algae biomass in a biorefinery based on a biochemical processing platform. In: Stuart PR, El-Halwagi MM, editors. *Integrated Biorefineries design, analysis, and optimization*. Boca Raton, FL: CRC Press; 2012. p. 707–26.
17. Bahadar A, Khan MB. Progress in energy from microalgae: a review. *Renew Sustain Energy Rev* 2013;**27**:128–48.
18. Singh A, Olsen SI. A critical review of biochemical conversion, sustainability and life cycle assessment of algal biofuels. *Appl Energy* 2011;**88**:3548–55.
19. Demirbas A. Political, economic and environmental impacts of biofuels: a review. *Appl Energy* 2009;**86**:S108–17.
20. Heiman K. Novel approaches to microalgal and cyanobacterial cultivation for bioenergy and biofuel production. *Curr Opin Biotechnol* 2016;**38**:183–9.
21. Razaghifard R. Algal biofuels. *Photosynth Res* 2013;**117**:207–19.
22. Dias MOS, Ensinas AV, Nebra SA, Filho RM, Rossell CEV, Maciel MRW. Production of bioethanol and other bio-based materials from sugarcane bagasse: integration to conventional bioethanol production process. *Chem Eng Res Des* 2009;**87**:1206–16.
23. Demirbas A, Demirbas MF. *Algae energy. Algae as a new source of biodiesel*. London: Springer-Verlag London Limited; 2010.
24. Lardon L, Hélias A, Sialve B, Steyer JP, Bernard O. Life-cycle assessment of biodiesel production from microalgae. *Environ Sci Technol* 2009;**43**(17):6475–81.
25. Hu Q, Sommerfeld M, Jarvis E, Ghirardi M, Posewitz M, Seibert M, et al. Microalgal triacylglycerols as feedstocks for biofuel production: perspectives and advances. *Plant J* 2008;**54**:621–39.
26. Tredici MR. Photobiology of microalgae mass cultures: understanding the tools for the next green revolution. *Biofuels* 2010;**1**:143–62.
27. Heasman M, Diemar J, O'connor W, Sushames T, Foulkes L. Development of extended shelf-life microalgae concentrate diets harvested by centrifugation for bivalve molluscs—a summary. *Aquac Res* 2000;**31**:637–59.
28. Bilad MR, Arafat HA, Vankelecom IFJ. Membrane technology in microalgae cultivation and harvesting: a review. *Biotechnol Adv* 2014;**32**:1283–300.
29. Scott SA, Davey MP, Dennis JS, Horst I, Howe CJ, Lea-Smith DJ, et al. Biodiesel from algae: challenges and prospects. *Curr Opin Biotechnol* 2010;**21**:277–86.
30. Baeyens J, Kang Q, Appels L, Dewil R, Lv Y, Tan T. Challenges and opportunities in improving the production of bio-ethanol. *Prog Energ Combust* 2015;**47**:60–88.
31. Gupta A, Verma JP. Sustainable bio-ethanol production from agro-residues: a review. *Renew Sustain Energy Rev* 2015;**41**:550–67.
32. Kumar AK, Sharma S. Recent updates on different methods of pretreatment of lignocellulosic feedstocks: a review. *Bioresour Bioprocess* 2017;**4**:7.
33. Mosier N, Wyman C, Dale B, Richard E, Lee YY, Holtzapple M, et al. Features of promising technologies for pretreatment of lignocellulosic biomass. *Bioresour Technol* 2005;**96**:673–86.
34. Kostas ET, White DA, Du C, Cook DJ. Selection of yeast strains for bioethanol production from UK seaweeds. *J Appl Phycol* 2016;**28**:1427–41.

35. Cho Y, kim H, Kim S. Bioethanol production from brown seaweed, *Undaria pinnatifida*, using NaCl acclimated yeast. *Bioproc Biosyst Eng* 2013;**36**:713–19.
36. Wargacki AJ, Leonard E, Win MN, Regitsky DD, Santos CNS, Kim PB, et al. An engineered microbial platform for direct biofuel production from brown macroalgae. *Science* 2012;**335**:308–13.
37. Sunwoo IY, Kwon JE, Nguyen TH, Ra CH, Jeong G, Kim S. Bioethanol production using waste seaweed obtained from gwangalli beach, busan, korea by co-culture of yeasts with adaptive evolution. *Appl Biochem Biotechnol* 2017;**183**:966–79.
38. Moreno-Garcia L, Adjallé K, Barnabé S, Raghavan GSV. Microalgae biomass production for a biorefinery system: recent advances and the way towards sustainability. *Renew Sustain Energy Rev* 2017;**76**:493–506.
39. Behera S, Singh R, Arora R, Sharma NK, Shukla M, Kumar S. Scope of algae as third generation biofuels. *Front Bioeng Biotechnol* 2015;**2**(90):1–13.
40. Kim KH, Choi IS, Kim HM, Wi SG, Bae HJ. Bioethanol production from the nutrient stress-induced microalga *Chlorella vulgaris* by enzymatic hydrolysis and immobilized yeast fermentation. *Bioresour Technol* 2014;**153**:47–54.
41. Kumar S, Gupta R, Kumar G, Sahoo D, Huhad RC. Bioethanol production from *Gracilaria verrucosa*, a red alga, in a biorefinery approach. *Bioresour Technol* 2013;**135**:150–6.
42. Wang X, Liu X, Wang G. Two-stage hydrolysis of invasive algal feedstock for ethanol fermentation. *J Integr Plant Biol* 2011;**53**(3):246–52.
43. Cho H, Ra CH, Kim SK. Ethanol production from the seaweed *Gelidium amansii*, using specific sugar acclimated yeasts. *J Microbiol Biotechnol* 2014;**24**(2):264–9.
44. Khambhaty Y, Mody K, Gandhi MR, Thampy S, Maiti P, Brahmbhatt H, et al. *Kappaphycus alvarezii* as a source of bioethanol. *Bioresour Technol* 2012;**103**:180–5.
45. Chen C-Y, Zhao X-Q, Yen H-W, Hod S-H, Cheng C-L, Lee D-J, et al. Microalgae-based carbohydrates for biofuel production. *Biochem Eng J* 2013;**78**:1–10.
46. Saravanan K, Duraisamy S, Ramasamy G, Kumarasamy A, Balakrishnan S. Evaluation of the saccharification and fermentation process of two different seaweeds for an ecofriendly bioethanol production. *Biocatal Agric Biotechnol* 2018;**14**:444–9.
47. Soliman RM, Younis SA, El-Gendy NS, Mostafa SSM, El-Temtamy SA, Hashim AI. Batch bioethanol production via the biological and chemical saccharification of some Egyptian marine macroalgae. *J Appl Microbiol* 2018;**125**(2):422–40.
48. Yanagisawa M, Nakamura K, Ariga O, Nakasaki K. Production of high concentrations of bioethanol from seaweeds that contain easily hydrolyzable polysaccharides. *Process Biochem* 2011;**46**:2111–16.
49. El-Dalatony MM, Kurade MB, Abou-shanab RAI, Kim H, Salama E, Jeon B. Long-term production of bioethanol in repeated-batch fermentation of microalgal biomass using immobilized *Saccharomyces cerevisiae*. *Bioresour Technol* 2016;**219**:98–105.
50. Tan IS, Lee KT. Enzymatic hydrolysis and fermentation of seaweed solidwastes for bioethanol production: an optimization study. *Energy* 2014;**78**:53–62.
51. Chew KW, Yap JY, Show PL, Suan NH, Juan JC, Ling TC, et al. Microalgae biorefinery: high value products perspectives. *Bioresour Technol* 2017;**229**:53–62.
52. Fermoso J, Coronado J, Serrano D, Pizarro P. Pyrolysis of microalgae for fuel production. In: Muñoz R, Gonzalez-Fernandez C, editors. *Microalgae-based biofuels and bioproducts*. Elsevier; 2018. p. 259–81.
53. Li K, Zhang L, Zhu L, Zhu X. Comparative study on pyrolysis of lignocellulosic and algal biomass using pyrolysis-gas chromatography/mass spectrometry. *Bioresour Technol* 2017;**234**:48–52.

54. Taylor G. Biofuels and the biorefinery concept. *Energy Policy* 2008;**36**:4406–9.
55. Huo Y-X, Cho KM, Rivera JGL, Monte E, Shen CR, Yan Y, et al. Conversion of proteins into biofuels by engineering nitrogen flux. *Nat Biotechnol* 2011;**29**(4):346–51.
56. Xu J, Dolan MC, Medrano G, Cramer CL, Weathers PJ. Green factory: plants as bioproduction platforms for recombinant proteins. *Biotechnol Adv* 2012;**30**(5):1171–84.
57. Suali E, Sarbatly R. Conversion of microalgae to biofuel. *Renew Sustain Energy Rev* 2012;**16**(6):4316–42.

Chapter 11

Nanobiocatalytic processes for producing biodiesel from algae

Tahereh Nematian[1] and Mohammad Barati[2]
[1]Department of Applied Chemistry, School of Chemistry, College of Science, University of Tehran, Tehran, Iran, [2]Department of Applied Chemistry, Faculty of Chemistry, University of Kashan, Kashan, Iran

11.1 Introduction

The daily increasing dependency of advanced countries on petroleum and exceeding consumption compared to production with drastic increases in oil prices have made industrial countries adopt novel approaches to energy. The reduction in fossil resources, environmental concerns, and the consequent exertion of strict standards have led researchers to take significant steps toward the conservation of the environment and sustainable development through taking advantage of various methods like making use of oil-replaced fuels know as clean fuels. The use of biofuels as an energy source has drawn attention, not only from bioenvironmental perspectives, but also due to other economic and social reasons as well as for their ease of use.[1] Biofuel is a term used to refer to fuels prepared from renewable resources. Biofuel is a type of fuel wherein energy is produced through the biological stabilization of carbon. These fuels are considered clean in that their constituent plant materials take in the carbon dioxide extant in the air and then give it off again after combustion. They are also considered as renewable because they can be repeatedly grown. Ethanol, methanol, biodiesel, and gaseous fuels like hydrogen and synthetic gas are among the most important biofuels.[2] Out of the preliminary resources from which biofuels can be produced, wood wastes, residues of agricultural crops like sugar cane, cereals, and plant and vegetable oils are considered as the most important. At present, diesel engines are considered as essential parts in supplying the propelling force for road and nonroad vehicles. On the other hand, nowadays, diesel engines are the main consumers of fossil fuels, hence they are envisioned as the most important factor giving rise to environmental pollutions.[1]

Sustainable Bioenergy. DOI: https://doi.org/10.1016/B978-0-12-817654-2.00011-3

Biodiesel is a fuel that consists of monoalkyl esters of long-chain fatty acids prepared from renewable resources like plant oils or animal fats. The addition of biodiesel to diesel oil of up to 30% of its weight causes an improvement in its knock rate. The lubrication of diesel engines can be increased by 20% by adding only 1% of biodiesel to diesel oil.[3] Among the most important advantages of biodiesel, its safeness, nontoxicity, renewability, and freeness of sulfur as well as its higher flash points than gasoline (over 130°C) and higher cetane number can be pointed out.[4] Having high oxygen content of its weight, biodiesel has more complete combustion. It also has better environmental compatibility compare with petroleum diesel because of no net carbon emission.[5] This type of biofuel is obtained from a wide range of resources and can be classified into three categories or generations. The first generation of biodiesel feedstock is composed of sugars, starches, and animal fat precursors that are produced using well-known processes or technologies. The second generation is prepared from nonedible products or agricultural wastes, especially lignocellulose biomasses and/or food industry and restaurant wastes. However, the third generation of biodiesels which is produced from microalgae have attracted a great deal of attention from researchers and industries.[6] Biofuels obtained from first generation feedstocks are produced from edible precursors and are hence regarded as inappropriate resources for fuel production. Biofuels prepared from second generation feedstocks are not of high quality. Microalgae with rapid growth rates and that produce large volumes of oil have been proposed as appropriate feedstocks for biodiesel production. As compared to oil-producing plants, microalgae produce greater amounts of oil and can be cultivated in any climate.[7]

In this chapter, nanobiocatalytic processes for biodiesel production from algal biomass have been described. The types of macro- and microalgae biomass for producing fatty acid methyl esters (FAMEs) in the presence of enzymes and in different reaction conditions have been reviewed. The efficiency of nanostructures such as nanoparticles, nanofibers, nanoporous carriers, polymeric nanocarriers, carbon nanotubes, and graphene oxide for the lipase immobilization of catalytic performance has been discussed in detail.

11.2 Microalgae as biodiesel feedstock

The high oxygen content of biodiesel (about 10%–11% of its overall weight) causes its complete combustion in engines. The amounts of the components present in diesel and biodiesel have been compared in Table 11.1.[8] Microalgae are third-generation feedstocks used for the production of biofuels such as bioethanol, biogas, and biodiesel. Due to their high capability for biofuel production, these renewable resources have drawn the attention of researchers and environmentalists. Their high oil content, high

TABLE 11.1 Ingredient and elemental analysis of diesel and biodiesel.[8]

Element/Compound	Content in biodiesel (wt.%)	Content in diesel (wt.%)
Carbon	79.6	86.4
Hydrogen	10.5	13.6
Oxygen	8.6	–
Nitrogen	1.3	–
C to H ratio	7.6	6.5
Aliphatics	15.2	67.4
Olefins	84.7	3.4
Aromatics	–	20.1
Naphthenes	–	9.1

rate of biomass production, high photosynthesis efficiency, minimum use of water, low land area requirement, ability to thrive without pesticides and fresh water, lack of competition with food resources, ease of production on a large scale, and ability to produce products with high biodegradability are important advantages of using microalgae for the production of biodiesel in contrast to other renewable plant resources.[1]

Algae belong to the "Protista" group of photoautotrophs, and can be classified as either multicellular organisms (macroalgae) or single-celled organisms (microalgae). In terms of the cellular structure of algae, two forms can generally be found: prokaryotic and eukaryotic. Blue-green algae are prokaryotic, whereas the other types of algae are eukaryotic. Single-celled algae or microalgae are deemed as more suitable than macroalgae for biodiesel production because over 60% of their weight is comprised of lipids and they are capable of absorbing more carbon dioxide.[8] Biofuels produced from microalgae include methane obtained via the anaerobic digestion of algae, biodiesel produced from algae oil, and biohydrogen. The annual oil output of algae is about 100 times that of corn oil, 40 times that of soybean oil, and 20 times that of sunflower seed oil.[9] Algae are fast growing organisms and complete their growth stage within 7–10 days, while the maturation period of most cereals is about 60–90 days. Algae can be cultivated on nonarable lands and even in salty waters. Microalgae need 6–8 Mt of nitrate per hectare per year. This amount is 55–111 times more than what is needed for other agricultural crops. The related costs can be considerably reduced through the use of urban sewage or industrial wastewater as a growth bed.[10]

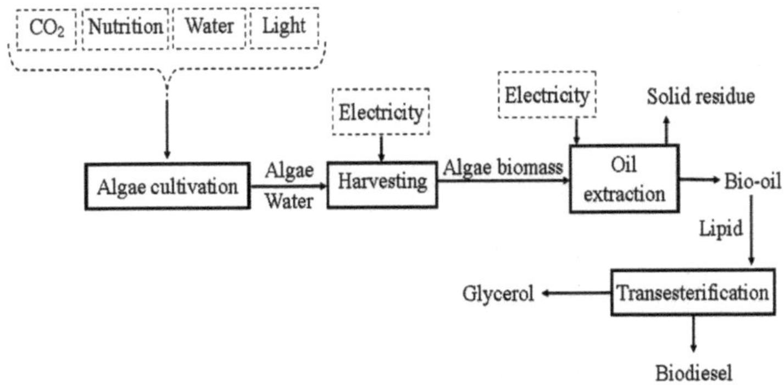

FIGURE 11.1 A schematic representation of biodiesel production from algae. *Adapted from Mata TM, Martins AA, Caetano NS. Microalgae for biodiesel production and other applications: a review.* Renew Sustain Energy Rev *2010;**14**:217–32, with copyright permission from Elsevier.*

For every kilogram of microalgae biomass dry weight produced, 1.8 kg of carbon dioxide is used.[11] An interesting and novel idea here is that the carbon dioxide emitted from industrial units can be used for algal growth. There have been numerous studies performed regarding the production of biodiesel from macro- and microalgae around the globe. In between these, some studies have investigated the growth and harvesting of algae in wastewater from factories. The remaining biomass after biodiesel production can be used as livestock feed. In addition, it can also be used as fertilizer in the form of compost after anaerobic digestion. Moreover, algae possess great potential for the synthesis of a variety of biodrugs like antiinflammatory and anticancer drugs, carotenoids, and pigments.[12]

11.3 Biodiesel production from algae

Biodiesel is obtained through different stages, including: cultivation, harvesting, oil extraction, transformation of oil into biodiesel via transesterification, and finally product separation. Fig. 11.1 illustrates a schematic representation of the entire process of biodiesel production from algae.[13]

11.3.1 Selection of algae

As previously mentioned, microalgae represent a potential and cost-effective feedstock source for biodiesel production for numerous reasons. The calorific value of thus produced biodiesel mainly depends on the algal biomass content. Normally, the calorific value of biodiesel produced from oilseeds like canola or soybean is 39.5 MJ/kg, whereas biodiesel obtained from algae features a calorific value equal to 41 MJ/kg.[14] According to reports, the oil production rates of microalgae are over 80% of their dry mass weight. Also, the amount of the oil extracted from microalgae is between 20% and 50% of their mass which is common.[8] The amount of produced oil is defined as the mass of oil produced from a certain volume of grown microalgae per day and depends on their growth rates and biomass oil contents. The most appropriate oil obtained from alga that can be used as a biodiesel feedstock is triglyceride containing a carbon chain length between 14 and 22 and the lowest saturation rate. Due to being rich in unsaturated fatty acids with structures containing more than four double bonds, microalgae oils are distinguished from plant oils.[15] Microalgae's cultivation media is an extremely important factor in lipid gains. For example, the amount of fat produced from *Chlorella protothecoides* will be higher when glucose or acetate is present in its cultivation media. Also, the use of cheaper feedstocks like ethanol, glycerol, or fructose can be used for the production of less costly products.[16] Generally, productivity and fat content are inversely correlated and stress conditions like a reduction in nitrogen or phosphate limit cell growth and increase fat content. Table 11.2 presents the amounts of fat extant in various microalgae species.[17]

TABLE 11.2 Lipid content of different algal biomasses.[17]

Algae	Lipid amount (dried biomass wt.%)
Botryococcus braunii	25–75
Chlorella sp.	28–32
Crypthecodinium cohnii	20–56
Cylindrotheca sp.	16–37
Dunaliella primolecta	23
Isochrysis sp.	25–33
Monallanthus salina	20
Isochrysis galbana	14.5
Nannochloris sp.	20–35
Nannochloropsis sp.	31–68
Neochloris oleoabundans	35–54
Phaeodactylum tricornutum	20–30
Schizochytrium sp.	50–77
Tetraselmis suecica	12–23

11.3.2 Lipid extraction from algae

The harvesting of algae and oil extraction can be performed simultaneously. These two stages are accomplished consecutively and are highly energy consuming. Oil extraction is usually carried out using physical or chemical processes or a combination of the two. Mechanical extraction, solvent-assisted extraction, supercritical fluid extraction, ultrasound extraction, chemical extraction, and biological extraction are also commonly applied for the extraction of oil from algae. The extraction method has to be scalable. It should be have acceptable gains and cause no damage to lipids.[11] Various kinds of lipids are obtained during extraction, but not all these lipids are convertible into biodiesel. In addition to lipids, other substances can also be produced during the process, but it is necessary to remove such substances before the conversion stage. Thus an extraction process should be selected that provides maximum selectivity in respect to lipids of interests. In many cases of oil extraction, there is a need of biomass pretreatment. Depending on the structure of the algal cell wall and the nature of the obtainable lipids, the use of such mechanical pretreatment methods as milling and exposition to ultrasound waves, and nonmechanical pretreatments like freezing, use of organic solvents, and osmotic shocks has been reported.[13,18–21]

11.3.3 Conversion of algal oil to biodiesel

The methods used for converting algal oil into biodiesel can be categorized into two types, namely chemical and biological. The method selection depends on the type of feedstock, the type of intended fuel, economic conditions, and the type of final product. Fig. 11.2 displays different kinds of these methods.[22,23]

11.3.3.1 Transesterification reaction

Pyrolysis, thermal cracking, microemulsification, distillation, and transesterification have been introduced as the most important methods of lowering viscosity during biodiesel production. Among these, transesterification is the best method for the production of high-quality biodiesel. It is a chemical reaction between triglycerides and short-chain alcohols. Monoesters and glycerin are produced form transesterification of long-branch triglycerides in the presence of appropriate catalysts. Fig. 11.3 exhibits a schematic representation of biodiesel production using the transesterification process.[10] In fatty acids' alkyl esters, the alkyl group is determined based on the acyl acceptor used in the reaction. The molecular weight of the FAME is one-third of the molecular weight of the plant oils used and its viscosity is one-eighth

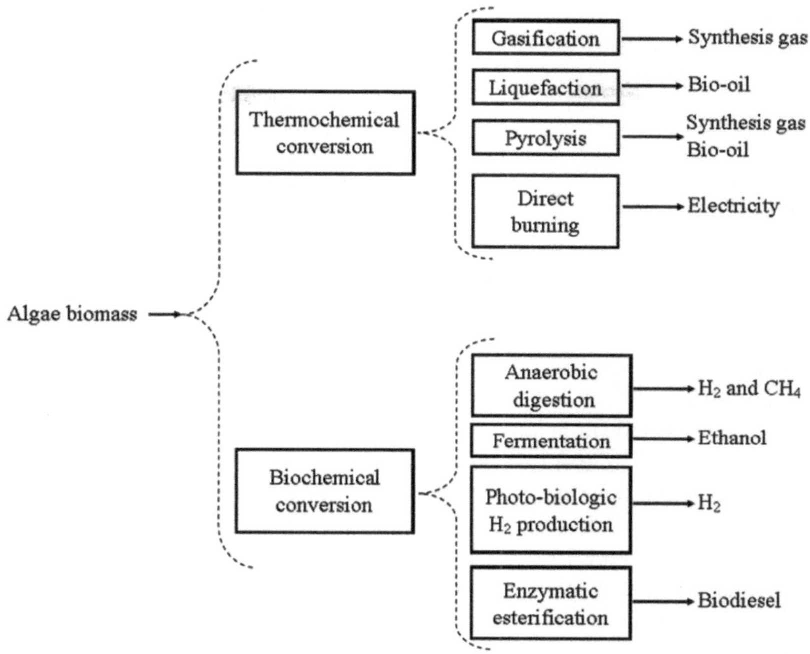

FIGURE 11.2 Methods for algae conversion into energy content. *Adapted and modified from Menon V, Rao M. Trends in bioconversion of lignocellulose: biofuels, platform chemicals & biorefinery concept.* Prog Energy Combust Sci *2012;**38**:522–50 , with copyright permission from Elsevier.*

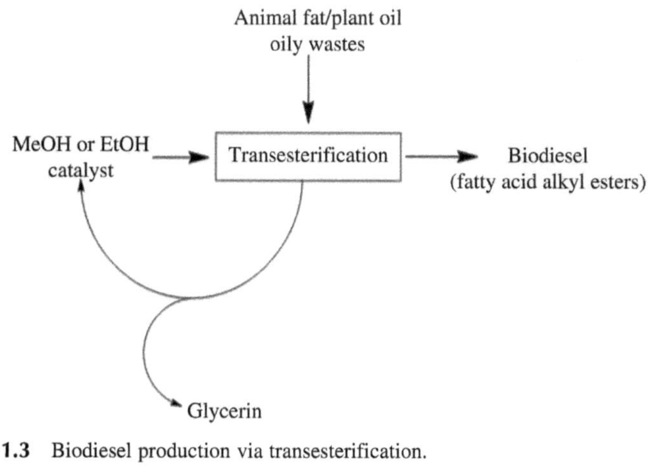

FIGURE 11.3 Biodiesel production via transesterification.

thereof. The cetane number of fatty acid alkyl ester is larger than that of plant oils. Cost-effectiveness is the most important factor in the diversity of biodiesel production methods.[24]

11.3.3.2 Chemocatalytic transesterification

Performing the transesterification process in the absence of a catalyst takes more time and requires higher temperatures and the yields are low. Catalysts used in transesterification reactions are divided into two general sets (homogeneous and heterogeneous) according to their consistency with the reaction environment and they are further grouped into three classes, namely acids, bases, and enzymatic or biocatalysts according to their chemical structure.[25]

The transesterification process can be catalyzed in the presence of both acids and bases. The reaction has a high output rate in the presence of acid catalysts, but the reaction rate is very low. H_2SO_4, HCl, BF_3, H_3PO_4, and some sulfonic acids can be pointed out as examples of acid catalysts. The maximum conversion rate reported using this type of catalyst is 99%. Using this process, a 99% biodiesel production yield has been obtained in the presence of one mole percent sulfonic acid (as the catalyst) with a 1:30 molar ratio of catalyst to feedstock in 51 hours of reaction time.[26] Another advantage of transesterification using an acid catalyst is the direct production of biodiesel from inexpensive and low content lipid feedstock such as kitchen oil wastes. The use of a soluble acid catalyst allows the reaction to take place in a homogeneous environment. It facilitates mass transfer onto the catalyst and as a result the reaction rate increases. However, the recycling and reuse of the catalyst in such reactions is rendered more difficult. Heterogeneous acid catalysts are responsible for decreasing the reaction rate and pose problems for glycerol recycling. However, they cause less corrosion in reactors and engines and their recycling is easier compared to homogenous acid catalysts.[27] The addition of acids leads to the protonation of the carbonyl ester groups and this is accompanied by carbocation creation which sets the foundation for the nucleophilic attack of alcohol. In such an interaction, a tetrahedral intermediate is obtained that forms a new ester by eliminating glycerol and the proton of the acid catalyst will be given back. According to this mechanism, the presence of water in reaction media should be prevented since water molecules also possess the capability of attacking the carbocation.[28]

Base catalysts have a greater level of activity than acid catalysts and their final products create less corrosion. The highest reported conversion rate for this type of catalyst in a transesterification process is 98% for a short time (30 minutes) and with a very small amount (0.5 mol.% of catalyst).[29] In this process, the presence of water also exerts a negative effect. Water in transesterification processes in the presence of base catalysts causes the acceleration of soap formation, reduces the yields, and increases the viscosity of the desired products. Investigations have shown that increasing the ratio of feedstock to catalyst to up to 6:1 can be effective in resolving the problem.[30] Alkaline catalysts are not usable for feedstocks that have free fatty acids of more than 5% of their weights. Thus acid catalysts are usually applied before making use of alkaline catalysts. Although the free fatty acids problem can be resolved using this method, there an additional purification stage is needed.[31] Generally, transesterification processes in the presence of base catalysts take place in three stages. In the first stage, the base reacts with alcohol and produces alkoxide. Then, alkoxide causes the formation of a tetrahedral intermediate by nucleophilically attacking the triglyceride carbonyl group. Afterwards, alkyl ester and diglyceride anions are formed from the separation of the formed intermediate. The optimum reaction temperature has been demonstrated to be between 20°C and 120°C depending on the type of feedstock.[28,32] Zeolites, alkali earth metal oxides, and hydrotalcites can be enumerated among the most important heterogeneous base catalysts. The advantages of this type of catalyst include their ease of use, recyclability, and the higher safety of reactions.[33]

In spite of the advantages recounted for heterogeneous catalysts such as their easier recycling and, resultantly, their higher consistency with sustainability, they have some disadvantages like the complexity and costliness of their treatment, low resistance to atmospheric carbon dioxide, low activity in low temperatures and pressures, and high alcohol needs. Metal oxides are the most important type of heterogeneous catalyst. Among metal oxides, mainly MgO and CaO based catalysts have drawn attention. These compounds are doped to other metals or metal oxides to moderate their chemical and physical characteristics, including acidity and mechanical resistance.[34] $CaO-Al_2O_3$ and $MgO-Al_2O_3$ catalysts were prepared based on the sol–gel method and applied in *Nannochloropsis oculata* algal oil conversion at 50°C with a methanol to oil ratio of 6:1. For this process, a biodiesel production yield of 97% was reported. However, the intended catalyst was found to be inactivated after two uses.[35] A mesoporous $CaO-Al_2O_3$ catalyst exhibited a 98% output in the conversion of palm oil into biodiesel.[36] The use of two metals, zirconium (Zr) and titanium (Ti), as metal oxide catalysts was also investigated. Zr and Ti cause an enhancement in the productivity of the catalyst by granting acidic characteristics and improving physical characteristics, respectively.[37] The use of Fe_3O_4 as part of a catalyst base for the recovery of the catalyst based on the exertion of a magnetic field has been reported to be highly useful. However, there was no report about the obvious effect of this substance on the transesterification reaction process.[38] Acidic metal oxides have been applied as feedstocks with abundant amounts of free fatty acids. For instance, an SnO_2-SiO_2 catalyst showed a 94.6% yield in biodiesel production reactions. SnO_2 amorphous crystals have been reported as active sites for both the esterification and transesterification reactions.[39] In a study, an $Fe_2O_4-SiO_2$ catalyst was reported to have converted feedstock into biodiesel with a 99% yield at 220°C and a methanol to oil molar ratio of 8:21 in 3 hours reaction time.[40] Different bimetal and trimetal oxides like $ZnFe_2O_4$ and $Mn_{3.5x}Zr_{0.5y}Al_xO_3$ have also been investigated in transesterification processes. The relatively high activity of these catalysts demonstrates the effect of bi- and trimetal oxides crystal structures on catalyst activity in transesterification.[41] The first point that has to be taken into account in biodiesel production is accomplishing the esterification and transesterification reactions in one step. Bi-agent base–acid metal oxide catalysts have been developed in order to perform the esterification reaction of fatty acids and the transesterification of triglycerides in one step. These catalysts have been prepared from transition metals and they simultaneously catalyze esterification and transesterification reactions under harsh conditions that need high temperature rates, high alcohol to oil ratios, and require longer reaction times.[42] Therefore these catalysts also maintain their activities in the presence of water; and due to this reason, they are extensively used in the production of biodiesel through esterification processes or single-stage esterification–transesterification processes. However, the results of these catalytic reactions are not always as good as they are expected to be, therefore they are not appropriate for large-scale industrial production.

11.4 Biocatalytic transesterification

The application of a biocatalyst (enzyme) in transesterification is more effective as compared to chemical catalysts. It has many advantages like the ready separation of product, requirement for less wastewater treatment, easy glycerol retrieval and the lack of producing side products. Over the past few decades, biocatalysts like lipases have attracted a great deal of attention due to their ability to catalyze a variety of precursors, their ability to act in different temperatures, pH ranges, and organic solvents, and their high rates of selectivity.[43] Lipases are enzymes that are capable of performing the reversible hydrolysis of animal fats and plant oils under cell conditions. Lipases catalyze the hydrolysis of triglycerides and other water insoluble esters. Plants, fungi, and bacteria (Table 11.3) synthesize these enzymes. The majority of lipases used in industry are produced from bacterial or fungal sources and they are widely applied in the pharmaceutical, dairy, detergent, cosmetic, and oil production industries.[44]

Also, the use of molecular methods such as recombinant DNA technology and protein engineering is highly effective in the large-scale production of lipases with modified characteristics and it can decrease enzyme preparation costs.[44] Lipases acquire the best position for transesterification in the water–oil interface; due to the same reason, enzyme has higher activity during the reaction in the aqueous–organic solvent interface. These biocatalysts have various mechanisms for performing hydrolysis, depending on their

TABLE 11.3 Natural resources of lipase.[44]

Bacteria	Fungus	Plant
Streptomyces sp.	*Mucor* sp.	*Lycopersicon esculentum*
Achromobacter sp.	*Rhizopus* sp.	*Carissa carandas fruit*
Alcaligenes sp.	*Candida* sp.	*Castor bean*
Arthrobacter sp.	*Geotrichum* sp.	*Cucumis melo*
Psuedomonas sp.	*Aspergillus* sp.	*Hibiscus cannabinus*
Bacillus sp.	*Pencillium* sp.	*Lycopersicon esculentum*
Staphylococcus sp.	*Humicola* sp.	*Moringa oleifera*
Chromobacterium sp.	–	*Rice bran*
Burkholderia sp.	–	*Triticum aestivum*

sources of production and their chemical structure. Intra- and extracellular lipases obtained from *Aspergillus niger* have shown greater selectivity for 1,3-triglycerides production.[45,46] Lipases obtained from *Candida*, except for *Candida deformans*, are unselective for glyceride ester bonds. However, lipase obtained from *Penicillium cyclopium* is more active in the conversion of di- and monoglycerides than in triglycerides. Lipases obtained from *Penicillium crustosum*, *Penicillium verrucosum*, and *P. cyclopium* have been found to be selective for butyric acid conversion. Lipases extracted from *Rhizopus* are more selective in the conversion of fatty acids with an average chain length of between 8 and 10 carbons. Moreover, other lipases from *Candida antarctica* and porcine pancreatic lipase tend toward the conversion of short-chain fatty acids, while lipases like those produced from *C. antarctica* B and *P. cyclopium* have specific activities toward the conversion of long-chain fatty acids.[47]

However, the use of enzyme as the biocatalyst in synthetic processes is accompanied by two problems; products being contamination with enzyme and high costs. To overcome these problems, enzyme immobilization is the general method of choice. Enzyme recovery is made feasible by its immobilization and it can be subsequently used in several reaction cycles. Enzyme immobilization means the entrapment or establishment of enzyme based on a solid material and/or a carrier matrix. The choice of carrier depends on several factors. Mechanical and microbial resistance, thermal stability, chemical durability, hydrophobic and hydrophilic nature, easy retrieval, high loading capability, and price are the most important factors involved in the selection of carrier type.[48]

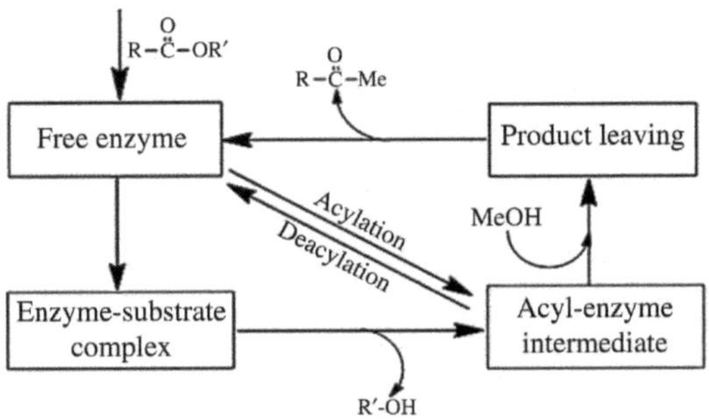

FIGURE 11.4 Conversion mechanism of triglycerides to FAMEs in the presence of lipase. *FAME*, Fatty acid methyl ester. *Adapted from context of Kobayashi S. Lipase-catalyzed polyester synthesis—a green polymer chemistry.* Proc Jpn Acad Ser B *2010;86:338–65,an open access article.*

Transesterification reaction using lipase follows the ping-pong-Bi-Bi mechanism that is an enzyme-catalyzed process involving multiple substrates/products.[49] The active site of lipase possesses two functional groups that are necessary for catalytic activity. The first is the hydroxyl group of serine amino acid that causes the formation of enzyme–substrate complex through a nucleophilic addition reaction. The second is the amine group of histidine amino acid that acts as a proton acceptor. Fig. 11.4 illustrates the mechanism of action of lipase enzyme in the transesterification process.[50] Moreover, Table 11.4 provides a list of various kinds of lipases that have been investigated using different feedstocks in transesterification reactions.[51–54]

The majority of common lipases that are used in the production of biodiesel have, most prevalently, bacterial, fermentative, and fungal sources.[55] However, all lipases are not responsible for high biodiesel yields. Porcine pancreatic (type II) lipase gives a below 20% yield of biodiesel due to its low stability.[56] Therefore the lipase type used has to be selected carefully so that an acceptable yield can be assured. The ideal enzyme for an efficient transesterification and conversion of biodiesel should have such characteristics as the ability to be used in interfaces (mono-, di-, and triglycerides) and free fatty acids, low inhibitory effect on products, high activity and productivity in nonaqueous environments, shorter reaction time, reusability, and resistance in alcohol media.[57]

The biodiesel production reaction is generally conducted in nonaqueous environments, therefore lipase resistance in such media is required. Nowadays, researchers are focused on obtaining lipases with high activities and stabilities in the presence of organic solvents.[58] It was demonstrated that the yield of biodiesel can be increased using lipases obtained from *Rhizopus oryzae* in transesterification reactions. This enzyme is stereoselective and

TABLE 11.4 **Transesterification in presence of enzymes from different sources.**[51–54]

Enzyme	Feed	Immobilization	Biodiesel yield (wt.%)
Burkholderia cepacia	Jatropha oil	Immobilized	90
Candida antartica lipase B	Olive oil	Immobilized	90
Candida sp.	Animal fat	Immobilized	90
Candida rugosa lipase	Colza oil	Immobilized	63.6
Photobacterium lipolyticum M37	Olive oil	Free	90
Aspergillus niger	Palm oil	Immobilized	90
Ralstonia sp.	Soybean/ palm oil	Free	90
Pseudomonas fluorescens	Soybean oil	Free	83.8
Chromobacterium viscosum	Jatropha oil	Free	62
		Immobilized	71
Thermomyces lanuginose	Palm oil	Immobilized	96.15
Rhizopus oryzae	Soybean oil	Immobilized	90.5
Penicillium expansum	Corn oil	Free	86

hydrolyzes the 1,3-tryglyceride site. The enzyme causes acyl migration during the transesterification reaction and enhances biodiesel production yields.[59] Due to the mentioned advantages, the enzyme can be a highly effective and considerable choice as a catalyst in the transesterification process. Table 11.5 compares the transesterification process in the presence of enzymatic and chemical catalysts.[29,60]

11.4.1 Enzyme immobilization

Many methods have been reported for enzyme immobilization over the past decade. Immobilization methods can be categorized into two general processes, namely reversible and irreversible, based on the interaction between the lipase and the support. Enzyme immobilization is carried out so as to deposit one species on a bed that is minimally influenced by other chemical compounds in the reaction environment.[61] Enzyme immobilization brings about an increase in thermal and chemical stability, facilitates enzyme

TABLE 11.5 Comparison of catalysts used for biodiesel production.[29,60]

Catalyst type	Catalyst example	Advantages	Disadvantages
Base-homogeneous	KOH, NaOH	High catalytic activity (4000 times higher than acid-homogenous), low cost, and mild operation conditions	Negative effect of high fatty acid in feed, negative sensitivity to water, soap formation, emulsion formation, and nonreusable
Acid-homogeneous	Sulfonic acid	Esterification and transesterification catalysis, no water sensitivity, no soap formation	Corrosion, high temperature need, high reaction time, low catalytic activity, nonreusable
Base-heterogeneous	CaO, MgO, $CaTiO_3$, KOH on Al_2O_3	Noncorrosive, reusable, biocompatible	High alcohol, temperature and pressure needed for reaction, high cost
Acid-heterogeneous	ZnO/ I_2ZrO_2/SO_4^{2-}	Esterification and transesterification catalysis, low corrosive, reusable	Low acid active sites, low porosity and substrate diffusion, high cost
Enzyme	Lipase, novozym 435	Reusable, biocompatible, no soap formation	High cost, enzyme denaturation

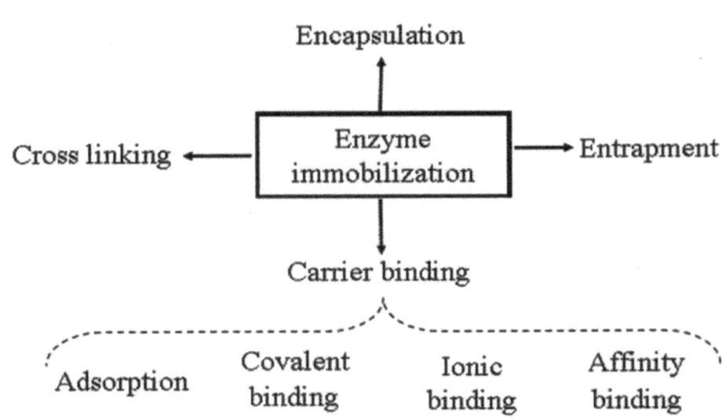

FIGURE 11.5 Different techniques for enzyme immobilization.

separation upon the termination of the reaction, and enhances the yield and activity. Common methods of enzyme immobilization are surface adsorption, covalent bonding, crosslinking, entrapping, and capsulation. A schematic comparison of these methods has been shown in Fig. 11.5.[62]

An old and common method of lipase immobilization is adsorption which is a reversible technique. In this method, lipase enzyme is attached to the support (acrylic and propylene resins and fabric membranes) through noncovalent adsorption (van Der Waals forces, hydrophilic interaction, hydrogen bonds, biophysical adsorption, electrostatic interaction, hydrophobic interaction, and adsorption tendency).[43] The advantages of this method mainly include direct performance and high enzyme activity. However, the reuse of enzyme is almost impossible because it is washed off from the support surface during the recycling process. Furthermore, there is a possibility of enzyme separation and enzyme activity reduction during reactions in water emulsion systems. Another constraint of this method is its sensitivity to temperature and ion concentration.[63]

The covalent bonding of enzyme onto the support surface is feasible through the use of various functional groups. Using multiple covalent bonds (enzyme attachment to the support at several points) is the best type of attachment in terms of thermal and functional stability for immobilization. Covalent attachment is extremely strong and makes enzyme separation from the support impossible.[64] Glutaraldehyde has been widely used in the covalent immobilization of enzyme on the support as an activator or chemical connector.[65] Porous glass, ceramic, sand, cellulose, synthetic polymers, and metal oxides are among the supports used for covalent enzyme immobilization.

Enzyme immobilization through trapping entails the confinement of the enzyme in a polymeric matrix allowing the substrate and product to pass through the matrix while the enzyme is confined therein. On the other hand, capsulation also entails enzyme entrapment, but in this method the enzyme is entrapped inside a polymeric gel capsule. As caused by this action, small molecules are enabled to permeate through a percolating membrane.[66] Enzyme trapping can be covalent or noncovalent. In comparison to adsorption, lipase immobilization using this method has shown more stability and can be more readily recycled. An important disadvantage of this method is mass transfer limitations that come about as a result of the adhesion to the carrier surface of the produced glycerol. Also, the lipase immobilization efficiency is low in this method.[67]

In the crosslinking method, there is no support for enzyme immobilization and as a result the production cost is decreased and productivity is increased. The method includes the crosslinking of enzyme molecules using one or two crosslinking agents.[68] The accumulation of attached enzyme produces insoluble enzyme masses providing relatively high stability and catalytic activity.[69] Lipase M37 obtained from Photobacterium was immobilized using this technique in a study, and the stability of the immobilized enzyme was reported to be higher than that of free enzyme.[70]

11.4.2 Nanotechnology for enzyme immobilization

During the past decade, enzyme immobilization using nanotechnology has drawn the attention of the scientific community. The invention and use of various methods for the synthesis of nanocatalysts that can be used in biodiesel production have opened new research avenues in this area. Many researchers have utilized nanostructured scaffolds for enzyme immobilization. Nanoscaffolds used for enzyme immobilization include nanofibers, nanoparticles, nanotubes, nanoplates, and nanocomposites. Important advantages of using nanomaterials for enzyme immobilization are their high surface area which increases the enzyme-loading capacity, lower resistance against mass transfer, high mechanical resistance, easy separation, and enzyme stability and activity enhancement.[71] Table 11.6 presents the results of a study pertaining to lipase immobilization on nanostructured supports for biodiesel production.[72,73]

11.4.2.1 *Nanofibers*

Nanofibers constitute a type of nanocarrier used in enzyme immobilization. Electrospinning is a common method of nanofiber production. Their high surface area and porosity are among the most important advantages of these nanocarriers which provide for a high enzyme-loading capability as well as homogeneous distribution in liquid phase.[74] Hybrid core–shell nanofibers have been utilized as enzyme carriers in order to investigate the effect of this

TABLE 11.6 Lipase immobilized on nanostructures for biodiesel production.[72,73]

Lipase source	Nanostructure support	Biocatalyst characteristics
Thermomyces lanuginose	Fe_3O_4 nanoparticles	High catalytic activity
Candida rugosa, Pseudomonas cepacia	Fe_3O_4 nanoparticles	High reaction rate
P. cepacia	Nanoporous Au	High activity, stability, and productivity
P. cepacia	Polyacrylonitrile electrospun nanofibers	High biodiesel yield
Burkholderia sp. *C20*	SiO_2–Fe_3O_4 nanocomposite	High productivity and reusability
B. cepacia	Chitosan–Fe_3O_4 core–shell nanoparticles	High activity and stability

type of carrier on enzyme performance. A nanofiber that could not be electrospun (such as the core) and a polymer that could be electrospun (such as the shell) were synthetized and used in a study on catalyst reactions in the presence of several enzymes, including lactate dehydrogenase, diaphorase, nicotinamide adenine dinucleotide (NADH) dehydrogenase, and 3-α-hydroxystroid dehydrogenase.[75] Despite the considerable advantages of the use of nanofibers, the nanofiber surface overlap in this method causes a limitation in enzyme penetration and hence a reduction in its activity during reaction. Since the most important advantage of nanofibers is their large surface area, studies have been carried out for the improvement of this feature through their functionalization by various oxidizers such as HNO_3 and H_2SO_4.[76]

11.4.2.2 Nanoporous carriers

To achieve more efficient enzyme immobilizations for improved productivity in enzymatic processes, mesoporous carbon and silica can be used as enzyme carriers.[77] Generally, porous structures possess large surface areas and are effective at encapsulating enzymes inside their cavities. Cavity size is of great importance in enzyme trapping when using these types of nanostructures; the closer the size of the cavities are to the hydrodynamic diameter of the enzyme the more the thermal stability and activity of the enzyme will be elevated.[78] In such a state, enzyme preserves its inherent conformation and the active sites of the enzyme can resultantly catalyze the reaction on their surfaces with no spatial restriction. If the size of the cavities of the porous nanostructures is smaller than that of the enzyme, the enzyme cannot enter the cavities and as a result its loading on the carrier will be limited. On the other hand, if the size of the cavities is larger, the activity of the enzyme will be decreased due to enzyme leaching and a reduction in its stability. Hybrid porous nanocarrier structures using other compounds have been reported to be extremely useful in modifying the characteristics and donation of new features. Many studies have been conducted on various magnetized silica supports.[79] One study investigated three structures for lipase immobilization, namely nanoporous, mesoporous, and silica-developed mesoporous structures, and all of these were surrounded in iron oxide nanoparticles. The mesoporous structure exhibited higher enzyme loading than the other structures, while the highest enzyme activity was reported in the third structure, but for its lower loading rate. The reason given for the reduction in enzyme loading was the lack of physical enzyme stability inside the cavities.[80]

11.4.2.3 Polymeric nanocarriers

Compared to other nanocarriers used for enzyme immobilization, materials with a polymeric nature can provide a surface containing a large number of adhering and biological agents having high stability and activity.[81] Polymers are appropriate supports for enzyme immobilization due to their adjustability of size and thickness. Nanostructured polymers additionally bring about increases in enzyme loading featuring modifiable surfaces.[82] In line with this, α-chymotrypsin was immobilized on a mesoporous magnetic nanogel functionalized with a carboxyl group through in situ polymerization, and it was reported that this biocatalyst possessed a higher stability in respect to temperature and pH variations in contrast to free enzyme and gave a 87% yield as well as maintaining its stability after being used 12 times. Reaction 1 indicates the catalyzed reaction using this biocatalyst[83]:

Reaction 1: Benzoyl-L-tyrosine ethyl ester + $H_2O \rightarrow$ *n*-benzoyl-l-tyrosine + ethanol

In another study on biodiesel extraction, lipase enzyme from *Candida rugosa*, was immobilized on Fe_3O_4—chitosan and its efficacy was evaluated. The results indicated that the biocatalyst showed a potential activity rate equal to 20 U/mg and that the biocatalyst still maintained 83% of its activity after being used 25 times.[84]

11.4.2.4 Carbon nanotubes

Carbon nanocarriers are extensively used as enzyme supports due to them being neutral, biodegradable, and thermally stable.[85] Among the nanostructures, carbon nanotubes, nanodiamonds, and graphene derivatives have been considerably taken into consideration for nanobiocatalysts production. Lipase from *Pseudomonas cepacia*, was immobilized for its use in biodiesel production on single-walled carbon nanotubes in two different solvents (buffer and ionic liquid). The produced biocatalyst in the medium of ionic liquid, exhibited better distribution and higher activity in the transesterification reaction, compare to the biocatalyst which produced in buffer.[86] In this context, in another study, iron oxide was deposited on a support made of carbon nanotubes based on an impregnation method and then lipase enzyme from *Yarrowia lipolytica* was immobilized thereon using covalent bonding. The enzyme stability was investigated through exposing it to ultrasound waves for 30 minutes and no change was reported in the enzyme structure.[87] Also, carbon nanotubes functionalized by certain compounds with ending groups such as carboxyl, amine, and ester were applied as lipase enzyme supports. In this study, 25% of the carbon nanotube weight was loaded with enzyme. This biocatalyst maintained 55% of its activity after 6 months under 4°C and 25% of its activity after 30 days under 60°C.[59]

11.4.2.5 Graphene oxide

Graphene oxide has also been considerably used for enzyme immobilization due to its unique characteristics and applications in various fields. These nanostructures with large surface areas, two-dimensional structures, and available functional groups, not only increase the enzyme loading capacity but also facilitate the mass transfer of substrate and product. The existence of oxygenated groups in graphene oxide renders it hydrophilic in a variety of polarized solvents. Its low production costs, modifiability, larger surface area (existence of two external surfaces, including upper and lower surfaces), absence of toxic metal particles, and higher biocompatibility are major advantages of graphene oxide as compared to carbon nanotubes as a nanocatalyst support.[88]

A. niger derived lipase was immobilized on graphene oxide functionalized with polyaniline-silver and glutaraldehyde as the lipase attachment agent and support, respectively. It was reported that enzyme loading on the support and enzyme activity were equal to 94% and 98.5%, respectively. This nanobiocatalyst is capable of tolerating solvent and mild pH changes.[89] In this regard, three different lipases, obtained from *Penicillium camemberti*, *C. rugosa*, and *R. oryzae*, were also immobilized on graphene oxide. The highest enzyme activity (73%) in the esterification reaction was shown by lipase recovered from *R. oryzae*.[90] In another study, the lipase enzyme, porcine pancreas, coupled with a propyl chlorosilane agent was attached to graphene oxide using mesoporous iron oxide nanoparticles coated with tetraethyl orthosilane (Fe_3O_4@MCM41) for which an efficiency rate equal to 98% was reported and it was shown to have maintained 87% of its activity after being used 10 times.[91] Three-dimensional hybrid graphene oxide composites with iron nanoparticles coated with chitosan have also been used as supports for *C. rugosa* lipase immobilization. A large surface area, the existence of amino and hydroxyl functional groups, as well as easy separation with a magnet are among the advantages of this composite. In this study, a graphene oxide–chitosan–iron oxide composite was functionalized using 1,6-hexadiene, following which two agents, namely glutaraldehyde and sodium chloroacetate, were used to covalently bind lipase onto the support. The results revealed a 0.05% superiority in the activity of the biocatalyst prepared using sodium chloroacetate.[92] Moreover, ethyl caprylate synthesis in the presence of *C. rugosa* on graphene oxide functionalized with (3-aminopropyl)triethoxysilane (APTES) was investigated for biodiesel production. A yield equal to 88% was reported for the biodiesel production of this biocatalyst and it was also found that graphene oxide has a higher loading rate without the use of any functional groups.[93] Table 11.7 provides an overview of several case studies on biodiesel production from algae in the presence of biocatalysts and nanobiocatalysts.[94–99]

11.4.2.6 Nanoparticles

Metal nanoparticles have been extensively utilized in nanobiocatalysts production. Various functional groups like amino, carboxylate, thiolate, and phosphate have been used for improving enzyme loading on metal nanoparticle supports. Gold nanoparticles have been introduced as one of the best metal supports due to them granting high stability and activity to the

TABLE 11.7 An overview of studies about biodiesel production from algae oil in presence of biocatalysts and nanobiocatalysts.

Feed (Microalgae)	Biocatalyst	Max. biodiesel conversion (wt.%)	Reuse (cycles)	Variable	Case study	Reference
Chlorella protothecoides	1. *Aspergillus niger* lipase 2. 2-Novozym 435 (immobilized lipase from *Candida antarctica*, type B)	97	5	✓ 3 Å molecular sieve ✓ Methanol to lipid: 5:1 ✓ Lipase dosage: 5% ✓ Temperature: 45°C	Two ways for biodiesels production: (1) in present of free lipase; (2) in present of immobilized lipase	94
Scenedesmus obliquus	1. *Pseudomonas fluorescens* lipases 2. *Candida rugosa* lipase	90.81	4	✓ Temperature: 35°C ✓ Methanol to oil: 3:1 ✓ 10% enzyme amount ✓ 2.5% water content	Two ways for biodiesels production: (1) in present of free lipase; (2) in present of immobilized lipase on Immobead 150 (sigma)	95
1. *Gracilaria edulis (G)* 2. *Enteromorpha compressa (E)* 3. *3-Ulva lactuca (U)*	Intracellular Cal A and Cal B lipase producing recombinant *Pichia pastoris* Immobilized *C. antarctica* lipase	*G*: 93*U*: 90*E*: 92 *G*: 88.5*U*: 87*E*: 89	–	✓ 10% Immobilized lipase ✓ 7% Whole cell ✓ Temperature: 40°C ✓ Oil to methanol: 1:3	Comparison of recombinant *Pichia pastoris* and immobilized *C. antarctica* lipase on calcium alginate	96

Nannochloropsis gaditana	*Rhizopus oryzae* lipase	83	3	✓ Lipase to lipid: 70 wt.% ✓ Methanol to lipid: 11:1 ✓ Butanol (mL) to lipid (g): 10 ✓ Reaction time: 72 h	Biodiesel composition investigation	97
S. obliquus	*A. niger* whole cell	80.97	2	✓ Temperature: 35°C ✓ Methanol to oil: 5:1 ✓ 2.5% water content	Investigation of catalytic performance of free whole cell and immobilized whole cell on polystyrene packaging	98
Chlorella sp.	Novozyme 435	293.82 mg from 1.0 g biomass	10	✓ Lipase to lipid: 50 wt.% ✓ 0.2% water content ✓ Temperature: 60°C ✓ Reaction time: 6 h	Investigation of dimethyl carbonate as reaction medium and acyl acceptor	99

enzyme.[100] Also, iron nanoparticles functionalized with carbodiimide have been used for cholesterol oxide immobilization, and gold nanoparticles and graphene oxide have also been utilized for catalase immobilization as well as enhancing enzyme activity in its free form.[101] Nanoparticles have also been applied as biocatalyst supports for biodiesel production. Iron nanoparticles functionalized with 1-ethyl-3-(3-diethylaminopropyl) carbonamide and *N*-hydroxysuccinimide have been used as supports for thermomyces lanuginosus lipase in para nitrophenyl palmitate and palm oil hydrolyses. In this study, a maximum biocatalyst activity of 97.2% was recorded.[102] Furthermore, iron oxide nanoparticles coated with SiO_2 were prepared based on the core–shell method and functionalized by (9-glycidyloxypropyl) trimethoxysilane for utilization in *C. antarctica* immobilization, for which a loading rate equal to 84% and a high efficiency, nearly 100%, were reported.[103] In another study, iron oxide nanoparticles were functionalized with various functional groups such as 2,2-(ethylenedioxy)bis(ethylamine), ethanolamine, and poly(ethylenimine) and attached to *C. rugosa* using glutaraldehyde. The highest activity was found for 2,2-(ethylendioxy)bis(ethylamine) because of the highest spherical radius of the catalyst and consequently, the lowest spatial restriction for substrate attachment. Lipase immobilized on iron oxide nanoparticles and functionalized with APTES and glutaraldehyde showed ability to converted 77% of olive oil to FAME.[104]

11.5 Conclusion

Algal oil conversion into biodiesel using enzymes is one of the most important and interesting methods to obtain biofuel. The high selectivity, low-cost purification of products, and working in mild and less expensive operational conditions make enzymatic processes more prominent. Even with the problem of high preliminary enzyme costs, it has been made clear that the use of high-performance retrieval methods can overcome this problem. Enzyme immobilization can improve thermal and chemical stability, facilitate enzyme separation after reactions, and enhance yields and activities. Enzyme immobilization on nanostructures, due to their high surface area available for staying enzymes, causes lower resistance to mass transfer onto the active sites of biocatalysts, better chemical and mechanical stability of enzymes, and hence higher activity of biocatalysts. Nanofibers, nanoparticles, carbon nanotubes, graphene oxide, and nanopolymers have been used as supports for enzyme immobilization in biodiesel production. An important and promising result of using nanotechnology in the synthesis of biocatalysts for biodiesel production from algal biomass is the significant improvement of the reusability of catalysts (up to 25 times). Reusability is highly effective at decreasing the cost of biocatalysts, while the cost of enzyme is the most important limitation in the use of biocatalysts for biodiesel production from algae oil.

References

1. Enamala MK, Enamala S, Chavali M, Donepudi J, Yadavalli R, Kolapalli B, et al. Production of biofuels from microalgae—a review on cultivation, harvesting, lipid extraction, and numerous applications of microalgae. *Renew Sustain Energy Rev* 2018;**94**:49–68.
2. Zuliani A, Ivars F, Luque R. Advances in nanocatalyst design for biofuel production. *ChemCatChem* 2018;**10**:1968–81.
3. Shahidi F, Zhong Y. Lipid oxidation: measurement methods. *Bailey's Ind Oil Fat Prod* 2005.
4. Sharma CY, Singh B, Madhu D, Liu Y, Yaakob Z. Fast synthesis of high quality biodiesel from 'waste fish oil'by single step transesterification. *Biofuel Res J* 2014;**1**:78–80.
5. Atabani AE, Silitonga AS, Badruddin IA, Mahlia T, Masjuki H, Mekhilef S. A comprehensive review on biodiesel as an alternative energy resource and its characteristics. *Renew Sustain Energy Rev* 2012;**16**:2070–93.
6. Haas MJ, Foglia T. *Alternate feedstocks and technologies for biodiesel production. The biodiesel handbook*. 2005. p. 42.
7. Kan A. General characteristics of waste management: a review. *Energy Educ Sci Technol A Energy Sci Res* 2009;**23**:55–69.
8. Chisti Y. Biodiesel from microalgae. *Biotechnol Adv* 2007;**25**:294–306.
9. Ahmad A, Yasin NM, Derek C, Lim J. Microalgae as a sustainable energy source for biodiesel production: a review. *Renew Sustain Energy Rev* 2011;**15**:584–93.
10. Feng Y, Li C, Zhang D. Lipid production of *Chlorella vulgaris* cultured in artificial wastewater medium. *Bioresour Technol* 2011;**102**:101–5.
11. Rawat I, Kumar RR, Mutanda T, Bux F. Dual role of microalgae: phycoremediation of domestic wastewater and biomass production for sustainable biofuels production. *Appl Energy* 2011;**88**:3411–24.
12. Markou G, Nerantzis E. Microalgae for high-value compounds and biofuels production: a review with focus on cultivation under stress conditions. *Biotechnol Adv* 2013;**31**:1532–42.
13. Mata TM, Martins AA, Caetano NS. Microalgae for biodiesel production and other applications: a review. *Renew Sustain Energy Rev* 2010;**14**:217–32.
14. Demirbas A. Importance of biodiesel as transportation fuel. *Energy Policy* 2007;**35**:4661–70.
15. Daroch M, Geng S, Wang G. Recent advances in liquid biofuel production from algal feedstocks. *Appl Energy* 2013;**102**:1371–81.
16. Amaro HM, Macedo ÂC, Malcata FX. Microalgae: an alternative as sustainable source of biofuels? *Energy* 2012;**44**:158–66.
17. Rodolfi L, Chini Zittelli G, Bassi N, Padovani G, Biondi N, Bonini G, et al. Microalgae for oil: strain selection, induction of lipid synthesis and outdoor mass cultivation in a low-cost photobioreactor. *Biotechnol Bioeng* 2009;**102**:100–12.
18. Braun T, Farag A. Cellular and foamed plastics as separation media: a new geometrical form of the solid phase in analytical liquid-solid contact. *Talanta* 1975;**22**:699–705.
19. Noraini M, Ong HC, Badrul MJ, Chong W. A review on potential enzymatic reaction for biofuel production from algae. *Renew Sustain Energy Rev* 2014;**39**:24–34.
20. Singh J, Gu S. Commercialization potential of microalgae for biofuels production. *Renew Sustain Energy Rev* 2010;**14**:2596–610.
21. Wiltshire KH, Boersma M, Möller A, Buhtz H. Extraction of pigments and fatty acids from the green alga *Scenedesmus obliquus* (*Chlorophyceae*). *Aquat Ecol* 2000;**34**:119–26.

22. Menon V, Rao M. Trends in bioconversion of lignocellulose: biofuels, platform chemicals & biorefinery concept. *Prog Energy Combust Sci* 2012;**38**:522–50.
23. Damartzis T, Zabaniotou A. Thermochemical conversion of biomass to second generation biofuels through integrated process design: a review. *Renew Sustain Energy Rev* 2011;**15**:366–78.
24. Baêsso RM, Oliveira PA, Moraes GC, Alvarenga AV, Costa-Félix RP. *Ultrasound as a metrological tool for monitoring transesterification kinetics. Advanced chemical kinetics.* Croatia: InTech Publisher; 2018.
25. Tran DT, Chang JS, Lee DJ. Recent insights into continuous-flow biodiesel production via catalytic and non-catalytic transesterification processes. *Appl Energy* 2017;**185**:376–409.
26. Marchetti JM, Miguel V, Errazu A. Possible methods for biodiesel production. *Renew Sustain Energy Rev* 2007;**11**:1300–11.
27. Borugadda VB, Goud VV. Biodiesel production from renewable feedstocks: status and opportunities. *Renew Sustain Energy Rev* 2012;**16**:4763–84.
28. Aransiola E, Ojumu T, Oyekola O, Madzimbamuto T, Ikhu-Omoregbe D. A review of current technology for biodiesel production: state of the art. *Biomass Bioenergy* 2014;**61**:276–97.
29. Lam MK, Lee KT, Mohamed AR. Homogeneous, heterogeneous and enzymatic catalysis for transesterification of high free fatty acid oil (waste cooking oil) to biodiesel: a review. *Biotechnol Adv* 2010;**28**:500–18.
30. Bharathiraja B, Chakravarthy M, Kumar RR, Yuvaraj D, Jayamuthunagai J, Kumar RP, et al. Biodiesel production using chemical and biological methods—a review of process, catalyst, acyl acceptor, source and process variables. *Renew Sustain Energy Rev* 2014;**38**:368–82.
31. Issariyakul T, Dalai AK. Biodiesel from vegetable oils. *Renew Sustain Energy Rev* 2014;**31**:446–71.
32. Shimada Y, Watanabe Y, Sugihara A, Tominaga Y. Enzymatic alcoholysis for biodiesel fuel production and application of the reaction to oil processing. *J Mol Catal B Enzym* 2002;**17**:133–42.
33. Lee DW, Park YM, Lee KY. Heterogeneous base catalysts for transesterification in biodiesel synthesis. *Catal Surv Asia* 2009;**13**:63–77.
34. Chang F, Zhou Q, Pan H, Liu XF, Zhang H, Xue W, et al. Solid mixed-metal-oxide catalysts for biodiesel production: a review. *Energy Technol* 2014;**2**:865–73.
35. Umdu ES, Tuncer M, Seker E. Transesterification of *Nannochloropsis oculata* microalga's lipid to biodiesel on Al_2O_3 supported CaO and MgO catalysts. *Bioresour Technol* 2009;**100**:2828–31.
36. Zabeti M, Daud WMAW, Aroua MK. Biodiesel production using alumina-supported calcium oxide: an optimization study. *Fuel Process Technol* 2010;**91**:243–8.
37. Amani H, Ahmad Z, Asif M, Hameed B. Transesterification of waste cooking palm oil by MnZr with supported alumina as a potential heterogeneous catalyst. *J Ind Eng Chem* 2014;**20**:4437–42.
38. Tang S, Wang L, Zhang Y, Li S, Tian S, Wang B. Study on preparation of $Ca/Al/Fe_3O_4$ magnetic composite solid catalyst and its application in biodiesel transesterification. *Fuel Process Technol* 2012;**95**:84–9.
39. Xie W, Wang H, Li H. Silica-supported tin oxides as heterogeneous acid catalysts for transesterification of soybean oil with methanol. *Ind Eng Chem Res* 2011;**51**:225–31.

40. Suzuta T, Toba M, Abe Y, Yoshimura Y. Iron oxide catalysts supported on porous silica for the production of biodiesel from crude Jatropha oil. *J Am Oil Chem Soc* 2012;**89**:1981–9.
41. Amani H, Ahmad Z, Hameed B. Synthesis of fatty acid methyl esters via the methanolysis of palm oil over $Ca_{3.5x}Zr_{0.5y}Al_xO_3$ mixed oxide catalyst. *Renew Energy* 2014;**66**:680–5.
42. Dimian AC, Srokol ZW, Mittelmeijer-Hazeleger MC, Rothenberg G. Interrelation of chemistry and process design in biodiesel manufacturing by heterogeneous catalysis. *Topics Catal* 2010;**53**:1197–201.
43. Al-Zuhair S, Ling FW, Jun LS. Proposed kinetic mechanism of the production of biodiesel from palm oil using lipase. *Process Biochem* 2007;**42**:951–60.
44. Houde A, Kademi A, Leblanc D. Lipases and their industrial applications. *Appl Biochem Biotechnol* 2004;**118**:155–70.
45. Amin HA, Ahmed HM, Mohamed SS, El-Bendary MA, Moharam ME, Abdel-Razek AG. Synthesis of biodiesel from used cooking oil using whole-cell lipase produced by solid state fermentation of *Aspergillus niger* Ky401431. *ARPN J Eng Appl Sci* 2018;**13**:75–88.
46. Utami TS, Hariyani I, Alamsyah G, Hermansyah H. Production of dry extract extracellular lipase from *Aspergillus niger* by solid state fermentation method to catalyze biodiesel synthesis. *Energy Procedia* 2017;**136**:41–6.
47. de Miranda AS, Miranda LS, de Souza RO. Lipases: valuable catalysts for dynamic kinetic resolutions. *Biotechnol Adv* 2015;**33**:372–93.
48. Karube I, Yugeta Y, Suzuki S. Electric field control of lipase membrane activity. *Biotechnol Bioeng* 1977;**19**:1493–501.
49. El Rassy H, Perrard A, Pierre AC. Application of lipase encapsulated in silica aerogels to a transesterification reaction in hydrophobic and hydrophilic solvents: Bi-Bi Ping-Pong kinetics. *J Mol Catal B Enzym* 2004;**30**:137–50.
50. Kobayashi S. Lipase-catalyzed polyester synthesis—a green polymer chemistry. *Proc Jpn Acad Ser B* 2010;**86**:338–65.
51. Kawakami K, Oda Y, Takahashi R. Application of a *Burkholderia cepacia* lipase-immobilized silica monolith to batch and continuous biodiesel production with a stoichiometric mixture of methanol and crude Jatropha oil. *Biotechnol Biofuels* 2011;**4**:42.
52. Lee KW, Min K, Park K, Yoo YJ. Development of an amphiphilic matrix for immobilization of *Candida antartica* lipase B for biodiesel production. *Biotechnol Bioproc Eng* 2010;**15**:603–7.
53. Shah S, Sharma S, Gupta M. Biodiesel preparation by lipase-catalyzed transesterification of Jatropha oil. *Energy Fuels* 2004;**18**:154–9.
54. Yang KS, Sohn JH, Kim HK. Catalytic properties of a lipase from *Photobacterium lipolyticum* for biodiesel production containing a high methanol concentration. *J Biosci Bioeng* 2009;**107**:599–604.
55. Salihu A, Alam MZ. Solvent tolerant lipases: a review. *Process Biochem* 2015;**50**:86–96.
56. Moreira AB, Perez VH, Zanin GM, de Castro HF. Biodiesel synthesis by enzymatic transesterification of palm oil with ethanol using lipases from several sources immobilized on silica–PVA composite. *Energy Fuels* 2007;**21**:3689–94.
57. Bajaj A, Lohan P, Jha PN, Mehrotra R. Biodiesel production through lipase catalyzed transesterification: an overview. *J Mol Catal B Enzym* 2010;**62**:9–14.
58. Bose A, Keharia H. Production, characterization and applications of organic solvent tolerant lipase by *Pseudomonas aeruginosa* AAU2. *Biocatal Agric Biotechnol* 2013;**2**:255–66.

59. Wang YD, Shen XY, Li ZL, Li X, Wang F, Nie XA, et al. Immobilized recombinant *Rhizopus oryzae* lipase for the production of biodiesel in solvent free system. *J Mol Catal B Enzym* 2010;**67**:45–51.
60. Vicente G, Martınez M, Aracil J. Integrated biodiesel production: a comparison of different homogeneous catalysts systems. *Bioresour Technol* 2004;**92**:297–305.
61. Garcia-Galan C, Berenguer-Murcia Á, Fernandez-Lafuente R, Rodrigues RC. Potential of different enzyme immobilization strategies to improve enzyme performance. *Adv Synth Catal* 2011;**353**:2885–904.
62. Jegannathan KR, Abang S, Poncelet D, Chan ES, Ravindra P. Production of biodiesel using immobilized lipase: a critical review. *Crit Rev Biotechnol* 2008;**28**:253–64.
63. Poppe JK, Fernandez-Lafuente R, Rodrigues RC, Ayub MAZ. Enzymatic reactors for biodiesel synthesis: present status and future prospects. *Biotechnol Adv* 2015;**33**:511–25.
64. Mendes AA, Giordano RC, de LC, Giordano R, de Castro HF. Immobilization and stabilization of microbial lipases by multipoint covalent attachment on aldehyde-resin affinity: application of the biocatalysts in biodiesel synthesis. *J Mol Catal B Enzym* 2011;**68**:109–15.
65. Barbosa O, Torres R, Ortiz C, Fernandez-Lafuente R. Versatility of glutaraldehyde to immobilize lipases: effect of the immobilization protocol on the properties of lipase B from *Candida antarctica. Process Biochem* 2012;**47**:1220–7.
66. De Souza MP, Petrus JCC, Gonçalves LAG, Viotto LA. Degumming of corn oil/hexane miscella using a ceramic membrane. *J Food Eng* 2008;**86**:557–64.
67. Ghaly A, Dave D, Brooks M, Budge S. Production of biodiesel by enzymatic transesterification. *Am J Biochem Biotechnol* 2010;**6**:54–76.
68. Sangeetha K, Abraham TE. Preparation and characterization of cross-linked enzyme aggregates (CLEA) of subtilisin for controlled release applications. *Int J Biol Macromol* 2008;**43**:314–19.
69. Matijošytė I, Arends IW, de Vries S, Sheldon RA. Preparation and use of cross-linked enzyme aggregates (CLEAs) of laccases. *J Mol Catal B Enzym* 2010;**62**:142–8.
70. Han JY, Kim HK. Transesterification using the cross-linked enzyme aggregate of *Photobacterium lipolyticum* lipase M37. *J Microbiol Biotechnol* 2011;**21**:1159–65.
71. Verma ML, Barrow CJ, Puri M. Nanobiotechnology as a novel paradigm for enzyme immobilisation and stabilisation with potential applications in biodiesel production. *Appl Microbiol Biotechnol* 2013;**97**:23–39.
72. Misson M, Zhang H, Jin B. Nanobiocatalyst advancements and bioprocessing applications. *J R Soc Interface* 2015;**12**:20140891.
73. Sankaran R, Show PL, Chang JS. Biodiesel production using immobilized lipase: feasibility and challenges. *Biofuel Bioprod Bior* 2016;**10**:896–916.
74. Wang ZG, Wan LS, Liu ZM, Huang XJ, Xu ZK. Enzyme immobilization on electrospun polymer nanofibers: an overview. *J Mol Catal B Enzym* 2009;**56**:189–95.
75. Feng C, Khulbe K, Matsuura T, Tabe S, Ismail A. Preparation and characterization of electro-spun nanofiber membranes and their possible applications in water treatment. *Sep Purif Technol* 2013;**102**:118–35.
76. Ros TG, Van Dillen AJ, Geus JW, Koningsberger DC. Surface oxidation of carbon nanofibres. *Chem Eur J* 2002;**8**:1151–62.
77. Ikemoto H, Chi Q, Ulstrup J. Stability and catalytic kinetics of horseradish peroxidase confined in nanoporous SBA-15. *J Phys Chem C* 2010;**114**:16174–80.
78. Wu L, Wu S, Xu Z, Qiu Y, Li S, Xu H. Modified nanoporous titanium dioxide as a novel carrier for enzyme immobilization. *Biosens Bioelectron* 2016;**80**:59–66.

79. Zhang YW, Tiwari MK, Jeya M, Lee JK. Covalent immobilization of recombinant Rhizobium etli CFN42 xylitol dehydrogenase onto modified silica nanoparticles. *Appl Microbiol Biotechnol* 2011;**90**:499–507.
80. Kalantari M, Kazemeini M, Arpanaei A. Evaluation of biodiesel production using lipase immobilized on magnetic silica nanocomposite particles of various structures. *Biochem Eng J* 2013;**79**:267–73.
81. Xu R, Zhou Q, Li F, Zhang B. Laccase immobilization on chitosan/poly (vinyl alcohol) composite nanofibrous membranes for 2,4-dichlorophenol removal. *Chem Eng J* 2013;**222**:321–9.
82. Neri DF, Balcão VM, Dourado FO, Oliveira JM, Carvalho Jr LB, Teixeira JA. Immobilized β-galactosidase onto magnetic particles coated with polyaniline: support characterization and galactooligosaccharides production. *J Mol Catal B Enzym* 2011;**70**:74–80.
83. Hong J, Xu D, Gong P, Yu J, Ma H, Yao S. Covalent-bonded immobilization of enzyme on hydrophilic polymer covering magnetic nanogels. *Micropor Mesopor Mater* 2008;**109**:470–7.
84. Kuo CH, Liu YC, Chang CMJ, Chen JH, Chang C, Shieh CJ. Optimum conditions for lipase immobilization on chitosan-coated Fe_3O_4 nanoparticles. *Carbohydr Pol* 2012;**87**:2538–45.
85. Guo L, Zeng S, Li J, Cui F, Cui X, Bu W, et al. An easy co-casting method to synthesize mesostructured carbon composites with high magnetic separability and acid resistance. *New J Chem* 2009;**33**:1926–31.
86. Lee HK, Lee JK, Kim MJ, Lee CJ. Immobilization of lipase on single walled carbon nanotubes in ionic liquid. *Bull Korean Chem Soc* 2010;**31**:650–2.
87. Tan H, Feng W, Ji P. Lipase immobilized on magnetic multi-walled carbon nanotubes. *Bioresour Technol* 2012;**115**:172–6.
88. Wang Y, Li Z, Wang J, Li J, Lin Y. Graphene and graphene oxide: biofunctionalization and applications in biotechnology. *Trends Biotechnol* 2011;**29**:205–12.
89. Asmat S, Husain Q, Azam A. Lipase immobilization on facile synthesized polyaniline-coated silver-functionalized graphene oxide nanocomposites as novel biocatalysts: stability and activity insights. *RSC Adv* 2017;**7**:5019–29.
90. Hermanová S, Zarevúcká M, Bouša D, Mikulics M, Sofer Z. Lipase enzymes on graphene oxide support for high-efficiency biocatalysis. *Appl Mater Today* 2016;**5**:200–8.
91. Shao Y, Jing T, Tian J, Zheng Y. Graphene oxide-based Fe_3O_4 nanoparticles as a novel scaffold for the immobilization of porcine pancreatic lipase. *RSC Adv* 2015;**5**:103943–55.
92. Wang J, Zhao G, Jing L, Peng X, Li Y. Facile self-assembly of magnetite nanoparticles on three-dimensional graphene oxide–chitosan composite for lipase immobilization. *Biochem Eng J* 2015;**98**:75–83.
93. Patel V, Gajera H, Gupta A, Manocha L, Madamwar D. Synthesis of ethyl caprylate in organic media using *Candida rugosa* lipase immobilized on exfoliated graphene oxide: process parameters and reusability studies. *Biochem Eng J* 2015;**95**:62–70.
94. Chen JZ, Wang S, Zhou B, Dai L, Liu D, Du W. A robust process for lipase-mediated biodiesel production from microalgae lipid. *RSC Adv* 2016;**6**:48515–22.
95. Guldhe A, Singh B, Rawat I, Permaul K, Bux F. Biocatalytic conversion of lipids from microalgae *Scenedesmus obliquus* to biodiesel using *Pseudomonas fluorescens* lipase. *Fuel* 2015;**147**:117–24.
96. Bharathiraja B, Kumar RR, PraveenKumar R, Chakravarthy M, Yogendran D, Jayamuthunagai J. Biodiesel production from different algal oil using immobilized pure lipase and tailor made rPichia pastoris with Cal A and Cal B genes. *Bioresour Technol* 2016;**213**:69–78.

97. López EN, Medina AR, Moreno PAG, Cerdán LE, Valverde LM, Grima EM. Biodiesel production from *Nannochloropsis gaditana* lipids through transesterification catalyzed by *Rhizopus oryzae* lipase. *Bioresour Technol* 2016;**203**:236–44.
98. Guldhe A, Singh P, Kumari S, Rawat I, Permaul K, Bux F. Biodiesel synthesis from microalgae using immobilized *Aspergillus niger* whole cell lipase biocatalyst. *Renew Energy* 2016;**85**:1002–10.
99. Lee OK, Kim YH, Na J-G, Oh Y-K, Lee EY. Highly efficient extraction and lipase-catalyzed transesterification of triglycerides from *Chlorella* sp. KR-1 for production of biodiesel. *Bioresour Technol* 2013;**147**:240–5.
100. Wang X, Liu X, Yan X, Zhao P, Ding Y, Xu P. Enzyme-nanoporous gold biocomposite: excellent biocatalyst with improved biocatalytic performance and stability. *PLoS One* 2011;**6**:e24207.
101. Kouassi GK, Irudayaraj J, McCarty G. Examination of cholesterol oxidase attachment to magnetic nanoparticles. *J Nanobiotechnol* 2005;**3**:1.
102. Raita M, Arnthong J, Champreda V, Laosiripojana N. Modification of magnetic nanoparticle lipase designs for biodiesel production from palm oil. *Fuel Process Technol* 2015;**134**:189–97.
103. Mehrasbi MR, Mohammadi J, Peyda M, Mohammadi M. Covalent immobilization of *Candida antarctica* lipase on core-shell magnetic nanoparticles for production of biodiesel from waste cooking oil. *Renew Energy* 2017;**101**:593–602.
104. Sarnoab M, Paciello L, Cirillob C, Parascandolaa P, Ciambelli P. Improvement of the lipase immobilization procedure on monodispersed Fe_3O_4 magnetic nanoparticles. *Chem Eng* 2016;49.

Chapter 12

Impacts of sustainable biofuels production from biomass

Shiv Prasad[1] and Avinash P. Ingle[2]

[1]Centre for Environmental Sciences and Climate Resilient Agriculture (CESCRA), ICAR-Indian Agricultural Research Institute, New Delhi, India, [2]Department of Biotechnology, Engineering School of Lorena, University of São Paulo, Lorena, Brazil

12.1 Introduction

The energy sector has played a vital role in the global economy as well as in socioeconomic development.[1,2] Energy consumption has been increasing steadily with the increasing population growth and industrial development. According to leading international body, the Intergovernmental Panel on Climate Change (IPCC), the burning of fossil fuels (coal, gas, and oil), which has increased the content of heat-trapping greenhouse gases (GHGs) in the atmosphere, represents the most significant anthropogenic source of climate change and air pollution.[2,3] About 98% of carbon emissions are generated from fossil fuel combustion. Traditional biomass burning is also known as a vital global source of gaseous emissions adding as much as 40% of the gross CO_2 and 38% of the O_3 in the troposphere.[3,4]

Reductions in open biomass burning and complete combustion of fossil fuels represent the most significant potential for the mitigation of expected catastrophic climate change effects, including lower emissions of harmful pollutants.[2,4,5] This can be achieved either by energy conservation or by a transition to clean, nonfossil fuel, energy sources like biofuels production from biomass. The most significant difference between biofuels and petroleum feedstocks is oxygen content. Liquid biofuels have oxygen levels of 10%–45%, while petroleum has none, making the chemical properties of biofuels very different from those of crude oil.[6] Extensive experience has been gained from using liquid renewable biofuels as pure fuel and by blending with gasoline.[7] For example, ethanol contains 34.7% oxygen by weight and adding oxygen to gasoline fuel results in complete fuel combustion, therefore offering a reduction in exhaust emission and petroleum use.[8,9]

Sustainable Bioenergy. DOI: https://doi.org/10.1016/B978-0-12-817654-2.00012-5

Biofuels have the potential to change the transport and agricultural sectors of decarbonizing societies.[10] Projections show that by 2050 biofuels could satisfy around one-third of worldwide primary energy demand, which is equivalent to approximately 250 EJ.[11] However, the close nexus between the progress of biofuels and the growing demand for food, feed, and fiber is a grave issue and requires the determination of the optimal utilization level of biofuels to ensure sustainable development. The modern forms of biomass energy include its potential conversion into liquid fuels (methanol and ethanol), gaseous fuels (biogas and producer gas), and electricity.[4] Second-generation biofuels include new cellulosic technologies, while firm policies have provided the greatest traction to ensure the commercial production of biofuel in a sustainable manner.[12] As these technologies become more mature, efficient, and economical, they could eventually replace traditional fossil fuels. Biofuels produced from biomass waste or from biomass grown on depleted land offer an excellent opportunity to reduce atmospheric CO_2 emissions and further make it more sustainable.[13]

Policy choices are influential in determining the direction of national as well as global biofuels production.[1] Biofuel policies aim to increase the production and use of biofuels made from biomass as well as other renewable fuels in the transport sector.[6] Biofuel policies concern job creation, higher efficiency in the general business situation, and the protection of the environment. Many nations have adopted policies to encourage liquid biofuel development, led by the United States, the European Union, Brazil, Canada, Australia, and Japan.[14] A growing number of developing countries such as China, India, Philippines, and Thailand have also started to introduce similar policies. Energy security and socioeconomic and environmental benefits can be achieved through the sustainable production of biofuels from biomass.[3,4]

12.2 The need for sustainable biofuel production

The growing energy demand along with the rapid depletion of petroleum and associated resources and concerns about climate and environmental deterioration have led to a resurgence in the development of sustainable and renewable energy alternatives.[3] However, concerns about the overall sustainability of biofuels have been raised, regarding competition with food production, water use and the use of other resources to produce biomass, the release of stored carbon, and impacts on biodiversity if land is cleared to grow energy crops.[1,3,15] In 2008 the Roundtable on Sustainable Biofuels[16,17] released its standards for sustainable biofuels, including: (1) Biofuel production shall follow international treaties and national laws regarding such things as air quality, water resources, agricultural practices, labor conditions, and more. (2) Biofuels projects need to be designed and operated in participatory processes that involve all relevant stakeholders in planning and monitoring. (3) Biofuels shall significantly reduce GHGs as compared to fossil fuels.

This seeks the establishment of a standard methodology for comparing GHG benefits. (4) Its production shall not violate human or labor rights and shall ensure decent work and the well-being of workers. (5) Its production shall contribute to the social and economic development of local, rural, and indigenous peoples and communities. (6) Its production shall not impair food security. (7) Its production shall avoid adverse impacts on biodiversity, ecosystems, and areas of high conservational value. (8) Its production shall promote practices that improve soil health and minimize degradation. (9) Surface and groundwater use should be optimized and contamination or depletion of water resources reduced. (10) Air pollution shall be declined along the supply chain. (11) Biofuels shall be produced most cost-effectively, with a commitment to improving production efficiency and socio-environmental performance in every stage of the biofuel value chain. (12) Biofuel production shall not violate land rights. Generally speaking, sustainability criteria can be used as a tool to safeguard sustainable production and sustainable products in a developing industrial sector.[17]

12.3 Biofuel from biomass

Biomass was the first fuel ever used by humankind and was the mainstay of the global fuel economy until the middle of the 18th century. Renewable energy sources are vital to the future of energy. Among all the renewable energy sources, biomass is the biggest, most diverse, and most readily exploitable resource, and is receiving increased attention as a renewable substitute for fossil fuels.[18,19] The potential contribution of modern biomass energy services to the new energy paradigm is indeed significant. Globally, biomass is the fourth most significant source of energy and supplies about 15% of the total energy use, while in developing countries, it accounts for 38% of the energy use. The world consumes about 400 EJ of energy per year. However, the equivalent of about 100 EJ of mostly remaining crop residues are generated all over the world,[20] which have the potential to produce an additional 180 EJ from energy dedicated grasses and trees.[21] Despite the rapid growth of commercial power, biomass remains the principal source of energy for rural households and traditional industries in developing countries. However, biomass waste burning is emerging as a major air pollution issue in many developing countries including India, now the conversion of abundant biomass to biofuels and associated techno-economics has recently changed global views of harnessing biomass resources as a valuable material rather than a waste.

12.3.1 Biomass feedstock for sustainable biofuel production

Biomass is the total mass of organic materials that originate from living organisms present on the Earth.[22,23] Biomass and its byproducts can be categorized, based on source, into primary, secondary, and tertiary sources. Primary biomass resources are produced by the direct use of solar energy through photosynthesis and are harvested directly from fields or land. This includes crops, perennial short-rotation woody plants, seeds of oil crops, and residues resulting from the harvesting of crops and trees (e.g., corn stover, wheat, and rice straw, tops, limbs, and bark from trees). Secondary biomass resources are produced from the processing of primary biomass either physically (e.g., sawdust in mills), chemically (e.g., waste from pulping processes), or biologically (e.g., cow-dung/slurry or derived from other animal feces). Tertiary biomass resources are postconsumer waste or residue streams, including packaging wastes, used cooking vegetable oils, animal fats, and grease from food-processing in-line grease traps, and construction and demolition debris.[23] Biomass is a complex resource that can be processed in many ways for food, feed, fiber, and fuel production. Annual and perennial species that are cultivated specially to produce solid, liquid, or gaseous energy feedstocks tend to be particularly efficient in biofuel conversion processes and are termed energy crops.

Ethanol production from biomass such as sugar-containing materials like sugarcane or sweet sorghum syrup/juice is the most straightforward route for producing biofuels.[24] Lignocellulosic biomass cannot be used directly to produce ethanol, but requires pretreatment and hydrolysis due to its heterogeneity and complex chemical nature.[22] The chemical composition of biomass feedstock is a crucial factor affecting the efficiency of biofuel production during conversion processes.[7,9] The structural and chemical composition of biomass is highly variable because of genetic and environmental influences and their interactions. The plant cell wall is constituted of mainly six components: (1) cellulose; (2) hemicellulose; (3) lignin; (4) water-soluble sugars, amino acids, and aliphatic acids; (5) ether and alcohol-soluble constituents (e.g., fats, oils, waxes, resin, and many pigments); and (6) proteins. The proportions of these constituents vary in different groups of plants and even in the same group. Collectively, lignocellulosic biomass consists of two essential components: carbohydrate polymers (cellulose and hemicelluloses), which can be transformed into sugars, and a nonfermentable fraction of lignin which can be burned for the production of electricity and the generation of heat.[4,22] Cellulose, hemicellulose, and lignin constitute 75%–95% of the total dry mass of lignocelluloses, of which around 70% is the polysaccharide fraction.[10,22]

The contents of cellulose, hemicellulose, and lignin in common biomass residues and wastes are presented in Table 12.1. Cellulosic biomass contains ~40%–50% cellulose, a glucose polymer; ~25%–35% hemicellulose, a sugar heteropolymer; and ~15%–20% lignin, a nonfermentable phenylpropene unit, along with lesser amounts of minerals, oils, soluble sugars, and other components that can be used for the production of fuel. In lignocelluloses, the cellulose and hemicelluloses are associated with lignin.[7,22] The lignin acts as a

TABLE 12.1 Cellulose, hemicellulose, and lignin content in common agricultural residues and wastes.[22]

Agricultural residue	Cellulose (%)	Hemicellulose (%)	Lignin (%)
Nutshells	25–30	25–30	30–40
Corn cob	45	35	15
Wheat straw	33–40	20–25	15–20
Rice straw	40	18	55
Leaves	15–20	80–85	0
Sorted refuse	60	20	20
Cottonseed hair	80–90	5–20	0
Solid cattle manure	1.6–4.7	1.4–3.3	2.7–5.7
Swine waste	6.0	28	
Primary wastewater solids	8–15	NA	24–29
Paper	85–99	0	0–15
Newspaper	40–55	25–40	18–30
Waste paper (chemical pulp)	60–70	10–20	5–10

TABLE 12.2 Cellulose, hemicellulose, and lignin content in dedicated energy crops.[22]

Dedicated energy crops	Cellulose (%)	Hemicellulose (%)	Lignin (%)
Grasses	25–40	35–50	10–30
Coastal Bermuda grass	25	35.7	6.4
Switchgrass	30–50	10–40	5–20

physical barrier and must be removed in order to make the carbohydrates available for further transformation.[25] Because lignin is believed to be a significant hindrance to enzymatic hydrolysis, its removal enhances cellulose digestion and also reduces the nonproductive binding of cellulose to lignin.[22]

Dedicated energy or solid crops such as *Miscanthus giganteus* (miscanthus), *Panicum vigratum* (switchgrass), *Hibiscus cannabinus* (kenaf), *Phalaris arundinacea* (reed canary grass), *Arundo donex* (giant reed), *Eucalyptus globulus* (eucalyptus) are considered as short-term crops for the production of biofuel.[22,23] The contents of cellulose, hemicellulose, and lignin in dedicated energy crops are presented in Table 12.2.

TABLE 12.3 Microalgae oil potential compared to other terrestrial oilseed crops.[27]

Crop	Oil content (%)	Oil yield (L/ha/year)
Corn	3.1–5.7	172
Soybean	18–22	446
Canola/Mustard	40–45	1190
Jatropha	20–40	1892
Coconut	17	2689
Oil palm	23	5950
Microalgae	30 (min. scenario)	58,700
Microalgae	70 (max. scenario)	136,900

Presently, major efforts are diverted towards the production of biodiesel, an alternative diesel fuel made from vegetable oil and animal fat. Biodiesel derived from different sources of vegetable oils, such as soybean oil, palm oil, sunflower oil, and rapeseed oil, are used in different parts of the world. The oil content in various edible oilseeds mostly ranges from 30% to 50%. Many algae are exceedingly rich in lipids or oil contents.[26] The average oil levels of microalgae are about 20%–50% (dry weight), while the oil contents themselves can be estimated to be 64.4% of the total lipid component.[27] Microalgal lipids are mostly neutral lipids with a low degree of unsaturation. That makes them potential feedstocks for biofuels production. Microalgae are capable of generating manifold more oil per unit area of land compared to conventional terrestrial oilseed crops. The oil content of some microalgae exceeds 70% of the dry weight of the algal biomass. Using microalgae to produce biodiesel would not compromise the production of food, fodder, and other products derived from crops. Microalgae have the highest oil yield among various plant oils. Microalgae can produce up to 100,000 L/ha/year of oil, whereas palm, coconut, castor, and sunflower produce up to 5950, 2689, 1413, and 952 L/ha/year of oil, respectively[27] (Table 12.3). The Aquatic Species Program (ASP) considered three main options for fuel production: methane gas, ethanol, and biodiesel, along with a fourth option, which entails the direct combustion of algal biomass for the production of steam or electricity; however, the ASP did not focus much on direct combustion.[28]

12.3.2 Biofuel production pathway from biomass

Biofuels derived from biomass may be in solid, liquid, or gaseous forms, of which liquid biofuels are the most imperative, given the current scenario in the world transportation sector and their ability to reduce/replace nonrenewable petroleum fuels. Liquid biofuels are now being produced from sugar, grain-starch, oilseed crops, and animal fats with first-generation technologies.[22,23] New second-generation biofuel production technologies are being used to produce ethanol and diesel from cellulosic biomass. The transformation of biomass-into-biofuels can be realized principally through chemical, biochemical, and thermochemical processes.[25] The biochemical conversion of biomass into ethanol happens in three steps: pretreatment; enzymatic hydrolysis; and fermentation. Pretreatment weakens the plant wall, then acid or enzymatic hydrolysis separates the cellulose into sugars, and lastly, fermentation converts the sugars into ethanol.[29] For the synthesis of essential biofuels such as biodiesel, cellulose needs to undergo thermochemical processes such as pyrolysis or gasification. Apart from the high-efficiency production of oil for biodiesel, microalgae are also well suited for the production of feedstocks for other biofuels. The development of technologies is also applicable to biohydrogen, biogas, bioethanol, and the biomass-to-liquid (BTL) approach using fast-growing algae. BTL, biohydrogen, and biomethane processes are discussed in this chapter as they are especially pertinent to the microalgal system. First and second-generation biofuel pathways are presented in Fig. 12.1.

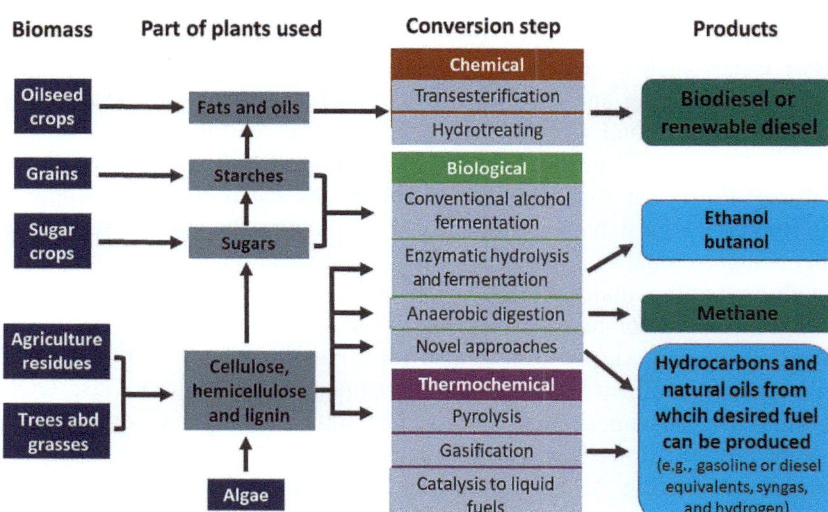

FIGURE 12.1 First and second-generation biofuel pathways. *Adapted from Pena N, Sheehan J. Biofuels for transportation. In:* CDM investment newsletter. *2007. Number 3/2007. <https://www.c2es.org/site/assets/uploads/2007/11/cdm-investment-newsletter-biofuels-transportation.pdf>; an open access article. Courtesy of the Center for Climate and Energy Solutions.*

12.4 Policies and standards to promote sustainable biofuels production

Several countries have introduced policies or adopted standards to promote sustainable biofuels production and use. India is one of the few countries to have a separate ministry for renewable energy (Ministry of New and Renewable Energy; MNRE) which addresses the development of renewables along with biofuels. MNRE consists a task force to assist in the development of a national program for the biomass-based cogeneration of energy sources. The government of India (GOI) is dedicated to expanding its contributions of sustainable energy (resources with low-carbon generation) in all of India's end-use sectors and undertakes policy and planning activities to that end.[30] The GOI through a notification in September 2003 made 5% ethanol blending mandatory in petrol in nine states and three union territories. In the next phase, the supply of ethanol-blended gasoline was extended to the whole country and efforts were made to increase the percentage of ethanol in the mixture with gasoline to 10%. In April 2003 the GOI launched the National Biodiesel Mission that identified *Jatropha curcas* as the most suitable tree-borne oilseed for the production of biodiesel and focused on promoting plantations of jatropha on wastelands.[31]

The GOI approved the National Biofuels Policy (NBP) on December 24, 2009. The NBP assures that the biofuel program would not compete with food security and that fertile farmlands would not be diverted for the plantation and production of biofuel crops.[31,32] The policy deals with many issues such as minimum support prices, subsidies for biofuel crop growers, marketing and subsidies for the biofuel industry, the mandatory blending of auto-fuel with biofuel, quality norms, and testing and certification of biofuels.[32] Further, the Union Cabinet of India has approved the National Policy on Biofuels—2018. The Policy expands on the scope of raw materials for ethanol production by allowing the use of sugarcane juice, sugar-containing materials like sugar beet and sweet sorghum, starch-containing materials like corn and cassava, as well as damaged foodgrains unfit for human consumption and rotten potatoes. The policy allows the use of surplus foodgrains for ethanol production for blending with petrol/gasoline.[33] Ministry of Petroleum and Natural Gas, GOI, assured the parliament that only grains unfit for human consumption would only be allowed on a conditional basis to be used as feedstock for ethanol production in an attempt to quell fears that deregulation of feedstock is allowing grains would potentially negatively impact food security. The policy also supports the establishment of supply chain mechanisms for biodiesel production from non-edible oilseeds, used cooking oil, and short gestation crops.[34]

12.5 Impacts of sustainable biofuels production

The sustainable use of biomass for biofuels production has the potential to address many environmental problems ranging from desertification to climate change and socioeconomic concerns, as well as to reduce demand for costly oil imports.[4] However, controversy now extends to the global impacts of national biofuels production as concerns have been raised regarding competition with food production, water use, and other resources to produce biomass and regarding the release of stored carbon stocks and the impacts on biodiversity if the land is cleared to grow energy crops. Second-generation biofuels technologies for commercial biofuel production from biomass waste or biomass grown on depleted land offer excellent opportunities to ensure energy security, and socioeconomic and environmental benefits.[9,11] Many countries are dedicated to expanding their contributions to sustainable energy (resources with low-carbon generation) in all of their end-use sectors and to undertaking policy and planning activities to that end.

12.5.1 The food-versus-fuel debate

The food-versus-fuel debate is a controversy embedded in this challenge, involving the trade-offs of utilizing grains and oilseeds for biofuels production versus animal feed and human food. This ethical debate has become known as the "food for fuel" debate.[35] Unlike producing ethanol from corn, cellulosic ethanol is produced from the inedible parts of plants or organic materials so there is no competition between food and fuel. However, crop residues protect the soil from erosion and maintain soil organic matter. The primary challenge within sustainable soil management is maintaining the balance between the residue removal rate and long-term soil health.[36] However, reports state that the removal of 20%–30% of crop residue is probably sustainable, although extra fertilizer would eventually be required to replace the nutrients removed. Soil erosion is affected by crop type and production practices. Increased bioenergy production increases erosion risk.[36,37]

The choice of plants is essential, especially if maize replaces grass and forages. Production practices such as winter cover crops, where appropriate, can mitigate the risk of erosion.[37] Prices in food markets, despite various interventions, are still affected substantially by fluctuations in supply and demand. However, only 6% of the total global foodgrain produced is used to generate ethanol.[38] Furthermore, biofuel coproducts contribute to the sustainability of food production because just 1%–2.5% of the overall energy efficiency is lost from transforming crops into biofuels and cattle feed, and about one-third of the corn used to produce ethanol is recovered as feed coproducts.[38,39] Second and third-generation biofuels are considered better alternatives to petrol-derived fuels when compared to food crop-derived first-generation biofuels on account of their ability to utilize wastes and other material with low food supply interactions.[39]

12.5.2 Changes in land use for sustainable biofuel production

The issue of indirect land use changes often arises from the perceived conflict between food production and biofuels production.[3,40] Food crops such as corn and wheat grains, sugarcane, and vegetable oils are some of the most commonly used energy feedstocks. There are several prospects to further reduce the conflict between food and fuel production, including increased use of agricultural wastes, logging residues, food scraps, municipal solid waste, and marginal lands. If grasslands, savannas, and scrublands with marginal productivities are considered for planting low-input high-diversity (LIHD) mixes of domestic perennials as energy crops, the total land availability can extend from 1107 to 1411 Mha, depending on whether pasturelands are discounted.[41] Growing second-generation biofuel feedstocks on abandoned and degraded land and LIHD perennials on grassland with marginal productivity may satisfy 26% − 55% of the current global liquid fuel consumption, without impacting on the use of land with regular productivity for conventional crops and without changing existing pastureland. Under the different land use scenarios, Africa may have more than one-third, and Africa and Brazil collectively, may hold more than half of the total land available for biofuel production. These estimates are based on physical conditions such as soil productivity, land slope, and climate.[41] The Energy Independence and Security Act of 2007 outlines standards for GHG reduction that must be met in order for a given fuel to be considered renewable. In December of 2010, the Environmental Protection Agency finalized a set of Renewable Fuel Standard (RFS2) regulations establishing the necessary level of GHG reduction as compared to the emissions of the 2005 baseline average for gasoline or diesel fuel that it replaces.[40] These reductions are calculated based on lifecycle analyses and include indirect emissions from land use changes associated with each fuel[42] (Table 12.4).

12.5.3 Use of biomass as mitigation option for climate change

The primary cause of climate change is the burning of fossil fuels.[43] At the Bologna, Italy, 2016 G7 summit, climate change was once again top of the agenda which led to several initiatives such as UNFCCC and the Kyoto Protocol. It has also motivated deep interest in all manner of renewable energy sources, biofuels among them.[2,43,44] With transport contributing to around 25% of global CO_2 emissions and with extremely few viable alternative fuels available, biofuels are seen as being instrumental in a shift to low-carbon fuels that would bring about sustainability in the transportation

TABLE 12.4 Reduction thresholds of various fuels.[42]

Fuel class	Lifecycle GHG reduction thresholds (%)	Examples
Renewable fuel	20	Corn-based ethanol
Advanced biofuel	50	Sugarcane ethanol
Biomass-based diesel	50	Soy-based biodiesel
Cellulosic biofuel	60	Cellulose-based ethanol

GHG, Greenhouse gas.

sector.[45] There is no question that when produced and used judiciously, biofuels can deliver substantially lower net GHG emissions than fuels derived from fossil sources. It seems virtually inevitable that biofuels will (and should) have a role in national and global strategies to address the threats of climate change. In the fourth climate change assessment report by the IPCC, the use biofuels was identified as a key mitigation strategy.[2,43]

The estimated range of reductions in GHG emissions per vehicle kilometer for rape methyl ester (RME) compared to conventional diesel fuel (for which RME can be substituted) according to Life-cycle assessment (LCA) study (Fig. 12.2) is 16%–63%; a range of a factor of four. This reduction indicated for soy methyl ester is 45%–75%.[46] Ethanol produced from wheat showed anywhere from a 38% GHG emissions benefit to a 10% penalty relative to gasoline. Conditioning current biofuel support measures to such criteria would ensure that government policies do not inadvertently lead to increases in GHG emissions in this area.

An LCA of Jatropha biodiesel regarding its environmental impact was carried out by researchers. More than 90% of GHG emissions during the lifecycle of both diesel and biodiesel are from the use phase.[47] The rate of total global warming potential from biodiesel production and use is just 23% that of diesel.[47] The energy efficiency of biodiesel conversion should be given priority for the improvement of the process as it is the main contributor to both energy use as well as GHG emissions.[47] Almost 60% of the energy consumption in the transesterification step is from steam. The alteration and maintenance of engines and more efficient biodiesel conversion technologies can also help in reducing energy consumption. However, this needs to be further investigated. Fig. 12.3 shows the comparative lifecycle GHG emissions of biodiesel and diesel.[47]

Biomass is the only renewable resource for producing carbon-bearing biofuels. Unused, discarded biomass residues from forestry, agricultural, and

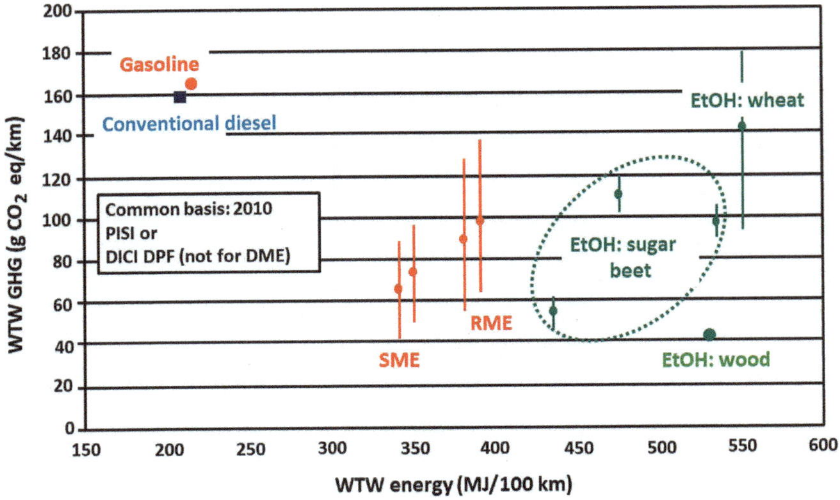

FIGURE 12.2 Well-to-wheels energy requirements and greenhouse gas emissions for conventional biofuel pathways compared with gasoline and diesel pathways, assuming 2010 vehicle technology.[46] *Notes: DICI DPF*, Direct injection compression ignition with diesel particulate filter; *EtOH*, ethyl alcohol (ethanol); *PISI*, port injection spark ignition; *RME*, rape methyl ester; *SME*, soy methyl ester. *Source:* From Biofuel production technologies: status, prospects and implications for trade and development, Originally adapted from CONCAWE (Oil Companies' European Association for Environment, Health and Safety in Refining and Distribution, but the acronym is derived from "Conservation of Clean Air and Water in Europe"), Joint Research Centre of the EU Commission, and European Council for Automotive R&D, 2004. "Well-to-Wheels Analysis of Future Automotive Fuels and Powertrains in the European Context," Version 1b, January, 60. https://iet.jrc.ec.europa.eu/about-jec/sites/iet.jrc.ec.europa.eu.about-jec/files/documents/wtw3_wtw_report_eurformat.pdf[46]*; an open access article.*

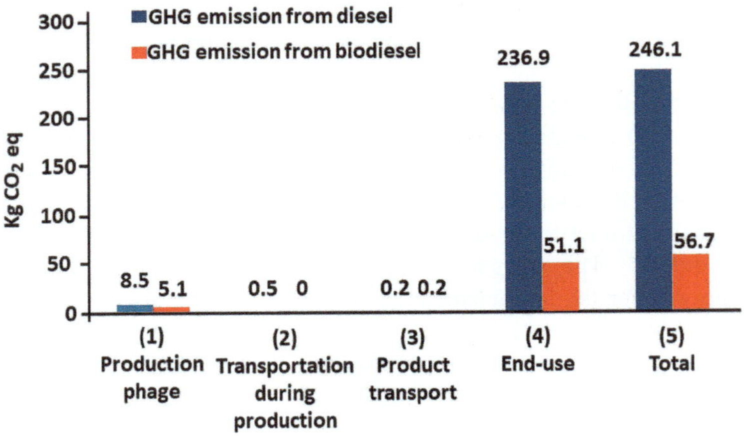

FIGURE 12.3 Lifecycle GHG emissions of biodiesel and diesel. *GHG*, Greenhouse gas. *Source: Adapted and modified from K. Prueksakorn and S.H. Gheewala*[47]*;* Energy and greenhouse gas implications of biodiesel production from *Jatropha curcas L, In: The 2nd joint international conference on sustainable energy and environment, 21–23 November, 2006, Bangkok, Thailand. an open access article.*

municipal sources are potential energy resources which at present are not well managed and thus pose significant environmental problems, especially GHG emissions. More effective use of this resource for the production of bioenergy and related byproducts could contribute to the displacement of fossil fuel emissions, stabilize CO_2 in terrestrial biomass again, and also achieve the direct mitigation of GHG emissions implicated in climate change.[48]

Most biofuels are sufficiently clean and environment-friendly for transportation.[40,47] Biofuels as alternatives to fossil fuels can help in the reduction of atmospheric CO_2 in three ways: (1) by avoiding the emissions associated with fossil fuels; (2) by allowing the CO_2 content in fossil fuels to remain in storage; and (3) by providing a mechanism for CO_2 absorption by growing new biomass for fuels. Due to their compatibility with the natural carbon cycle, biofuels offer the most beneficial alternative for reducing GHG emissions from the transportation sector.[49]

The Paris Climate Agreement took effect on November 4, 2016. On October 2, 2016, many countries including India ratified the Paris agreement on climate change. As part of its initial commitments to the Paris agreement, over the next 15 years, India plans to reduce its carbon emission per unit of gross domestic product (GDP) by 33% from its 2005 levels, and aims to use nonfossil fuels to produce 40% of its installed electricity capacity by 2030.[50] This development would not only reduce the use of fossil fuel, but would also mitigate the threat of climate change.

12.5.4 Sequestering carbon in biomass

Terrestrial and aquatic carbon sinks such as plant biomass and microalgae effectively capture and dispose of atmospheric CO_2 by converting it, through photosynthesis, into carbon. For instance, forests in the northern hemisphere have been estimated to sequester up to 0.7 GT of C annually which accounts for almost 10% of current global fossil fuel C emissions.[51] Furthermore, through the use of carbon-sequestering biomass as feedstock for biofuel production, emissions from the production and combustion of fossil fuels can be offset within the transportation and energy sectors. Promising results have been observed for algal biofuel production and carbon dioxide fixation and sequestration into aquifers or depleted oil and gas wells. Commercial interest in large-scale algal cultivation systems could be obtained by placing algae plants near coal power plants, sewage treatment facilities, or any industry that emits large quantities of carbon dioxide into the atmosphere. Much work has been done on the effect of different flue gas constituents on microalgal growth rates and carbon dioxide fixation. Flue gases emitted from coal-based power plants have carbon dioxide levels ranging from 10% to 15% (4% for natural gas-fired ones). At average carbon dioxide percentages in the atmosphere (0.038%), microalgae show no signs of significant growth inhibition.

Furthermore, various studies have shown that microalgae respond better to increased carbon dioxide concentrations, outgrowing (on a biomass basis) microalgae exposed only to ambient air. This strategy of large-scale algal cultivation system provides not only raw materials for producing greater biodiesel but also converting captured CO_2 and nutrient wastes into valuable resources.[27,32]

12.5.5 Sustainability and air pollution abatement

Air quality is an essential aspect of sustainability because the location and magnitude of air pollutants can have harmful effects on human health and the environment. Emissions from gasoline engines contain SO_x, NO_x, CO, CO_2, volatile organic compounds (VOCs), and particulate matter (PM) which cause pollution.[52] Biofuels by nature are oxygenated fuels and have oxygen levels ranging between 10% and 45%, while petroleum has none, essentially making the chemical properties of biofuels extremely different from those of petroleum.[6] Ethanol and biodiesel contain 34.7% and 11% oxygen contents by weight, respectively. The addition of oxygen to fuel results in the complete combustion of the fuel, and therefore contributes to a reduction in exhaust emissions and petroleum use.[5,53,54] The effects of using low biofuel blends on air quality are mostly positive. For example, ethanol and biodiesel blends reduce carbon monoxide emissions by 25%–50%.[53,55]

The burning of crop residues during the wheat and rice harvesting seasons in the Indo-Gangetic plains releases many pollutants into the atmosphere. Punjab and Haryana burn almost 30 MT of crop residues annually. These two states contribute to 48% of the total emission across India due to paddy burning. According to Gupta et al.,[56] the burning of 1 t of straw releases PM as well as 60 kg of CO, 1460 kg of CO_2, 199 kg of ash, and 2 kg of SO_2. This has adverse impacts on the environment and the economy as this straw is wasted as well as having grave implications for human health and society due to the smoke and fumes produced.[55] It is predicted that if crop residues and wasted crops are used to produce second-generation biofuels, it can reduce carbon emissions by 90%, and by 2040 these could replace up to 40% of all conventional fossil fuels.[57] Thus the utilization of these crop biomasses not only reduces air pollution, but also provides an alternative for Indian farmers to the burning of biomass in fields. Thus this option could support biofuels production and sustainable transportation.

12.5.6 Energy services to the local community

One of the direct benefits that biofuel projects can provide to rural communities, especially in the developing countries, is the availability of fuel itself which can generate different energy services.[19,58] The amount and quality of available energy significantly affects the living conditions in rural areas which is currently a significant limitation in many developing countries.[59] The relatively uneven distribution and use of modern energy sources (e.g., electricity, petroleum products, etc.) in rural areas lead to important issues in economics, equity, and quality of life.[60] Nearly half the world's population still relies on solid fuels for household use, resulting in severe impacts on health, particularly for young children and women.[61] Worldwide, almost two million deaths occur annually from pneumonia, chronic lung disease, and lung cancer which have been linked to exposure to indoor air pollution resulting from cooking with biomass and coal.[62]

Since traditional bioenergy sources are unable to provide clean, cheap, and efficient energy (such as electricity and ethanol fuel) and given their potential to pose serious environmental and health risks, it is essential to address the mentioned energy-related problems from an environmental, energy sustainability, and economic perspective.[59] Access to energy services is closely linked to social and economic development and human welfare.[63,64] Energy services can significantly reduce the amount of time and effort that rural women and children spend collecting fuelwood and performing household activities. The extra time saved can be used on more productive as well as social activities, including education and income-generating activities. People cannot perform efficiently or produce goods if much of their time is being spent searching for fuel or if much of their income is utilized to pay for inefficient power.[60] It has been argued that access to high quality energy sources can assist in the improvement of a variety of human development activities such as education, health, poverty alleviation, and creating excellent local environments.[65,66]

12.5.7 Job creation and rural development

The biofuel industry is a significant employer and could potentially generate thousands of new jobs.[67] It is estimated that the effect of producing 2 billion liters of ethanol in the rural economy would create 6645 jobs in rural Canada. According to International Energy Renewable Energy Agency, employment in the renewable energy sector expanded by 5% in 2015 to 8.1 million jobs. China is the most significant global renewable energy sector employer with 44% (3.4 million) of the world's jobs, followed by Brazil with more than 0.9 million jobs, the United States (0.7 million), India (0.4 million), and Germany (0.3 million). The emergence of biofuels in India has been an uphill struggle that is slowly getting easier. The Bloomberg analysis estimates that the Indian cellulosic ethanol industry could create up to 1 million jobs in predominantly rural areas by 2020. The number of people employed in the renewable energy sector across the globe could increase to 24 million by 2030.[67]

12.6 Conclusion

Biofuels, renewable energy sources produced by living organisms or their metabolic products, have excellent future prospects for ethanol production in developing countries. Since the advancement of society, biofuels have been the only energy resource used as a substitute for fossil fuels for cooking, lighting, heating, and later for powering diesel or gasoline engines. The modern forms of biomass energy such as liquid fuels (methanol and ethanol), gaseous fuels (biogas and producer gas), and electrical power from biomass gasification offer the significant advantage of being carbon neutral. However, the close nexus between biofuels progress and the growing demand for food, feed, and fiber is a grave issue and requires the determination of the optimal utilization level of biofuels to ensure sustainable development. The use of biofuel, primarily for transport purposes would be more economical and environmentally friendly. The efficient use of biomass for sustainable biofuels production has the potential to create jobs and economic growth in developing countries, as well as to reduce the demand for costly oil imports, improve rural livelihoods, and address many environmental problems ranging from desertification to climate change.

Acknowledgment

The authors are grateful to the Centre for Environmental Sciences and Climate Resilient Agriculture (CESCRA), ICAR-Indian Agricultural Research Institute (IARI), New Delhi, and the Indian Council of Agricultural Research for providing facilities and financial support to undertake these investigations. There are no conflicts of interest. API is highly thankful to the Research Council of the State of Sao Paulo, Fundação de Amparo à Pesquisa do Estado de São Paulo (FAPESP), Brazil for providing financial assistance (Process Number-2016/22086-2) in the form of Post-Doctoral Fellowship.

References

1. Prasad S, Dhanya MS, Gupta N, Kumar A. Biofuels from biomass: a sustainable alternative to energy and environment. *Biochem Cell Arch* 2012;**12**(2):255–60.
2. IPCC. *The fourth climate change assessment report, mitigation of climate change: technical summary*. Geneva: IPCC; 2007.
3. Prasad S, Dhanya MS. Biofuel production potential, processing and utilization. In: Haggerty AP, editor. *Biomass crops: production, energy and the environment.* NOVA Publisher; 2011. p. 165–87.
4. Levine JS. *Global biomass burning: atmospheric, climatic, and biospheric implications.* Cambridge, MA: The MIT Press; 2011. p. 25–30.
5. Prasad S, Dhanya MS. Air quality and biofuels <http://www.intechopen.com/books/environmental-impact-of-biofuels/air-quality-and-biofuels/> (accessed 11.08.18) In: Dos Santos Bernardes MA, editor. *Environmental impact of biofuels.* InTech; 2011.
6. Demirbas A. Political, economic and environmental impacts of biofuels—a review. *Appl Energy* 2009;**86**:S108–17.

7. Wyman CE. Ethanol from lignocellulosic biomass: technology, economics, and opportunities. *Bioresour Technol* 1994;**50**:3–15.
8. Huang HJ, Ramaswamy S, Tschirner UW, Ramarao BV. A review of separation technologies in current and future biorefineries. *Sep Purif Technol* 2008;**62**:1–21.
9. RFA. *Ethanol: clean air, clean water, clean fuel.* Washington, DC: Renewable Fuels Association; 2001.
10. Araújo K, Devinder M, Ryan K, Marcelo S. Global biofuels at the crossroads: an overview of technical, policy, and investment complexities in the sustainability of biofuel development. *Agriculture* 2017;**7**:32. Available from: https://doi.org/10.3390/agriculture7040032.
11. WEC. *World energy resources.* Bioenergy World Energy Council (WEC); 2013. <https://www.worldenergy.org/wpcontent/uploads/2013/10/WER_2013_7_Bioenergy.pdf/>.
12. Nguyen Q, Jim B. *Global production of second-generation biofuels: trends and influences.* <http://www.dovetailinc.org/report_pdfs/2017/dovetailbiofuels0117.pdf/>; 2017 [accessed 13.08.18].
13. Cao Y, Pawłowski L. Effect of biofuels on the environment and sustainable development. *Ecol Chem Eng Sci* 2013;**20**(4):799–804.
14. Kemausuor F, George YO, Abeeku BH, Alfred D. A review of trends, policies and plans for increasing energy access in Ghana. *Renew Sustain Energy Rev* 2011;**15**:5143–54.
15. ETIP Bioenergy-SABS. *ETIP biofuels and sustainability issues.* <http://www.etipbioenergy.eu/?option = com_content&view = article&id = 268/>; 2018 [accessed 13.07.18].
16. Pavlovskaia E. Using sustainability criteria for biomass. *J Multidiscip Engine Sci Technol* 2015;**2**(1):192–201.
17. Roundtable on Sustainable Biofuels. *Global principles & criteria for sustainable biofuels production, version 0. 5.* Lausanne, Switzerland: Swiss Federal Institute of Technology. < http://cgse.epfl.ch/page65660-en.html/ > ; 2009 [accessed 19.07.18].
18. P.R. Shukla. *Biomass energy in India: policies and prospects.* <http://www.decisioncraft.com/energy/papers/ecc/re/biomass/bpi.pdf/>; 2010 [accessed 13.07.18].
19. Sheelanere P, Kulshreshtha S. Sustainable biofuel production: opportunities for rural development. *Int J Environ Resour* 2013;**2**(1):1–13.
20. Woods J, Hall DO. Bioenergy for development - Technical and environmental dimensions - FAO Environment and energy paper 13, FAO - Food and Agriculture Organization of the United Nations, FAO- Fiat/Panis Rome, 1994. <http://www.agrienvarchive.ca/bioenergy/download/bioenergy_fao.pdf> [accessed 21.02.19].
21. IPCC. Revised *guidelines for national greenhouse gas inventories 2006.* < https://www.ipcc-nggip.iges.or.jp/public/gl/invs1.html> [accessed 21.02.19].
22. Prasad S, Singh A, Joshi HC. Ethanol as an alternative fuel from agricultural, industrial and urban residues. *Resour Conserv Recycle* 2007;**50**:1–39.
23. ESMAP. *Sustainable bioenergy development in UEMOA member countries, UNF biomass conversion technologies.* <https://www.coursehero.com/file/13604319/UNF-Bioenergy-5/#!/>; 2005 [accessed 15.08.18].
24. Prasad S, Singh A, Jain N, Joshi HC. Ethanol production from sweet sorghum syrup for utilization as automotive fuel in India. *Energy Fuels* 2007;**21**(4):2415–20.
25. Prasad S, Malav MK, Kumar S, Singh A, Pant D, Radhakrishnan S. Enhancement of bio-ethanol production potential of wheat straw by reducing furfural and 5-hydroxymethylfurfural (HMF). *Bioresour Technol Rep* 2018;**4**:50–6.
26. Veillette M, Chamoumi M, Nikiema J, Faucheux N, Heitz M. Production of biodiesel from microalgae. *Adv Chem Eng* 2012;**2**:245–68.
27. Chisti Y. Biodiesel from microalgae. *Biotechnol Adv* 2007;**25**:294–306.

28. Sheehan J, Dunahay T, Benemann J, Roessler P. *A look back at the US Department of Energy's Aquatic Species Program: biodiesel from algae*. Golden: National Renewable Energy Laboratory; 1998.
29. Pena N, Sheehan J. Biofuels for transportation. In: *CDM investment newsletter*. 2007. <https://www.c2es.org/site/assets/uploads/2007/11/cdm-investment-newsletter-biofuels-transportation.pdf>.
30. Planning Commission. *Commission report of the committee on development of biofuel*. Govt. of India. <http://planningcommission.nic.in/reports/genrep/cmtt_bio.pdf/>; 2003 [accessed 13.07.18].
31. MoPNG. Energizing the nation. In: *Annual report 2013-14*. GOI; 2014. Available at <http://petroleum.nic.in/docs/Annual_Report/AR13-14.pdf/>.
32. Prasad S, Kumar A, Muralikrishna KS. Biofuels production: a sustainable solution to combat climate change. *Indian J Agric Sci* 2014;**84**(12):1443–52.
33. The Hindu. *Cabinet GOI approves new biofuels policy*. The Hindu. <https://www.thehindu.com/news/national/cabinet-approves-new-biofuels-policy/arti/>; 2018 [accessed 13.07.18].
34. Biofuel Digest. *India will only allow grain unfit for human consumption for biofuel feedstock*. <http://www.biofuelsdigest.com/bdigest/2018/07/25/india-will-only-allow-grain-unfit-for-human-consumption-for-biofuel-feedstock/>; 2018 [accessed 13.07.18].
35. Ajanovic A. Biofuels versus food production: does biofuels production increase food prices? *Energy* 2011;**36**:2070–6.
36. Williams PRD, Inman D, Aden A, Heath GA. Environmental & sustainability factors associated with next-generation biofuels in the US: what do we really know. *Environ Sci Technol* 2009;**43**(13):4763–75.
37. Kim S, Dale BE. Life cycle assessment of various cropping systems utilized for producing biofuels-bioethanol and biodiesel. *Biomass Bioenergy* 2005;**29**(6):426–39.
38. Shurson GC. The role of biofuels co-products in feeding the world sustainably. *Annu Rev Anim Biosci* 2017;**5**(5):1–5.
39. Sikarwar VS, Zhao M, Paul SF, Shah M, Anthony EJ. Progress in biofuel production from gasification. *Prog Energy Combust Sci* 2017;**61**:189–248.
40. De-Wit M, Junginger M, Faaij A. Learning in dedicated wood production systems-past trends, future outlook and implications for bioenergy. *Renew Sustain Energy Rev* 2013;**19**:417–32.
41. Cai X, Zhang X, Wang D. Land availability for biofuel production. *Environ Sci Technol* 2011;**45**(1):334–9.
42. EPA (United States Environmental Protection Agency). *EPA lifecycle analysis of greenhouse gas emissions from renewable fuels: regulatory announcement*. Available at <https://nepis.epa.gov/>; 2010 [accessed 31.10.18].
43. IPCC. *Climate change, fifth assessment report of the intergovernmental panel on climate change*. Geneva, Switzerland: IPCC; 2014. p. 151.
44. Lee RA, Lavoie JM. From first- to third-generation biofuels: challenges of producing a commodity from biomass of increasing complexity. *Anim Front* 2013;**3**(2):6–11.
45. IRENA. *Renewable energy and jobs. Annual review*. <https://www.irena.org/DocumentDownloads/Publications/IRENA_RE_Jobs_Annual_Review_2016.pdf/>; 2016 [accessed 13.07.18].
46. CONCAWE (Oil Companies' European Association for Environment, Health and Safety in Refining and Distribution, but the acronym is derived from "Conservation of Clean Air and Water in Europe"), Joint Research Centre of the EU Commission, and European Council

for Automotive R&D. *Well-to-wheels analysis of future automotive fuels and powertrains in the European context, version 1b*. 2004. p. 60.

47. Prueksakorn K, Gheewala SH. Energy and greenhouse gas implications of biodiesel production from *Jatropha curcas* L, In: *The 2nd joint international conference on sustainable energy and environment, 21–23 November, 2006, Bangkok, Thailand.*
48. IEA. *Answers to ten frequently asked questions about bioenergy, carbon sinks and their role in global climate change.* <http://www.joanneum.ac.at/iea-bioenergy-task38/pub/>; 2001 [accessed 13.07.18].
49. Ferrell J, Glassner D, Cushman J. Biofuels- a solution for climate change. U.S. Department of Energy Office of Energy Efficiency and Renewable Energy Office of Transportation Technologies, DOE/GO-10098-580, Revised September 1999 <https://www.nrel.gov/docs/fy99osti/24052.pdf> [accessed 21.02.19].
50. Kumar SS. *Hindustan Times, India ratifies Paris climate treaty: here's all you need to know.* <https://www.hindustantimes.com/india-news/what-signing-the-paris-climate-change-treaty-means-for-india/>; 2018 [accessed 13.08.18].
51. Goodale CL, Michael JA, Richard AB, Christopher BF, Linda SH, Richard AH, et al. Forest carbon sinks in the Northern Hemisphere. *Ecol Appl* 2002;**12**:891–9.
52. Gaffney JS, Marley NA. The impacts of combustion emissions on air quality and climate: from coal to biofuels and beyond. *Atmos Environ* 2009;**43**:23–36.
53. Yang HH, Shu-Mei C, Mei-Yu L, Chi-Wei LJ, Wen-Chang L, Yong-Yuan K. Effects of biodiesel on emissions of regulated air pollutants and polycyclic aromatic hydrocarbons under engine durability testing. *Atmos Envon* 2007;**41**:7232–40.
54. Lopes S, Geng P. Estimation of the elemental composition of diesel fuel containing biodiesel. *SAE Int J Fuels Lubr* 2013;**6**(3):668–76.
55. IEA. *International Energy Agency. Biofuels for transport.* Paris, France: OECD and the International Energy Agency; 2005.
56. Gupta PK, Sahai A, Singh N, Dixit CK, Singh DP, Sharma C, et al. Residue burning in the rice-wheat cropping system, causes, and implications. *Curr Sci* 2004;**87**(12):1713–17.
57. Krisztina U, Scarpete D, Panait T, Marcel D. Thermo economic performance criteria in using biofuels for internal combustion engines. *Adv Energy Plann* 2010;81–6.
58. WWI. *Biofuels for transport—global potential and implications for sustainable energy and agriculture*. Earthscan: World Watch Institute; 2007.
59. Rosillo-Calle F. The role of biomass energy in rural development. In: *Paper presented at the 3. Encontro de Energia no Meio Rural, Campinas, Brazil*; 2000.
60. Barnes DF, Floor WF. Rural energy in developing countries: a challenge for economic development. *Annu Rev Energy Environ* 1996;**21**:497–530.
61. UNDP-WHO. *The energy access situation in developing countries*. New York: United Nations Development Program and the World Health Organization; 2009.
62. Teske S, Morris T, Nagrath K. *100% Renewable energy for Tanzania: access to renewable and affordable energy for all within one generation*. Report prepared by ISF for Bread for theWorld; 2017.
63. Goldemberg J, Suani TC. Renewable energy: traditional biomass vs. modern biomass. *Energy Policy* 2004;**32**:711–14.
64. WEHAB. *A framework for action on energy*. UN-Department of Economic and Social Affairs, United Nations Development Programme; 2002.
65. Kanagawa M, Nakata T. Assessment of access to electricity and socio-economic impacts in rural areas of developing countries. *Energy Policy* 2008;**36**:2016–29.

66. Cabraal RA, Douglas FB, Agarwal SG. Productive uses of energy for rural development. *Annu Rev Energy Environ Resour* 2005;**30**:117–44.
67. IRENA. *Renewable energy and jobs. Annual review.* <http://www.irena.org/DocumentDownloads/Publications/IRENA_RE_Jobs_Annual_Review_2017.pdf/>; 2017 [accessed 16.07.18].

Chapter 13

Socioeconomic impacts of biofuel production from lignocellulosic biomass

Avinash P. Ingle[1], Pramod Ingle[2], Indarchand Gupta[3] and Mahendra Rai[2]

[1]*Department of Biotechnology, Engineering School of Lorena, University of Sao Paulo, Lorena, SP, Brazil,* [2]*Nanobiotechnology Lab, Department of Biotechnology, Sant Gadge Baba Amravati University, Amravati, Maharashtra, India,* [3]*Department of Biotechnology, Institute of Science, Aurangabad, Maharashtra, India*

13.1 Introduction

Energy is not only considered as an important national resource, but is also among the key building blocks that strengthen the foundation for the social and economic development of any country.[1] It is a well-known fact that over the past few decades there has been tremendous depletion in various energy resources including coal, oil, and other fossil fuels. In addition, different problems associated with environmental pollution generated by the burning of fossil fuels have serious impacts on global development.[2] Considering these major issues, many countries including both developing and developed countries, demand the development of novel renewable energy sources to overcome energy crises and environmental pollution problems.[3,4]

In this context, various attempts have been made toward the production of different biofuels from various biomass feedstocks. However, the utilization of renewable feedstocks like lignocellulosic biomass for the production of bioethanol is considered as most promising renewable material in the present scenario of energy crises.[5–8] Lignocellulosic biomass is amongst the cheapest, most economically compatible, and abundant feedstocks on the planet.[9] There are many sources from which lignocellulosic feedstocks can be generated, for example, agricultural waste, forest residues, grasses, aquatic plants, waste from the paper and pulp industries, municipal solid waste, etc. All these lignocellulosic feedstocks can be used for the production of biofuels like bioethanol due to the presence of large amounts of

Sustainable Bioenergy. DOI: https://doi.org/10.1016/B978-0-12-817654-2.00013-7

lignocellulose.[10] Lignocellulose, the key component of plant cell walls, is mainly composed of cellulose, hemicellulose, and lignin. Moreover, it is estimated that the percentage of these components may vary depending on the origin of the lignocellulosic materials.[11,12] However, biodiesel can be used as a sustainable substitute for mineral diesel and is generally produced from biological materials such as vegetable oil or animal fats through a chemical process known as transesterification.[13,14]

It is globally accepted that biofuels have many advantages as well as the potential to contribute to a wide range of policy objectives, mainly reducing dependency on fossil fuels, thereby improving energy security, overcoming issues related greenhouse gas (GHG) emissions, renewability, cleanliness, and increasing rural employment and farmers' income.[4,15] In addition, it also provides economic efficiency and reduces ecological degradation. Considering these benefits both developed and developing countries have approved these policies to support biofuel production. However, issues and conflicts triggered due to biofuels production continue to emerge and it increases the uncertainty about the advantages of promoting biofuels. Moreover, various policies adapted for the production of biofuel also show some drawbacks and potential risks. Although it is widely regarded that biofuels have some positive impacts, many studies have proven that biofuels have significant negative impacts on the environment as well as social and economic aspects.[16,17]

Considering the worldwide acceptance of biofuels as an alternative transportation fuel and their benefits, the majority of studies have focused on the positive impacts of biofuels production. However, as discussed earlier, the production of biofuels has significant negative impacts on many factors. Therefore the present chapter is mainly focused on the environmental, economic, and social impacts of biofuels production. In addition, a brief overview of biofuels and worldwide developments in its production have been also discussed.

13.2 Worldwide development in biofuel production

The history of biofuels as a transport fuel goes back to the early 19th century. However, in the past few years biofuels have caught the imagination of the industrialized world. The global desire for biofuels expanded significantly around 2003 and 2004 because the European Union and the United States passed some bold energy initiatives to replace billions of liters of fossil fuel with biofuels like biodiesel and ethanol.[18] Currently, numerous developments in the biofuels sector are observed due to low oil prices and increased concerns about their effects especially in developing countries. The extent to which the biofuel policies of the European Union might have contributed to rising food prices, reduced availability, pressure on agricultural land, and other adverse effects has not been fully measured. Due to the mentioned advantages of biofuels, biorefineries have attracted a great deal of attention from all over the word. As a result, a rising trend in the production of biofuels has been observed globally since 2000 in both developed and developing countries.

According to the International Energy Agency (IEA) the production of biofuels grew tremendously between 2000 and 2011, that is, from 16 billion liters to more than 100 billion.[19] It is estimated that nearly 3% of total transportation fuel energy usage is from biofuels, although in some countries biofuel accounts for higher shares. In 2008 about 21% of road transport fuel demand was met by biofuels in Brazil which is now much higher. Similarly, in the European Union and United States, this share was about 3% and 4%, respectively. Further, it was reported that global biofuel production has jumped from 10,021,000 t of oil in 2001 to 58,457,000 t of oil in 2010, an increase of nearly 500%.[20] Currently, even Thailand is using bioethanol as an alternative fuel for passenger cars as its production has rapidly increased from 0.4 million liters per day in 2006 to 1.1 million liters a day in 2010. Considering the increasing demand for biofuel it is expected that its production will increase to 9 million liters per day by 2022, according to the alternative energy development plan (2008–22) adopted by Thailand.[21] Moreover, Fig. 13.1 represents the worldwide distribution of ethanol production and its projected use until 2025.[22] However, Fig. 13.2 shows the yearwise global production of ethanol in the past 10 years.[23] The worldwide increase in the production of biofuels using various feedstocks raises some concerns related to the environment and economic and social issues.

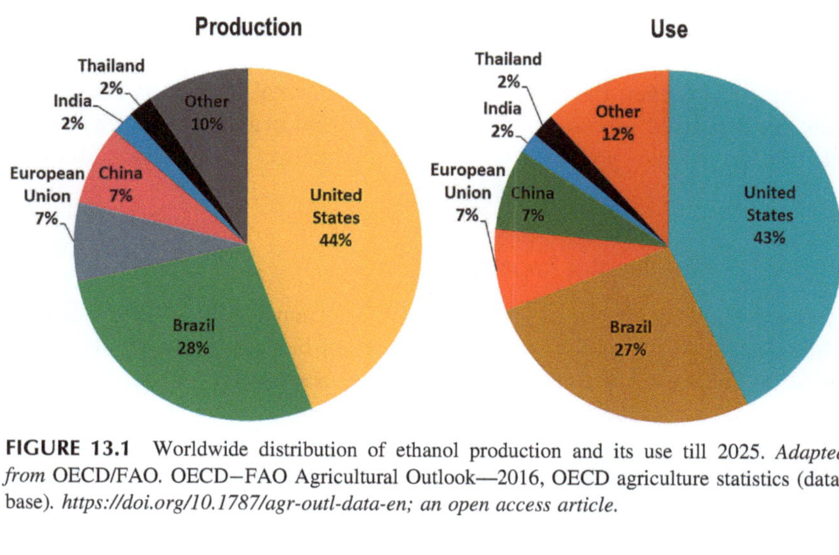

FIGURE 13.1 Worldwide distribution of ethanol production and its use till 2025. *Adapted from* OECD/FAO. OECD–FAO Agricultural Outlook—2016, OECD agriculture statistics (database). *https://doi.org/10.1787/agr-outl-data-en; an open access article.*

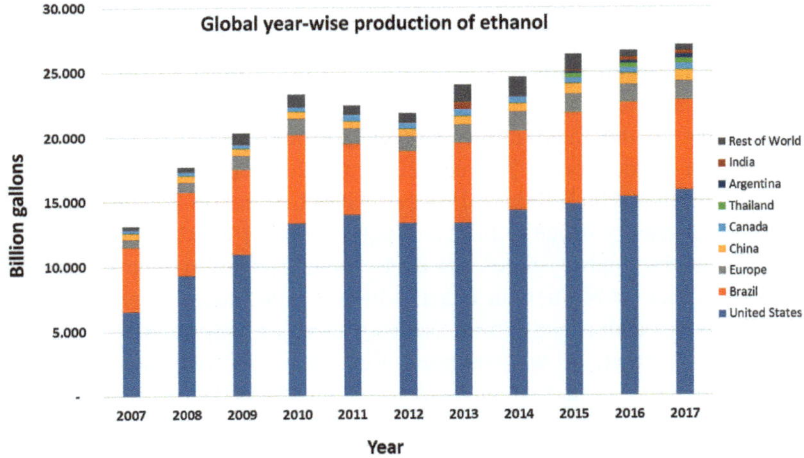

FIGURE 13.2 Year wise global production of ethanol between 2007 and 2017. *Data source: Renewable Fuels Association.* World fuel ethanol production. *Available at <http://www.ethanolrfa.org/resources/industry/statistics/#1454098996479-8715d404-e546> [accessed 12.11.18].*

13.3 Impact of biofuel production on various factors

Various socioeconomic sustainability criteria and indicators were proposed for the assessment of the socioeconomic and environmental impacts of biofuel production. They cover a wide range of aspects related to socioeconomic sustainability, including contribution to the local economy, working rights and conditions, health and safety, gender, land rights and conflicts, food security, and different environmental impacts that could affect local communities.[24] Moreover, the assessment of socioeconomic impacts includes systematic analyses to identify and evaluate the potential socioeconomic and cultural impacts of the utilization of biomass feedstock and biofuel production on the lives and circumstances of people, their families, and their communities.[25] However, the assessment of environmental impacts is considered to be an organized process to identify, predict, and evaluate the environmental effects of any proposed actions and projects, for instance, biofuel production.

13.3.1 Environmental impact

Environmental impact is an astonishingly important factor to be considered while designing guidelines for the assessment of biofuel sector impacts. This includes all factors such as water quality and resources; biodiversity, flora, fauna, and landscapes; soil quality and resources; and land use. The Renewable Energy Directive 2009/28/EC of the European Union targets a 10% biofuel use by 2020, inclusive of first-generation biofuels (8.5%), second-generation biofuels (1%), and renewable electricity (1%).[16,26–28]

In developing countries, the biofuel sector is rising at a commendable rate because of its low cost and the increasing concern on the environment.[29] Presently 30 Mha of land produces a considerable number of byproducts, contributing less than 1% of total biofuels. Alongside land acquisition by Western companies, African third states with much less compensation lead to the environmental and social impact of land use change.[19] The uncontrolled use of fossil fuels is responsible for issues such as the degradation of land, water, biodiversity and natural resources, as well as GHG emissions. In addition, the globally distributed lignocellulosic biofuels production system also has impacts on access to different resources which are summarized in Table 13.1.

Therefore it is extremely important to ensure the transition to sustainable practices that implement the reduced use of fossil fuels in the food and agricultural sectors. Both opportunities and challenges are presented for the economy and the environment by developing bioenergy projects. The environmental impacts are dependent on the bioclimatic conditions of the site utilized. Agronomists have recognized that it is necessary to respect the fundamentals of the equilibrium and positive synergy between soil–plant–air for sustainable economic, social, and environmental policies.[30]

13.3.1.1 Impact on soil quality

The Bioenergy and Food Security Criteria and Indicator Project of the FAO has divided agricultural practices into three groups for ensuring sustainability, namely: agricultural management approaches, integrated sustainable agricultural and forestry management systems, and field-level agricultural and forestry practices. Biofuel cultivation can thus be beneficial in many ways like preventing soil erosion caused by afforestation, physically stabilizing soil through root systems and leaf littering, and reducing water runoff and sediment loss. The adverse effects of eutrophication and groundwater contamination can be minimized using good agricultural practices and judicious application of agrochemicals.[31]

Woody trees and bioenergy crops can help in nitrogen fixation directly or indirectly, improve soil texture by adding organic matter, and increase water and nutrient holding capacity. Shelterbelts and agroforestry systems can prevent soil loss due to wind erosion. Also, the rotational harvesting of forests or biofuel crops can help in reducing erosion on mountains.[32,33] A comparative account of soil management and soil protection was reported by The Center for International Forestry Research (CIFOR). The report illustrates criteria and indicators like lowering the use of agrochemicals, soil quality and fertility maintenance, and sustainable use of resources.[34] The plantation of jatropha and other perennial biofuel crops on degraded or marginal land can improve soil quality and help in rehabilitating soil not favorable for other food crops.[35,36]

TABLE 13.1 Potential impacts of a globally distributed lignocellulosic biofuel production system on access to resources.

Resource	Perennial crops	Crop residues	Forestry residues	Processing and conversion	Distribution of impacts
Food	(+) If this helps replace current use of food crops for fuel (−) If crops were grown on land of value for food	No direct competition with food: (+) if this helps replace use of food crops for fuel; and (−) If residues are part of animal feed (e.g., straw)	No direct competition; indirect impacts not assessed	No direct impact	(−) Impacts are globally distributed, disproportionately affecting lower income populations relying on the market to access food; (−) Local or regional impacts where biomass is sourced by displacing people from common property
Water	Could be (−) in current conditions in the case of miscanthus which requires high water inputs. Can be (−) if crops are grown in areas requiring irrigation Technological improvements may compensate if crops are grown at smaller scale (+) Impact of water retention due to year-round cover from perennial crops	Not widely studied, but residue cover reduces evaporation of water from soil, conserving moisture. Potential (−) impacts of residue removal on water, therefore, cited as a risk, but manageable with careful choice of sites, where residues become commodities in their own right; how water use is allocated between the crop and coproducts may be at stake. Overall: (nil/ −)	Similar issues regarding residue cover conserving moisture are relevant. Overall: (nil/ −)	Based on current processes, lignocellulosic conversion is more water-intensive than first-generation biomass due to added conversion steps; hence, (−) Provided improvements occur (e.g., highlight modern plants, improving recycling practices), impacts could be (nil) in future Overall: (nil/ −)	Likely to be local, affecting specific areas where biomass is sourced

Energy	(+) Impacts for fuel users if it helps meet transport energy needs as expected, but this is constrained by the low energy density of biomass and the "blend" wall (in vehicles as currently designed) (−) impact if biomass competes with other bioenergy options	(+) Impacts for fuel users if it helps meet transport energy needs as expected. But this is constrained by the low energy -density of biomass and the "blend" wall (in vehicles as currently designed) (−) impact if the biomass competes with use for other bioenergy options	(−) Impacts if residues are sourced from areas where they fulfill the energy needs of forest dwellers	(−) Impacts in current conditions where energy inputs for processing and conversion are high Overturning (−) Impacts are possible through technological improvements	Effects are likely to be distributed globally as well as locally and regionally, for example, (−) impacts where residues use local fuel in subsistence farming Contribution to energy security of poorer countries may therefore be nil

Source: Adapted and modified from Raman S, Mohr A, Helliwell R, Ribeiro B, Shortall O, Smith R, et al. Integrating social and value dimensions into sustainability assessment of lignocellulosic biofuels. *Biomass Bioenergy* 2015;**82**:49–62; an open access article.

13.3.1.2 Impact of monoculture plantations and biodiversity

In the case of afforestation in Kenya, monocultures of leguminous biofuel crops/trees covered over 200 ha which helped in nitrogen fixation and reduced soil erosion. This ultimately solved the problem of bare lands covered with thickets. For jatropha cultivation using indigenous trees, monocultures assisted in uprooting ecological imbalances,[16] thus improving the microclimate and creating a new carbon sink. However, continuous intensive monoculture systems are responsible for pathogen and pest buildup, declined soil fertility, biodiversity loss, and degradation of land and natural resources.[37,38] Degraded lands can also be used for biofuel plantation, but it is not always sustainable because these marginal lands harbor some biodiversity and constitute biological corridors.[39] In India, it has been seen that the availability of extra biomass due to biofuel plantation reduces the pressure on forests.[40]

13.3.1.3 Impact on household level

The impacts of biofuel at a local level, that is, at household level and for small-scale farmers, involves their social, economic, ecological, and geopolitical benefits. Also, the cost of commodities, business models, and landscapes poses a profound effect on the benefits to people and the environment such as can be seen in Brazil.[41] There are different frameworks accepted worldwide such as the Convention on Wetlands (Ramsar Convention), the Convention on Biological Diversity, the Rotterdam Convention on pesticides and industrial chemicals, and the Stockholm Convention on persistent organic pollutants, for the assessment of the impacts of biofuel on the environment.[29] In these frameworks, due respect is given to ecological connectivity.[42]

13.3.1.4 Impact on greenhouse gas emissions

The FAO reported a GHG saving of about 1.8 t/ha/year of CO_2 from bioethanol and a GHG saving of 8.6 t/ha/year generated from switch-grass. For this purpose, bare lands are being used for biofuel cultivation in Brazil, Indonesia, Malaysia, and the United States; thus, covering open spaces will help in minimizing soil erosion and water runoff. This will reduce the carbon dioxide emission 17-times over as compared to fossil fuels.[39] CIFOR reported a study on the comparative emission of GHG and carbon stock changes between fossil fuel consumption and carbon emission after converting forest land into biofuel plantations.[34] One case study in Nigeria, Argentina, and Kenya showed that multiple cropping systems and crop rotations have a remarkable impact on carbon sequestration.[43] Studies have reported that a reduction of 1 t of CO_2 by biofuel costs only US$960–US $1700.[44] A study in Thailand on the combined use of ethanol and gasoline reported only a nominal GHG emission.[45,46] Moreover, Table 13.2 summarizes the potential environmental impacts of a globally distributed lignocellulosic biofuel production system.

TABLE 13.2 Potential environmental impacts of a globally distributed lignocellulosic biofuel production system.

Environmental outcome	Perennial crops (incl. processing and conversion)	Crop residues (incl. processing and conversion)	Forestry residues (incl. processing and conversion)	Distribution of impacts
Water, soil, and air quality	Contrary (+) and (−) results reported across studies. Review suggests more research is needed, especially on under-researched aspects (e.g., impacts of enzymes and catalysts used in conversion)	Contrary (+) and (−) results across LCAs Dependent on allocation of impacts between crop and coproduct Frequently cited concern of (−) impact on soil structure and nutrient balance from the removal of residues (vs. ploughing back to soil) Difficulties with scientific measurement of impacts on soil	(−) Impact on soil quality and structure from the removal of residues cited as a concern; difficulties with scientific measurement of the impacts on soil quality	Local impacts experienced in and around sites of biomass sourcing and conversion
Biodiversity	Some evidence of (+) impact on biodiversity. Depends on indicators used for biodiversity, and which perennial crops were compared: arable crops, grassland, or leaving the land fallow; when compared with fossil fuel extraction, high (+) impact was observed	According to IEA, it depends on whether these are "primary" or "secondary" residues (−) impact for primary residues which are likely to be left as cover or ploughed back into soil with benefits for microflora. (+) Impact for secondary residues which would be treated as waste (−) Impacts may be ameliorated depending on how much residue is removed versus left behind	Similar concerns about (−) impact due to role of forest residue cover in enhancing soil biodiversity. It is also suggested that even thinning the forest cover (as opposed to large-scale deforestation) can be harmful	Locally specific impacts experienced in sites of biomass sourcing

(*Continued*)

TABLE 13.2 (Continued)

Environmental outcome	Perennial crops (incl. processing and conversion)	Crop residues (incl. processing and conversion)	Forestry residues (incl. processing and conversion)	Distribution of impacts
GHG emissions	Most studies report (+) impact, namely GHG savings compared to fossil fuels; but actual impact on the ground depends on choice of land use	Most studies report (+) impact, namely GHG savings compared to fossil fuels; but including the (−) impact of residue removal on soil carbon would change the assessment	Studies report (+) impact, namely GHG savings; but challenged by other reports of high potential for (−) impact, that is, increase in GHG emissions from iLUC (due to diversion of forestry residues from current uses in furniture and paper/pulp industries) or from total increase in use of forestry resources. Impact of residue removal on soil carbon also cited as (−)	Impact on national carbon saving targets Physical impacts of GHG changes will be felt on a global level

GHG, Greenhouse gas; *LCA's*, Life-cycle assessments; *iLUC*, Indirect land-use change.
Source: Adapted and modified from Raman S, Mohr A, Helliwell R, Ribeiro B, Shortall O, Smith R, et al. Integrating social and value dimensions into sustainability assessment of lignocellulosic biofuels. *Biomass Bioenergy* 2015;**82**:49–62; an open access article.

13.3.2 Economic impacts

Bioethanol and biodiesel are biofuels obtained from biomass feedstocks which harbor energy from the sun and which represent considerable renewable fuel sources. These are organic, nontoxic, and clean alternatives that thus contribute less to environmental damage and GHG emissions. These diverse sources of energy help the environment and the economy.[28,47,48] Although the feasibility of some of the present biofuel programs is considerable, the increasing use of land for biofuel cultivation imparts huge effects on social and environmental aspects. This also raises competition between food production and biofuel cultivation.[49] There are over six million vehicles in use that can accept mixed ethanol/gasoline of up to 85% in Brazil, Sweden, and the United States. Of these biofuels, ethanol contributes 90% and it represent the most publicized first-generation biofuels,[50] promoting the production of bioethanol from specific crops. The production of biodiesel is accompanied by the formation of byproducts like fatty acids, oilseed meal, fertilizers, and glycerin which are important for the food and cosmetic industries and are also considered safe by the US FDA (United States Food and Drug Administration), thus promoting the establishment of other industries at the same time.[51] The biofuel industry has been continuously increasing over the past two decades.

The IEA[19] reported that biofuel production increased from 1 billion liters in 2000 to 100 billion liters in 2011. Arndt et al.[52] investigated the impact of biofuel production on the economy using a recursive-dynamic computable general equilibrium model. Many studies have emphasized the feasibility of the implementation of biofuels in the European Union toward achieving a 20% reduction in GHG emissions by 2020.[52,53] Taheripour et al.[54] studied the impact of increased renewable energy production and consumption on the economy and the environment by applying a global CGE model. Lifecycle assessment studies on the biofuel industry in Brazil have demonstrated direct positive impacts on the economy and the environment, while simultaneously reducing the negative effects of the overuse of fossil fuels.[54,55]

The combined use of conventional fuels and lignocellulosic ethanol from sugarcane bagasse, as used in Canada for example, reduces the pressure on gasoline use and GHG emissions along with promoting the commercialization of the biofuel industry, ultimately contributing to the economy.[17,56] Macroeconomic studies on the effects of the biofuel sector on the Canadian economy in 2008, considering the "input-output model," provide data indicating direct employment generated by the biofuel sector of 0.04% of the whole Canadian employment market. The studies indicate that the biofuel industry has a direct impact on the agricultural sector as the raw materials required for biofuel production come from agriculture. Also, the mining and manufacturing industries are considerably affected by the entry of biofuels into the economic system.[29,57]

Brazil is the world's largest producer of ethanol from sugarcane, as well as being the largest producer of biodiesel. Brazil is emerging internationally as a contributor to the global economy. Gradually, Brazil is sharing leadership with the United States in terms of biodiesel production.[58,59] The Bioenergy and Food Security Analytical Framework of the FAO stated that food security is affected through competition for resources between food production and biofuel manufacture. According to the FAO, 14% of the Asian population is undernourished.[60,61] Sometimes farmers compromise actual food crops for commercial biofuel crops in an attempt to gain financial benefits; however, inadequate compensation or low costing eventually lowers the economic condition of farmers.[62,63]

According to USAID, in 2009, Brazilian sugarcane was shown to be the most economically viable feedstock.[64] In 2008 the OECD reported that the biodiesel industry nurtures the ideology of expanding the market share and value of local export industries.[44] Biofuel plays an increasingly critical role in the economy by supporting the stimulation of energy demand of nations. Also, the efforts taken by government to develop production strategies and biofuel chains (like production, transportation, and exports) will ensure increased employment and increased incomes, particularly in developing countries like Africa.[65–68] The increasing demand for biodiesel will increase the cost of the relevant agricultural raw materials used in biofuel production, contributing farmers' incomes in developing countries.[69] Studies in South American countries confirm the direct and close relationship between biomass energy consumption and economic growth, except in Argentina and Jamaica.[70,71] Gohin[72] used a computable general equilibrium model for the EU15 economy, and estimated the impact of European biofuel policies on the agricultural sector, indicating more than 43,000 job opportunities and economic increases of €3.2 billion.[29,72] Food security is also gaining much importance as the biofuel sector continues to develop. Studies have been performed based on framework assessment of the impact of biofuel on food security. Conceptual frameworks at national and international levels include the development of Global Bioenergy Partnership Sustainability Indicators for bioenergy[71,73] and High-Level Panel of Experts on Food Security and Nutrition Review.[74] On the contrary, increased biofuel production increases competition among producers, feedstock and food production, as well as land, water, and labor; thus, reducing soil organic carbon losses and hampering the quality life of people.[29,75–77]

Biofuel production causes the deterioration of food security for urban consumers in net food importer countries, while net exporter countries show increased food security.[78] However, at the same time, the displacement of agricultural land by biofuel companies without sufficient compensation, and continuous increases in food costs ultimately cause a decrease in peoples' incomes.[73,79] The economic cost of biofuel generation depends on the cost of feedstock, the cost of conversion, and the cost land. This makes the cost of biofuel higher than that of fossil fuel.[80,81] Along with production and transportation, the supply chain of biofuel also provides a huge number of job opportunities. However, at the same time, the supply chain presents many challenges, for instance, multiscale challenges, multi-player challenges, and multiobjective and sustainability challenges.[82,83] Other positive impacts of biofuel production and industry establishment include multiple impacts on the environment and the surrounding communities. It provides jobs as employees in biofuel projects and as laborers in processing plants. Sources of income are abundant from the production and sale of feedstocks which they can grow on their farms or agricultural lands. Additional sources of income are provided through the setting up of mechanical shops (like automobile and machinery repair, etc.). Conversion plants can use feedstock produce from local farmers, thus improving peoples' lives.[43,84]

13.3.3 Social impacts

Remarkably little attention has been given to the social aspects in scientific publications. Although biofuel production to meet global demand requires cheaper and high-throughput technology, comparable attention is also needed on the social challenges associated with the production and use of biofuels. This mainly includes health risks and the requirements of water, land, jobs, and equality. The European Commission has established certain mandatory directives for biofuels, biomass, and bioliquids; however, they did not set any clear criteria for social sustainability.[85] Hence, the social impacts of biofuel need to be carefully evaluated prior to public support of biofuel on a large scale.

The food versus fuel issue is always highly complex. Therefore the production of biofuel from food or agricultural products needs to be carefully examined.[86] Nevertheless, it has received wide attention after the world food price crisis in 2007–08 which impelled persistent high food prices for over a decade as food production and supplies were displaced by biofuel production, especially in developing countries that are net food importers.[87] Therefore rising food prices will also lead to biofuel technology being expensive. For a growing population if it is already hard to meet rising food demands and as such it becomes harder to meet biofuel demands.[88]

It is a well-known fact that the production of biofuels from lignocellulosic material depends on plant waste or excess materials. However, land officially designated as marginal or degraded land may still meet the requirements of rural poor people for their livelihood, food, and fuel. Moreover, companies producing biofuel may target lands that are already in use for agricultural production, forests, irrigation, or those producing high crop yields.[88] In countries with weak social governance, land acquisition is always a major challenge due to discrepancies between large companies and local people.[89] It is a matter of a great concern that global land has been heavily used for crops required for food and biofuel production. Hence, the production of biomass needed for biofuel may require deforestation. Deforestation may force local people to migrate as well as reduce the available land for foodstuff, feed, and fiber.[90]

As mentioned earlier, biofuels have been considered as a potential source of renewable energy and can capture a major portion of the energy market. However, some of the present biofuel programs, although viable, include great social costs. Bioenergy programs at rural level can be socially beneficial in that they can create employment which would provide a source of income for people living in rural areas. Gender issues are highly significant in the biofuel industry, in terms of who is working, how much money they have been paid for their assigned workload, and changes in their social status. For instance, the Tanzanian government owns all the agricultural land. The rural citizens of Tanzania do not have any legal right to their land. Moreover, most of the production workload is given to women. This is because women laborers are cheaper to hire as is the case in many developing countries and get less remuneration as compared to their male counterparts. Additionally, they are submissive and dependable and can therefore be more exploited.[91]

Supporters of the lignocellulosic biofuel industry are concerned about its social impacts on rich and poor populations. There is always a chance that not everyone will benefit equally from this industry. People owning larger lands will benefit more than others. Similarly, those getting greater financial benefits can invest elsewhere to get more money whereas others cannot. Therefore, it is of prime importance that this technology should also ensure that every component of this chain should benefit equally. It is also important to note that the development of such technology should not invite the destruction of social structures, nor violate the rights of traditional users, and most importantly there should not be any kind of large displacement of local populations. On the contrary, the involvement of local people should be given priority in terms of employability with decent wages. Such approaches will definitely make biofuel a "social fuel" that contributes to the development of local and global communities. Fig. 13.3 shows a schematic representation of various impacts of biofuel production.

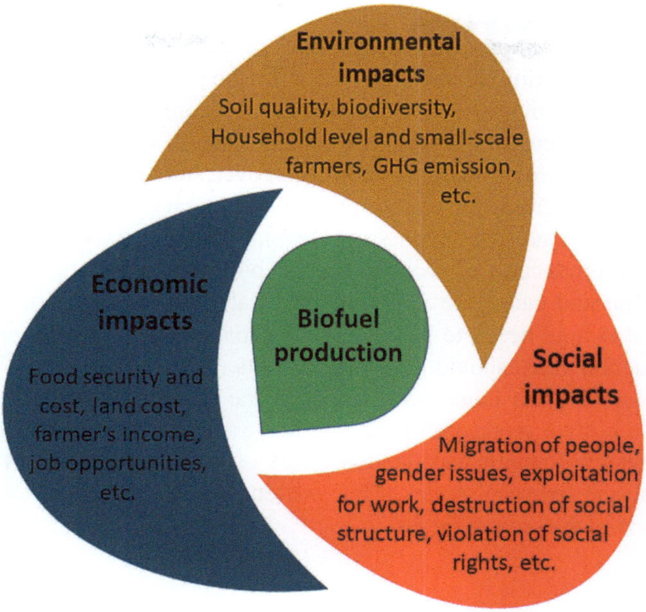

FIGURE 13.3 Schematic representation of various impacts of biofuel production.

13.4 Conclusion

Biofuels are considered as promising alternative renewable energy sources and are widely accepted due to their potential to mitigate the major concerns associated with fossil fuels. Biofuels are green fuels that release negligible amounts of GHG emissions as compared to fossil fuels, thereby aiding in climate change relief and avoiding environmental pollution. Due to a large number of benefits, worldwide production of biofuels has significantly increased. Different countries including both developed and developing countries have adopted various policies and programs for biofuel production and have set targets. Out of such programs, only a few are considered feasible; however, many of these have not been found to be environment-friendly and they also involve high social costs. The process of biofuel production is inherently interconnected with various other important factors such as the environment, the economy, and society. Therefore it is necessary to understand the emerging widespread impacts of biofuel production on economics, the environment, and human welfare as available reports have projected that the widespread adoption and commercialization of biofuel production programs may adversely affect the environment, the economy, and society.

Biofuel is highly dependent on agricultural production. Hence, it can have considerable impacts on the contents of soil due to deforestation. The continuous production of the same crop in the same field may affect the plantation and biodiversity associated with it. Economically it may help to minimize the expenses involved in the production of fuel as many countries can reduce their fossil fuel imports. Moreover, the production of biofuel locally will generate job opportunities for local people. It will help strengthen the economy at a grassroots level. On the contrary, if more land is used for the production of crops useful for biofuel production, it may invite deforestation and cause the migration of local people. Therefore the social aspect of this field needs to be evaluated so that each component of the biofuel production chain should get equal benefits.

Considering these concerns, there is prime need to design and develop sustainable biofuel production programs that are contingent on collaborations across research domains and the evaluation of potential impacts over multiple spatial and temporal scales. This will ensure the all-round development of local and global communities.

References

1. Prasad S, Dhanya MS, Gupta N, Kumar A. Biofuels from biomass: a sustainable alternative to energy and environment. *Biochem Cell Arch* 2012;**12**(2):255–60.
2. Owusu PA, Asumadu-Sarkodie S. A review of renewable energy sources, sustainability issues and climate change mitigation. *Cogent Eng* 2016;**3**:1167990 Available from: https://doi.org/10.1080/23311916.2016.1167990.
3. Russi D. Biofuels: solution for energy crisis? *Econ Pol Wkly* 2007;**42**(19):1664–7.
4. Ližbetin J, Hlatká M, Bartuška L. Issues concerning declared energy consumption and greenhouse gas emissions of FAME biofuels. *Sustainability* 2018;**10**:3025 Available from: https://doi.org/10.3390/su10093025.
5. Limayem A, Ricke SC. Lignocellulosic biomass for bioethanol production: current perspectives, potential issues and future prospects. *Prog Energy Combust Sci* 2012;**38**(4):449–67.
6. Zabed H, Sahu NJ, Boyce AN, Faruq G. Fuel ethanol production from lignocellulosic biomass: an overview on feedstocks and technological approaches. *Renewable Sustainable Energy Rev* 2016;**66**:751–74.
7. Zhou Z, Lei F, Li P, Jiang J. Lignocellulosic biomass to biofuels and biochemicals: a comprehensive review with a focus on ethanol organosolv pretreatment technology. *Biotechnol Bioeng* 2018;**115**(11):2683–702.
8. Robak K, Balcerek M. Review of second generation bioethanol production from residual biomass. *Food Technol Biotechnol* 2018;**56**(2):174–87.
9. Ingle AP, Rathod J, Pandit R, da Silva SS, Rai M. Comparative evaluation of free and immobilized cellulase for enzymatic hydrolysis of lignocellulosic biomass for sustainable bioethanol production. *Cellulose* 2017;**24**:5529–40.
10. Fatma S, Hameed A, Noman M, Ahmed T, Shahid M, Tariq M, et al. Lignocellulosic biomass: a sustainable bioenergy source for the future. *Protein Pept Lett* 2018;**25**(2):148–63.
11. Sorek N, Yeats TH, Szemenyei H, Youngs H, Somerville CR. The implications of lignocellulosic biomass chemical composition for the production of advanced biofuels. *BioScience* 2014;**64**(3):192–201.

12. Krasznai DJ, Hartley RC, Roy HM, Champagne P, Cunningham MF. Compositional analysis of lignocellulosic biomass: conventional methodologies and future outlook. *Crit Rev Biotechnol* 2018;**38**(2):199–217.
13. Demirbas A, Karslioglu S. Biodiesel production facilities from vegetable oils and animal fats. *Energy Sources, A: Recover Util Environ Eff* 2007;**29**(2):133–41.
14. Kim DS, Hanifzadeh M, Kumar A. Trend of biodiesel feedstock and its impact on biodiesel emission characteristics. *Environ Prog Sustain Energy* 2018;**37**(1):7–19.
15. Payam F, Taheri A. Challenge of fossil energy and importance of investment in clean energy in Iran. *J Energy Manag Technol* 2018;**2**(1):1–8.
16. Ji X, Long X. A review of the ecological and socioeconomic effects of biofuel and energy policy recommendations. *Renewable Sustainable Energy Rev* 2016;**61**:41–52.
17. Raman S, Mohr A, Helliwell R, Ribeiro B, Shortall O, Smith R, et al. Integrating social and value dimensions into sustainability assessment of lignocellulosic biofuels. *Biomass Bioenergy* 2015;**82**:49–62.
18. Hubbard P. Biofuels and the developing world*SAIS Eur J Global Aff* 2008. Available at Available from: http://www.saisjournal.org/posts/biofuels-and-the-developing-world.
19. *IEA. Technology roadmap: bioenergy for heat and power.* Available at <https://webstore.iea.org/technology-roadmap-bioenergy-for-heat-and-power>; 2012.
20. *European Biofuels. Global biofuels—an overview.* <http://www.biofuelstp.eu/global_overview.html>; 2014 [accessed 04.11.18].
21. Silalertruksa T, Gheewala SH. The environmental and socio-economic impacts of bioethanol production in Thailand. *Energy Procedia* 2011;**9**:35–43.
22. *OECD/FAO. OECD–FAO Agricultural Outlook—2016, OECD agriculture statistics (database).* https://doi.org/10.1787/agr-outl-data-en.
23. *Renewable Fuels Association. World fuel ethanol production-2018.* Available at <http://www.ethanolrfa.org/resources/industry/statistics/#1454098996479-8715d404-e546> [accessed 12.11.18].
24. Rutz D, Janssen R. *Socio-economic impacts of bioenergy production.* Switzerland: Springer International Publishing; 2014. p. 312.
25. *MVEIRB (Mackenzie Valley Environmental Impact Review Board). Issues and recommendations for social and economic impacts assessment in the Mackenzie Valley.* MVEIRB: Yellowknife, Canada; 2007. Available at <http://reviewboard.ca/upload/ref_library/SEIA_paper.pdf> [accessed 04.11.18].
26. *European Commission. Press release from Brussels, 2012.* Available at <http://europa.eu/rapid/press-release_IP-12-1112_en.htm?locale = en> [accessed 11.11.18].
27. *European Commission. Impact assessment guidelines.* Available at: <http://ec.europa.eu/smart-regulation/impact/commission_guidelines/docs/iag_2009_en.pdf>; 2009 (accessed 12-11-2018).
28. *European Commission. Final report of The European Union's Framework Contract Commission (N° 2012/299193 FWC COM 2011), entitled assessing the impact of biofuels production on developing countries from the point of view of policy coherence for development.* Available at <https://ec.europa.eu/europeaid/sites/devco/files/study-impact-assesment-biofuels-production-on-development-pcd-201302_en_2.pdf> [accessed 12.11.18].
29. Gheewala SH, Damen B, Shi X. Biofuels: economic, environmental and social benefits and costs for developing countries in Asia. *WIREs Clim Change* 2013;**4**:497–511.
30. *CFS (Committee on World Food Security), Price volatility and food security, Committee on World Food Security High Level Panel of Experts on Food Security and Nutrition.* 2011.
31. Neumann K, Verburg PH, Stehfest E, Muller C. The yield gap of global grain production: a spatial analysis. *Agric Syst* 2010;**103**:316–26.

32. Cherubin MR, Oliveira DMS, Feigl BJ, Pimentel LG, Lisboa IP, Gmach MR, et al. Crop residue harvest for bioenergy production and its implications on soil functioning and plant growth: a review. *Sci Agric* 2018;**75**(3):255–72.
33. *European Parliament. Impact of EU bioenergy policy on developing countries.* Available at <https://www.ecologic.eu/4644>; 2011 [accessed 12.11.18].
34. Guariguata MR, Masera OR, Johnson FX, von Maltitz G, Bird N, Tella P, et al. A review of environmental issues in the context of biofuel sustainability frameworks. In: *Occasional Paper 69*. Bogor, Indonesia: CIFOR; 2011.
35. Chum H, Faaij A, Moreira J, Berndes G, Dhamija P, Dong H, et al. Bioenergy. In: Edenhofer O, Pichs-Madruga R, Sokona Y, Seyboth K, Matschoss P, Kadner S, et al., editors. *IPCC special report on renewable energy sources and climate change mitigation*. Cambridge and New York: Cambridge University Press; 2011.
36. Achten W, Verchot L, Franken Y, Mathijs E, Singh V, Aerts R, et al. Jatropha bio-diesel production and use. *Biomass Bioenergy* 2008;**32**:1063–84.
37. FAO, 2008, *From subsistence farming to sugar-cane monoculture: impacts on agrobiodiversity, local knowledge and food security*, 2008. Available at http://www.fao.org/3/aj042e/aj042e.pdf [accessed 12.11.18].
38. Daley E, Park CMY. *The gender and equity implications of land-related investments on land access and labour and income-generating opportunities: a case study of selected agricultural investments in Northern Tanzania.* FAO; 2012. Available at <http://www.fao.org/docrep/016/ap433e/ap433e.pdf> [accessed 12.11.18].
39. Bringezu S, Schütz H, O'Brien M, Kauppi L, Howarth RW, McNeely J. *Towards sustainable production and use of resources: assessing biofuels. A report by United Nations Environment Programme (UNEL) produced by International Panel for Sustainable Resource Management.* Available at <http://www.unep.fr/shared/publications/pdf/WEBx0149xPA-AssessingBiofuelsSummary.pdf>; 2009 [accessed 12.11.18].
40. *IEA Bioenergy. Bioenergy—a sustainable and reliable energy source: a review of status and prospects.* Available at <https://www.ieabioenergy.com/wp-content/uploads/2013/10/MAIN-REPORT-Bioenergy-a-sustainable-and-reliable-energy-source.-A-review-of-status-and-prospects.pdf>; 2009 [accessed 12.11.18].
41. FAO. *Climate change adaptation and mitigation: challenges and opportunities for food security.* 2008.
42. Olukoye GA, Mukanga H. A review of multilateral environmental agreements and their implications for environmental governance in Kenya. In: *A guide for higher education in Kenya*. Kenyatta University Press, Kenya; 2007. p. 101–120.
43. *FAO. Carbon sequestration in dryland soils.* Available at <http://www.fao.org/3/a-Y5738e.pdf>; 2004 [accessed 12.11.18].
44. *OECD. Biofuels support policies: an economic assessment.* Paris: Organization for Economic Cooperation and Development; 2008.
45. Nguyen TLT, Gheewala SH, Garivait S. Energy balance and GHG-abatement cost of cassava utilization for fuel ethanol in Thailand. *Energy Policy* 2007;**35**:4585–96.
46. Nguyen TLT, Gheewala SH, Garivait S. Fossil energy savings and GHG mitigation potentials of ethanol as a gasoline substitute in Thailand. *Energy Policy* 2007;**35**:5195–205.
47. Hansen AC, Gratton MR, Yuan W. Diesel engine performance and NOX emissions from oxygenated biofuels and blends with diesel fuel. *Trans ASABE* 2006;**49**(3):589–95.
48. Benavides A, Benjumea P, Pashova V. El biodiesel de aceite de higuerilla como combustible alternativo para motores diesel. *Revista Dyna Redaly* 2007;**74**(153):141–50 2007.
49. Hernández M, Hernández J. Verdades y mitos de los biocombustibles. *Elementos* 2008;**71**:15–18.

50. Chauvet MY, González R. Biofuels and biopharmaceutical crops: opportunities or threats? (original article is in Portuguese). *Cotidiano* 2008;**23**(147):51–61.
51. Bruso J. *Foods containing glycerin-2015.* <http://www.livestrong.com/article/333686-foods-containing-glycerin/> [accessed 25.10.18].
52. Arndt C, Pauw K, Thurlow J. Biofuels and economic development: a computable general equilibrium analysis for Tanzania. *Energy Econ* 2012;**34**(6):1922–30.
53. De Lucia C, Bartlett M. Implementing a biofuel economy in the EU: lessons from the SUSTOIL project and future perspectives for next generation biofuels. *Renewable Sustainable Energy Rev* 2014;**29**(0):22–30.
54. Taheripour F, Hertel TW, Tyner WE, Beckman JF, Birur DK. Biofuels and their by-products: global economic and environmental implications. *Biomass Bioenergy* 2010;**34**(3):278–89.
55. Souza SP, Seabra JEA. Integrated production of sugarcane ethanol and soybean biodiesel: environmental and economic implications of fossil diesel displacement. *Energy Convers Manage* 2014;**87**(0):1170–9.
56. Wang L, Quiceno R, Price C, Malpas R, Woods J. Economic and GHG emissions analyses for sugarcane ethanol in Brazil: looking forward. *Renewable Sustainable Energy Rev* 2014;**40**(0):571–82.
57. Xi C. *Economic and environmental impacts of biofuel policy in Canada: an application of input-output modelling* [Master thesis]. Montreal: Department of Agricultural Economics, McGill University; 2015.
58. Gorter G, Just D. The social costs and benefits of biofuels: the intersection of environmental, energy and agricultural policy. *Appl Econ Perspect Policy* 2010;**32**(1):4–32.
59. Hill J, Nelson E, Tilman D, Polasky S, Tiffany D. Environmental, economic, and energetic costs and benefits of biodiesel and ethanol biofuels. *Proc Nat Acad Sci USA* 2006;**103**(30):11206–10.
60. FAO. *Bioenergy and food security—the BEFS analytical framework.* Rome: FAO; 2011.
61. FAO. *Trends and impacts of foreign investment in developing country agriculture, evidence from case studies.* Rome: Food and Agriculture Organization of the United Nations; 2012.
62. Schott C. *Social-economic dynamics of biofuel development in Asia Pacific.* Jakarta: Friedrich Ebert Stiftung; 2009. p. 186.
63. Gouret M, Hought J, Petersen JD, Thormose S. *Field report—impacts of agricultural commercialization on land use and livelihood in Banteay Chhmar Commune, Cambodia.* Copenhagen: University of Copenhagen; 2009.
64. USAID. *Biofuels in Asia: an analysis of sustainability options.* Washington, DC: USAID; 2009. p. 151.
65. Openshaw K. Biomass energy: employment generation and its contribution to poverty alleviation. *Biomass Bioenergy* 2010;**34**(3):365–78.
66. Ajanovic A. Biofuels versus food production: does biofuels production increase food prices? *Energy* 2011;**36**(4):2070–6.
67. Havlík P, Schneider UA, Schmid E, Böttcher H, Fritz S, Skalský R, et al. Global land use implications of first- and second-generation biofuel targets. *Energy Policy* 2011;**39**(10):5690–702.
68. Amigun B, Musango JK, Stafford W. Biofuels and sustainability in Africa. *Renewable Sustainable Energy Rev* 2011;**15**(2):1360–72.
69. FAO. *From subsistence farming to sugar-cane monoculture: impacts on agrobiodiversity, local knowledge and food security.* Available at <http://www.fao.org/docrep/pdf/010/aj042e/aj042e.pdf>; 2008 [accessed 12.11.18].

70. Bildirici ME. The relationship between economic growth and biomass energy consumption. *J. Renewable Sustainable Energy* 2012;**4**(2):023113.
71. GBEP Secretariat. *The Global Bioenergy Partnership Sustainability Indicators for Bioenergy*. 2011.
72. Gohin A. Impacts of the European biofuel policy on the farm sector: a general equilibrium assessment. *Appl Econ Perspect Policy* 2008;**30**(4):623–41.
73. Locke A, Henley G. *A review of the literature on biofuels and food security at a local level*. London: Overseas Development Institute (ODI); 2014.
74. Nelson G, Cai Z, Hassan R, Santos M, Swaminathan H. *Food security and climate change. A report by the High-Level Panel of Experts (HLPE) on Food Security and Nutrition of the Committee on World Food Security*. 2012.
75. Popp J, Lakner, Harangi-Rákos M, Fári M. The effect of bioenergy expansion: food, energy, and environment. *Renewable Sustainable Energy Rev* 2014;**32**:559–78.
76. Popp J, Pető K, Nagy J. *Impact of pesticide productivity on food security. Sustainable agriculture reviews*. Springer International Publishing; 2014. p. 19–33.
77. Anderson Teixeira KJ, Davis SC, Masters MD, Delucia EH. Changes in soil organic carbon under biofuel crops. *GCB Bioenergy* 2009;**1**(1):75–6.
78. Verpoorten M, Arora A, Stoop N, Swinnen J. Self-reported food insecurity in Africa during the food price crisis. *Food Policy* 2013;**39**:51–63.
79. De Gorter H, Drabik D, Just DR. The perverse effects of biofuel public-sector policies. *Ann Rev Resour Econ* 2013;**5**(1):463–83.
80. Duer H, Christensen PO. Socio-economic aspects of different biofuel development pathways. *Biomass Bioenergy* 2010;**34**(2):237–43.
81. Khanna M, Chen X, Huang H, Önal H. Supply of cellulosic biofuel feedstocks and regional production pattern. *Am J Agric Econ* 2011;**93**(2):473–80.
82. Zhang Z, Lohr L, Escalante C, Wetzstein M. Food versus fuel: what do prices tell us? *Energy Policy* 2010;**38**(1):445–51.
83. Garcia DJ, You F. Supply chain design and optimization: challenges and opportunities. *Comput Chem Eng* 2015;**81**(4):153–70.
84. *FAO. The state of food and agriculture*. Available at <http://www.fao.org/3/a-i0100e.pdf>; 2008 (accessed 12.11.18).
85. European Committee. *Directive 2009/28/EC of the European Parliament and of the Council of 23 April 2009 on the promotion of the use of energy from renewable sources and amending and subsequently repealing Directives 2001/77/EC and 2003/39/EC*. 2009.
86. Hall DO. Biomass energy. *Energy Policy* 1991;**19**(8):711–37.
87. OECD/FAO. *OECD-FAO Agricultural Outlook 2007–2016*. Paris: OECD Publications; 2007.
88. Searchinger T. How biofuels contribute to the food crisis. *Washington Post*. February 10, 2011.
89. van der Horst D, Vermeylen S. Spatial scale and social impacts of biofuel production. *Biomass Bioenergy* 2011;**35**:2435–43.
90. McKone TE, Nazaroff WW, Berck P, Auffhammer M, Lipman T, Torn MS, et al. Grand challenges for life-cycle assessment of biofuels. *Environ Sci Technol* 2011;**45**(5):1751–6.
91. Hurst P, Termine P, Karl M. *Agricultural workers and their contribution to sustainable agriculture and rural development*. Geneva, Switzerland; 2007. ISBN: 978-92-2-118709-7. Available at <https://www.ilo.org/wcmsp5/groups/public/---ed_dialogue/---actrav/documents/publication/wcms_113732.pdf> [accessed 12.11.18].

Chapter 14

Regulatory challenges in nanotechnology for sustainable production of biofuel in Brazil

Wilson Engelmann and Raquel Von Hohendorff
University of the Vale do Rio dos Sinos—UNISINOS, São Leopoldo, Brazil

14.1 Introduction

Biofuels are substances derived from renewable biomass, for instance, biodiesel and ethanol. A biofuel may partially or completely replace compounds of fossil origin in engines or other types of energy generating processes. Since biofuels are biodegradable and practically free of sulfur and aromatic compounds, they do not cause high impacts on the environment.[1] The transportation sector was the second largest contributor to carbon dioxide (CO_2) emissions in Brazil in 2016. However, the agricultural and livestock sector was rated as the largest, according to a report by the Weather Observatory.[2] Therefore it is essential to develop policies for the production of biofuels which maintain economic development and reduce environmental pollution.

Currently, in Brazil biofuels represent 38% of the energy matrix, being mostly from the sugar–energy industry (alcohol, sugar, and biomass from sugarcane) or originating from oilseeds like soya, palm oil, sunflower, peanuts, babassu, papaya, and physic-nuts. However, other fuels including biodiesel (commonly produced from the residues of the food industry) and biomethane have been discretely gaining space. According to the Science, Technology, Innovation and Communication Ministry, Brazil, residual biomass like agricultural residues, slaughtering residues, and urban solid residues have great potential for the production of biofuels.[2] Such residues which would otherwise be disposed of, causing pollution, can be harnessed for the generation of biofuels.

Apart from the traditional sources and approaches of biofuel production, nanotechnology-based approaches which involve the use of various nanomaterials have attracted great attention. Fig. 14.1 shows important nanomaterials used in the production of biofuels and their comparison with some

Sustainable Bioenergy. DOI: https://doi.org/10.1016/B978-0-12-817654-2.00014-9

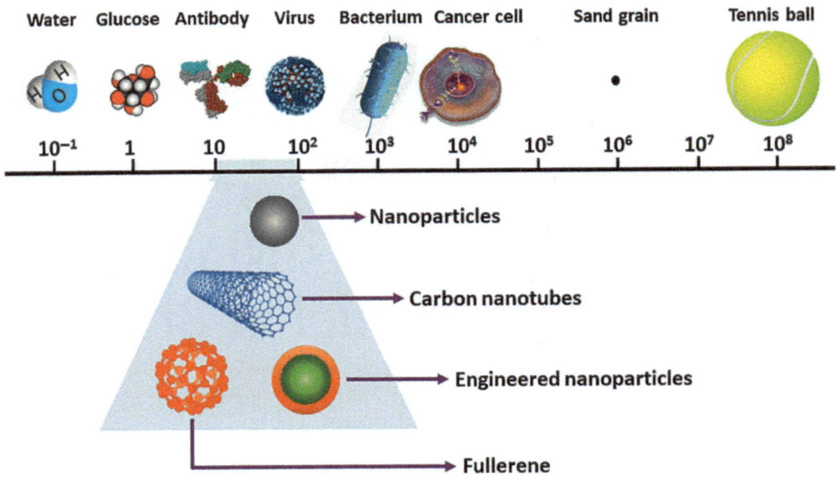

FIGURE 14.1 Important nanomaterials used in the production of biofuels and their comparison to some natural materials.

natural materials. Many studies have confirmed the promising role, directly and indirectly, of nanotechnology in biofuel production. Wegner and Jones[3] proposed that nanotechnology can be applied in biofuel production in various ways which mainly include: (1) manipulation of nanoscale cell wall structures (i.e., nanofibrils) within trees and plant materials which facilitates their easier disassembly into constitutive materials for biofuels production (either through fermentation, gasification, or catalysis); (2) direct application of nanocatalysts for the degradation of cellulose which is a major component of biomass; (3) application of engineered nanoscale enzymes or a system of enzymes (e.g., glycol hydrolases, lignin-degrading enzymes) to improve the conversion efficiency of cellulose into sugars; and (4) the development of novel symbiotic nanoscale biological systems which would work together to create ethanol or other biofuels.

Although nanotechnology has promising advantages in biofuel production, it has raised many toxicological concerns and adverse impacts on the economy and environment. According to Renn,[4] one of the most important concerns is the lack of knowledge about the potential risks and negative effects of nanotechnological developments on living organisms (including ecosystems and humans), as many nondegradable nanomaterials entering the environment might persist in it forever.[5] Nanomaterials deposited in soils, can cross several layers of soil and may reach groundwater sources (aquifers).[6] Moreover, nanomaterials like nanoparticles if released into the environment may enter into humans and animals in various ways like through inhalation or skin pores. Having entered the body, there is a chance that nanomaterials may enter the bloodstream due to their small size, and as a consequence, they may have many toxicological effects on the human body.[7] Therefore extensive studies involving the evaluation of the toxicological concerns of nanotechnology in any kind of application including biofuels production are required.[8,9] Similarly, the implementation of strict policy regulations is needed for the establishment of well-grounded oversight and monitoring procedures.[10,11]

Moreover, considering the eventual risks posed by the application of nanomaterials, it is necessary to follow and understand the importance of the precautionary principle proposed by the Sustainable Development Goals (SDGs) and the Millennium Development Goals (MDGs) organized by the United Nations (UN) in the 2030 Agenda. In the same line, a contribution will be given so that Brazil can implement the various SDGs to assure the quality of life and help save the environment. Moreover, the proposed SDGs are in agreement with the Article of Law 225 of the Brazilian Constitution and may be crossed with Annexes of the European Union's (EU) chemical legislation with regard to nanomaterials which were reviewed at the beginning of 2018. With this plural gathering of "sources," it is intended, through the dialog between them, to propose a specific regulatory framework to deal with the risks that may be generated by the utilization of nanomaterials in biofuels in Brazil, considering the existence of an exclusive regulation for this sector.

The present chapter is focused on legal issues necessary to regulate the application of nanotechnology in biofuels. Considering the current Brazilian legislation, there is a need to design a regulatory model that fills the gap of a specific regulation for biofuels produced from different nanoscale materials.

14.2 Challenges and possibilities of nanomaterial usage in biofuels

Klaus Schwab in his book entitled "The Fourth Industrial Revolution" mentions the beginning of a new period in the development of human civilization. According to him, the period before and up to the 20th century is historic; however, after that new horizons and possibilities have ushered in an industrial revolution in the form of artificial intelligence, robotics, autonomous vehicles, 3D printers, nanotechnology, biotechnology, materials science, storage of energy, and quantic computation, etc.[12]

The rapid changes generated by the Fourth Industrial Revolution are different from all previous times in the history of humanity. In the case of the fuel and energy sector, traditional fossil fuels have a substitution alternative for renewable energy sources. Emerging technologies for sustainable energy production can represent a reduction in investment costs, and therefore, it will be beneficial for businesses and consumers, while at the same time it will help to reduce industrial emissions which negatively affect the environment.[13] Despite this positive scenario, there is already an evident risk; the risk of the slow transition to clean energy. Previous transitions in energy systems combined science, infrastructure, norms and products of ecosystems; these systems arose over generations because of the extensive deadline for the implementation of materially intensive technologies.[13] However, the present situation is already concerning, therefore, there is not much time to mitigate environmental issues, particularly those caused by newer fuel technologies including nanotechnology.

The adequate utilization of residual biomass may be beneficial for biofuels production. However, the production of biofuels from residual biomass is accompanied by local environmental issues that need to be addressed. For example, stillage (the liquid residue from distillation in the sugarcane industry) used for biofuel production has a high chemical and biological oxygen demand and requires appropriate processing before final disposal. From a global climate change perspective, properly designed and managed policies are required for biofuel production which should contribute minimally to greenhouse gas (GHG) emissions. However, the ground reality is that many biofuel production systems in the world are contributing significantly to GHG emissions.[14] The production of biofuels is eco-friendly approach because it is said to be a solution to combat global warming and reduce pollution levels in large cities around the world.

Nowadays, biofuel production technologies are more intensified with the utilization of nanotechnology. According to the 229 report made by the Technical Committee of the International Organization for Standardization, nanotechnology is the "comprehension and control of matter and processes in nanoscale, which are typically, but not exclusively, situated under 100 nanometers in one or more dimensions, in which the appearing of the phenomenon depending on size usually allows new applications." Nanomaterials possess unique properties that are different from the properties of atoms, molecules, and matter in individual mass and such nanomaterials can be used to create materials, devices, and improved systems that explore these unique properties.[15]

Currently, a number of nanotechnology-based products are already available on the market. According to data collected on September 15, 2018, from the "STATNANO: Nano Science, Technology and Industry Information," website more than 8000 products that contain some type of nanoparticle are produced from around 2000 factories and these products are globally distributed in over 50 countries. Moreover, refining of the search with keywords like "petroleum" and "renewable energies" also showed various products (Fig. 14.2) on the same website.[16] This data clearly indicates the wide-spread application of various nanomaterials in different products and their worldwide use.

As discussed earlier, the small size of nanomaterials generate risks due to the physicochemical changes that occur at this scale. In spite of many new fundamental advances, nanotechnology is confronted with critical challenges

FIGURE 14.2 Information about a number of nanobased products, the companies producing them, and the countries where these products are distributed. *Prepared from the data available at Nanotechnology Products Database (NPD): <http://product.statnano.com>; 2018 [accessed 04.11.18].*

and there is still work to be done to translate the new fundamental advances reported in scientific literature into tangible technological applications that may be appreciated by laypersons.[17] Therefore the use of nanoscale materials in the production of biofuels is developing under an ambivalent panorama. On one hand, there are many technological promises, offering several possibilities to improve environmental conditions with the usage of this fuel category; on the other hand, the questions relating to the risks that can be generated are still greatly unanswered and deserve further study. The questions that arise in this context are characterized by the term "world metamorphosis." In spite of changing, metamorphosis destabilizes the certainties of modern society; it includes events and processes that provoke a fundamental shock. Metamorphosis means that what was impossible yesterday is real and possible today.[18] Access to the nanotechnology and nanoscale materials is one of the engines that drive the metamorphosis of the world, in which biofuels are found. It is a panorama of risks, many yet unknown. Although biofuels generated with and from nanomaterials are already a reality.

To a certain extent, the advances enabled by nanotechnologies create a manufactured ignorance, meaning that knowledge of global risks is highly dependent on scientists and specialists. For instance, industries and nuclear specialists occupy a double position, they are at the same time creators and evaluators of the risk, exploiting their positions of power based on the relations of the definition of risk.[18] In the case of nanomaterials, the situation is not different. The people who decide whether they are safe or not or if they could offer risks are the same scientists who research and use them. Similar doubts arise about whether nanomaterials including nanoparticles used for biofuels production are effectively safe and environmentally friendly. Or do they have the potentiality to generate more harm than benefits? Those and other questions are still without an appropriate answer, since a safe response will probably be difficult to obtain. According to Beck,[18] metamorphosis is profoundly connected to the idea of the unknown, creating a paradox. It emphasizes the inherent limitations of knowledge, particularly the reality that knowledge is cognizant or does not attract a disposition to know the nanotechnology, bioengineering, and other types of emerging technology having knowable and unknown risks. It offers a range of fundamental limitations to the capacity of society to perceive and govern these risks.[18]

In spite of having little awareness about the concerns, the utilization of nanomaterials grows constantly. This scenario does not cause preoccupation in the production sector which uses nanomaterials, constructing a reality of diverse superimposed levels of risk consequently empowering them. Antunes et al.[19] proposed the necessity of discussing issues related to the safety of nanomaterials used in biofuel/bioenergy production as many such nanomaterials have the ability to generate diseases in human beings and cause severe environmental problems. Further, they also mentioned that the increased use of nanoparticles for biofuel applications and the effects of exposure in human beings and on the environment is obvious. Therefore for the safe use of nanomaterials, a safety assessment is of utmost importance. In the present scenario, various approaches are in practice for assessing the toxicity of nanoparticles. Most studies involve in vitro evaluation of nanotoxicity. However, in-depth studies that focus on the in vivo interactions of nanoparticles, particularly those used for biofuel and bioenergy production, are necessary.[19]

Klaus Schwab rightly said in his book (The Fourth Industrial Revolution) that in spite of the many advances, human beings will still need to make a choice about whether we will continue on the path of the first three industrial revolutions, leaving environmental considerations at the edge of our priorities or take control and seize the "Fourth Industrial Revolution" to solve environmental problems through deliberated choices and collaborations among interested partners.[13] The set of technologies that are integrated in the new industrial revolution enables possibilities to support human needs. Therefore the effects of new technologies, positive or negative, must be socialized in order to assure a better life for all the members of society globally. The privatization of positive results and the publication of the eventual negative effects will be unacceptable. Herein lies the challenge, since the decision-making is in the hands of humans. Moreover, it will be important to utilize the technologies that are integrated in the structure of the fourth industrial revolution to improve the environment and planet Earth as a whole. In this context, the law has an interesting contribution to offer, by promoting the development of normative structures with this objective. An example of this practice is the SDGs of the MDGs proposed by the UN as a plan for 2030. It is intended that the performance of these goals in terms of their contribution to qualitatively assuring the wellness of human beings and the preservation of the environment be observed by present and future generations.

14.3 Sustainable Development Goals proposed in Agenda 2030 by the United Nations

In September 2015 193 countries at the UN Summit (SDGs of Agenda 2030 of the UN) adopted the worldwide recognized Agenda 2030 which is an action plan with 17 global goals in the form of SDGs.[20] It mainly involves various broad themes such as sustainability, education, and human rights that are to be developed throughout the span of 15 years (2015–30) to eradicate poverty, promote peace and equality, boost inclusive development, and most importantly to protect the environment.[21,22] The SDGs and objectives are integrated and indivisible, global by nature, and universally applicable; they take into account the different realities, capabilities, and levels of national development, respecting the national politics and priorities of the various countries. The goals are defined as global ideas, each government defines its own national targets, guided by the global level of ambition but considering national circumstances. Each government has decided how these ideal and global goals must be incorporated into national processes, politics, and strategies of planning.[22] In the documents on the SDGs, sustainable development is defined as development that seeks to satisfy the needs of the current generation without compromising the ability of future generations to satisfy their own needs. Consequently, it requires a group effort for the formation of an inclusive, resilient, and sustainable future for all people around the planet, and in order to be achieved the harmonization of three key elements is necessary, namely economic growth, social inclusion, and the protection of the environment. These elements are interconnected and fundamental to the wellness of the individuals of society.[22]

Different from the preceding Development Goals of the Millennium, the SDGs explicitly summon all organizations to use their creativity and innovation to solve the challenges associated with sustainable development. These goals can be understood as opportunities for the development and implementation of solutions and corporate technologies to treat the biggest worldwide challenges when it comes to sustainable development. The SDGs elaborate the global agenda for the development of the society, and hence they help organization leaders to advance sustainable development, minimizing the negative impacts and maximizing the positive impacts on people and on the planet.[23]

The SDGs particularly include these aspects: (1) the eradication of hunger and misery; (2) quality basic education for all; (3) promoting equality between genders and the autonomy of women; (4) reducing child mortality; (5) improving the health of pregnant women; (6) combating AIDS, malaria, and other diseases; (7) guaranteeing life quality and respect to the environment; and (8) establishing partnerships in development (website given at no. 18). To be more precise and considering the focus of the present chapter, item (7) will be fundamental, that is, the regulatory frameworks must assure life quality and respect for the environment.

14.4 Sustainable development, nanotechnology, and legislation

Considering the regulation of nanomaterials, it has been emphasized that European member states have approved the European Commission's draft regulation revising the Annexes of the EU's chemical legislation regarding nanomaterials. The revised Annexes provided a great opportunity for the European Commission to clarify the EU's chemical registration duties for nanomaterials and to resolve the regulatory uncertainty that has plagued the development and commercialization of nanomaterials since the EU's new chemical legislation was adopted in 2006.[24] The 10 regulatory issues are a concrete indication that the regulation of the production and commercialization of nanomaterials, despite being determined by the EU, can help qualify normative structures in other countries around the globe.[24] This group of recommendations reveals a great preoccupation with the risks that nanomaterials may generate. For this reason it is necessary to fill out forms, addressed to the public authorities, to monitor the production of nanomaterials, generating quantitative and qualitative information to support future regulatory decisions. This initiative is essential to learn more about nanomaterials, especially about the different possibilities of their interactions and consequences, with respect to the human and the environment.

The focus of EU is mainly on the sresponsible development in nanotechnology which can be characterized by only maintaining balance between efforts to maximize the positive impacts and minimize their negative consequences. This implies that there is a commitment to develop and use technology for attending to the most urgent human needs as well as to make reasonable efforts to anticipate and mitigate adverse implications and unintentional consequences.[25] There is also concern with the life quality of the environment as well as it receiving due respect which represents one of the SDGs and is also found in the base of the legislation review on chemical products made by the EU. The EU chemical legislation currently includes 17 annexures that specify many technical and scientific details. It also describe the criteria for when, how, and what different actors such as manufacturers, importers, downstream users, and regulatory authorities should deal with chemical safety of nanomaterials.[24] The regulatory indications are destined for all those who find themselves involved in the lifecycle of nanomaterials and that must together take care of the health and security of people and the preservation of the environment by controlling nanotoxicological issues.

Considering the various actors related to the review of the Annexes of the legislation on chemical products, it is clear that there is also a preoccupation with the conjecture of responsible research and innovation. There is a movement of transparency and engagement of all the actors involved and the clarification of doubts, through the responsible circulation of information for the structuring of the debate. This conjunct, in turn, will orientate decision makers so as to be careful not to interrupt the flux of ideas. Here there is also an advantage related to the insertion of the SDGs and the legal pluralism. Teubner[26] claims that global law can only be appropriately interpreted through the legal pluralism theory and the theory of the sources of the law, correspondingly conceived in pluralist terms. The structure of legal pluralism and the possibility to recognize the normative activity of other actors (distant from normative state production) are proposed by Teubner[26]. According to him, there is necessity to rethink on a theory of the law sources, and concentrate on the spontaneous processes of law formation.

14.5 Brazilian regulatory structures on biofuels and possible applications of the use of nanomaterials in this type of fuel

In Brazil, biofuels have been regulated by Bill 13,576 since December 26, 2017, creating the National Policy of Biofuels (RenovaBio) (website on no. 22).[27] RenovaBio is a state policy that aims to draw up a joint strategy to recognize the strategic role of all types of biofuels in the Brazilian energy matrix. Although the bill does not mention anything about nanoparticles in biofuels, article three of this bill states the principles of RenovaBio, emphasizing the redaction of section VI which establishes the need for technological development and innovation, envisages the consolidation of the technological base, and aims to increase the competitiveness of biofuels in the national energetic matrix as well as the of speed development and commercial insertion of advanced biofuels and new biofuels. There is a necessity to make the amendments about the use of nanotechnology in existing law on biofuels-2017 in Brazil.[27]

Besides these principles, RenovaBio still presents the following starting points regarding structures:

1. Predictability in the participation of biofuels, emphasizing the sustainability of the industry of biofuels and on safety.
2. Protection of the interests of consumers regarding price, quality, and choice of products.
3. Effectiveness of biofuels to contribute to the effective mitigation of GHG emissions which exacerbates the greenhouse effect and local pollution.
4. Potential to contribute to the biofuel market toward the generation of jobs and incomes and to regional development as well as to promote value chains related to the sustainable bioeconomy.
5. Advancement of energetic efficiency with the usage of biofuels in vehicles, machines, and equipment (Website no. 22). These principles are in harmony with many of the principles of the SDGs projected by the UN in the 2030 Agenda which may serve as a guide to the safe installment and environment-friendly production of biofuels using nanoparticles in Brazil.[27]

This law (Bill 13,576 since December 26, 2017) is important in the context of use of nanotechnology for biofuels production, because it establishes in its article 1st: “It is instituted the National Policy of Biofuels (RenovaBio), integrating the National Energetic Policy with the following goals: I – contribute to the attendance of the commitments of the Country on the ambit of the Paris Agreement under the United Nations Framework Convention on Climate Change.” Here the global commitment of Brazil is verified with a reduction of the effects that cause climate change, being parallel with the SDGs, especially regarding to the preservation of the environment. Therefore this Brazilian law can orientate researches and the development of biofuels regarding the usage of nanoparticles.[27]

Brazilian law, with this redaction, intends to offer the same competitiveness to biofuels when compared to the other forms of fuel available on the national market. The law emphasizes the commitment of Brazil to the advancement of the production of fuels from different fossil matrices and opens space for the entry of nanotechnology in the energy sector.

The normative framework brought into play by Bill 13,576, on December 26, 2017 (Website no. 22), is aligned with the guidance that descends from the Federal Constitution of 1998 which consecrates a new structure focused on the preservation of the environment in Article 225 of this document. It states that every person has the right to an ecologically balanced environment, of common usage for all people and that is essential to a quality and healthy life. It is the duty of the public to collectively defend and preserve the environment for present and future generations (Website no. 23). Therefore nanomaterials used in the production of biofuels must observe the limited presumptions that are stated in this article of the Brazilian Constitution.[27]

It is worth saying that it should not bring risks to the environment nor disease to humans for present or future generations. Therefore although there is no reference to nanotechnologies, nanoparticles, or nanomaterials, the duty to preserve the environment and assure a healthy quality of life are general guidelines to any modality of scientific or technological advancement in Brazil.

Extending this preoccupation, the Federal Constitution of Brazil still requires the presentation of an environmental impact report which will become public and will be presented upon the installment of construction or activity that could potentially cause significant degradation to the environment (Article 225, § 1st, IV) (Website no. 23).[28] In cases where research, production, commercialization, and the use of nanotechnology-based biofuels portray any potential to cause degradation to the environment, an environmental impact report must be provided. Moreover, in order to guarantee the legal and social efficiency of the content inserted in article 225, in its “caput”, the Brazilian authorities must control the production, commercialization and use of techniques, methods and substances that present a risk to life, quality of life and the environment.[28] This constitutional article will also serve as a parameter for the introduction of nanotechnology in biofuel production. This constitutional article presents, at least implicitly, a precautionary principle in the Federal Constitution of Brazil. The precautionary principle requires a special level of risk, i.e. the possibility of a significant risk that can be considered in the presence of scientific uncertainty and the likelihood of damage to life, life quality, and the environment.[29] Being the structuring elements of the precautionary principle, it can and must always be summoned when there are scientific doubts regarding the risk of damage or regarding types of damage associated to a specific category of residues or to a certain management operation. The fundamental assumption of the precautionary principle is, naturally, that the risks are real and the possible damages are grave and irreversible. This principle reflects on the precautionary finalities of environmental law and allows for the adoption of anticipatory measures (namely the suspension of the operation) before the existence of concrete scientific proof against the possible environmental damages or health damages, in accordance with the maximum: "if there is any doubt, it should decide in favor of the environment."[30] It may be noted that within the concept of the "precautionary principle" there are elements that must be structured on the basis of their advances in the use of nanomaterials in biofuels. The characterization of the precautionary principle reinforces the preoccupation which is brought by the Brazilian Constitution regarding the protection of the health of people in present and future generations, in addition to the preservation of the environment.

When in doubt about the occurrence of damages, the preservation of the environment must be prioritized. Consequently, scientific and technological progress and economic development must yield before life, the quality of life, and the environment. Here, a potential the path to be followed by biofuels is outlined. It does not strictly relate to Brazil, but rather to the global scenario, considering that the precautionary principle is aligned with the construction of modern content relating to human rights and that it is found at the center of the SDGs, as elaborated by the UN.

14.6 The way forward

The combination of the different elements examined so far will require the preparation of an environmental impact report and the promotion of individual or collective registration, outlining the Chemical Safety Assessment procedures to be followed and, finally, clarifying when specific information requirements should be addressed to nanomaterials and when they can be dispensed with.[24] Furthermore, the following steps must be considered:

1. Manufacturers and importers have to decide whether they have nanomaterials that need to be registered including detailed information about

their particle size distribution, surface functionalization or treatment, and shape and aspect ratio. To determine whether a nanomaterial is produced or imported, the number-based particle size distribution has to be provided; however, generally approved methods and technical standards to measure particle size distribution are not yet available.[31]

2. The second key decision to be made by manufacturers and importers is whether and how to group different nanomaterials as this decision can be of paramount importance to how much information they subsequently have to provide in the registration dossiers and in the chemical safety report.[32]
3. Many of the test-specific requirements are hung on vague terms such as "poorly soluble," "high insolubility," and "low" versus "high" dissolution rate. This is the third most important aspect that manufacturers and importers have to be made aware of. It is well-established that nanomaterials do not go into solution, but rather are dispersed and as these terms are not clearly defined in the Annexes, whether specific test requirements have to be met will depend on how the registrant interprets such terms.[24]

Although these steps are required by the EU, it is suggested that in Brazil, biofuels containing nanomaterials also fulfill this normative path, since it will be a trajectory toward the collection of data and information about the usage of nanotechnology from this productive sector. With these normative steps, the principles found in Bill 13,576 from December 26, 2017, will be attended. In addition to this, it will normalize the content of Article 225 and its complements in the Brazilian Constitution, making feasible the concretization of the SDGs, especially the one relating to guaranteeing the quality of life and respect for the environment. Therefore the dialog between the mentioned sources of the law is an alternative, constructed from legal pluralism which allows for the organization of a normative structure that is creative and flexible in order to support the law when it comes to the usage of nanoparticles in biofuels.

14.7 Conclusion

The advances presented by human access to nanomaterials are important and promising for improving the quality of life of people on planet Earth. However, since they exhibit extremely small particle sizes in the range of 1–100 nm and have unique physicochemical characteristics, many concerns have arisen which are as yet unexplained. For this reason, it is necessary to exercise caution regarding their large-scale utilization. The regulation of the use of nanoparticles and nanomaterials is not yet existing in many countries including Brazil. There are regulatory possibilities, through structures that should be considered normative, such as the framework, decision trees, and other models, organized from the precautionary principle and ODS, which do not originate in the Legislative Branch, the traditional place for the production of legislation.

The novelties presented by the fourth industrial revolution, offer space to the law to mobilize different normative structures produced by various national and international actors, characterizing legal pluralism, according to Günther Teubner. In this context, the methodology of the "dialog between the various sources and actors of the law" can be applied, seeking to standardize the diverse stages of the lifecycle of nanoparticles and nanomaterials which will be added into the production process of biofuels.

It was proposed that the elements found in the base of the precautionary principle are of significant risk (verisimilar risks). There is scientific uncertainty regarding the probability of damage to quality of life, and the environment. The concrete scientific proof on possible damages to the environment or human health needs to be presented or "if there is any doubt, it should decide in favor of the environment," the possible damages must be grave and irreversible.

Acknowledgment

The authors are highly thankful to the Foundation for Research Support of the State of Rio Grande do Sul (FAPERGS) (Announcement No. 02/2017) and the National Council for Scientific and Technological Development (CNPq) (Announcement No. 12/2017) for providing financial assistance in the form of a research project and research productivity scholarships, respectively.

References

1. NAOGB (National Agency for Oil, Gas and Biofuels). *Production and supply of biofuels (original article is in Portuguese).* Available at: <http://www.anp.gov.br/producao-de-biocombustiveis>; 2018 [accessed 09.09.18].
2. Brazilian Agency. *Research seeks to reduce environmental impact of biofuels (original article is in Portuguese).* Available at: <https://www.inovacaotecnologica.com.br/noticias/noticia.php?artigo = diminuir-impacto-ambiental-biocombustiveis&id = 010175170830#.W5VVQK3OqYV>; 2017 [accessed 30.10.18].
3. Wegner TH, Jones EP. A fundamental review of the relationships between nanotechnology and lignocellulosic biomass. In: Lucia LA, Rojas OJ, editors. *The nanoscience and technology of renewable biomaterials.* West Sussex, UK: Blackwell Publishing Ltd/Wiley; 2009.
4. Renn O. Nanotechnology: frame 1 policy implications, In: *IRGC symposium Zürich, Stuttgart University and DIALOGIK GmbH, 6–7 July 2006.*
5. Colvin VL. Research vision—sustainability for nanotechnology. *Scientist* 2004;**18** (16):234–45.
6. Alargova RG, Tsujii K. Stable colloidal dispersions of fullerenes in polar organic solvents. *J Am Chem Soc* 2001;**123**:10460–7.
7. Santos CAD, Seckler MM, Ingle AP, Gupta I, Galdiero S, Galdiero M, et al. Silver nanoparticles: therapeutical uses, toxicity, and safety issues. *J Pharm Sci* 2014;**103**:1931–44.

8. Justo-Hanani R, Dayan T. European risk governance of nanotechnology: explaining the emerging regulatory policy. *Res Policy* 2015;**44**:1527–36.
9. Hull MS, Bowman DM. Nanotechnology environmental health and safety what we've learned and where we're (potentially) heading. In: Hull MS, Bowman DM, editors. *Nanotechnology environmental health and safety*. 2nd ed. UK: Elsevier; 2014. p. 3–8.
10. Colvin VL. The potential environmental impact of engineered nanomaterials. *Nat Biotechnol* 2003;**21**(10):1166–70.
11. Molins R. Opportunities and threats from nanotechnology in health, food, agriculture and the environment, Comunica magazine- Fourth Year-Second Phase, 2008, IICA Headquarters, 38–53.
12. Schwab K. *The fourth industrial revolution (original book is in Portuguese)* [Miranda DM, Trans. in English]. Sao Paulo, Brazil: Edipro Publication; 2016. p. 269–80.
13. Schwab K. *Applying the fourth industrial revolution (original book is in Portuguese)* [Miranda DM, Trans. in English]. Sao Paulo, Brazil: Edipro Publication; 2018. p. 11–3.
14. Trindade SC. Nanotech biofuels and fuel additives. In: Bernardes MADS, editor. *Biofuel's engineering process technology*. Croatia: InTech Open Publisher; 2011. p. 103–14.
15. ISO (International Standardization Organization). *ISO/TC-229; nanotechnologies, an technical committee report*. Available at: <http://www.iso.org/iso/iso_technical_committee?commid = 381983>; 2005 [accessed 14.09.18].
16. Nanotechnology Products Database (NPD). Available at: <http://product.statnano.com>; 2018 [accessed 04.11.18].
17. Editorial. A matter of scale. *Nat Nanotechnol* 2016;**11**:733 Available from: https://doi.org/10.1038/nnano.2016.180.
18. Beck U. translated by Borges MLXA *The metamorphosis of the world—new concepts for a new reality (original book is in Portuguese)*. São Paulo, Brazil: Zahar Publication; 2018. p. 11–12.
19. Antunes FAF, Gaikwad S, Ingle AP, Pandit R, dos Santos JC, Rai M, et al. In: Rai M, Da Silva SS, editors. *Nanotechnology for bioenergy and biofuel production: green chemistry and sustainable technology*. Switzerland: Springer International Publishing; 2017. p. 3–18.
20. *Portal Brazil. The sustainable development goals*. Available at: <http://www.odmbrasil.gov.br/os-objetivos-de-desenvolvimento-do-milenio> 2018 [accessed 15.09.18].
21. Ministry of Foreign Affairs, Brazil. *Sustainable development goals*. 2016. p. 1–18. Available at: <http://www.itamaraty.gov.br/images/ed_desenvsust/ODSportugues12fev2016.pdf>. [accessed 01.09.18].
22. United Nations. *Sustainable development goals*. Available at: <http://www.un.org/sustainabledevelopment/sustainabledevelopment-goals>; 2017 [accessed 01.09.18].
23. SDG Compass. *Guidelines for implementing ODS in business strategy. [S.l.]*. Available at: <https://sdgcompass.org/wp-content/uploads/2016/04/SDG_Compass_Portuguese.pdf>; 2016 [accessed 01.09.18].
24. Clausen LPW, Hansen SF. The ten decrees of nanomaterials regulations. *Nat Nanotechnol* 2018;**13**:766–8.
25. Grunwald A. *The hermeneutic side of responsible research and innovation*, Vol. 5. London: Wiley/ISTE; 2016.
26. Teubner GA. Bukowina global sobre a emergência de um pluralismo jurídico transnacional. *Impulso,* 2003;**14**(33):11–18.
27. *Brazil Bill no. 13,576*. Available at: <http://www2.camara.leg.br/legin/fed/lei/2017/lei-13576-26-dezembro-2017-786013-publicacaooriginal-154631-pl.html>; 2017 [accessed 15.09.18].

28. Federal Constitution of Brazil. *Article 225, § 1st, V*. Available at: <http://www.planalto.gov.br/ccivil_03/Constituicao/Constituicao.htm> 2016, [accessed 15.09.18].

29. Machado PAL. Comment of the article 225, § 1°, IV e V. In: Canotilho JJG, et al., editors. *Comentários à Constituição do Brasil*. Saraiva; Almedina: Instituto Brasileiro de Direito Público, São Paulo; 2018. p. 2187–9.

30. Aragão A. *Princípios fundamentais do direito dos resíduos. Direito dos resíduos. In: Miranda J, et al. Scientific coord. Entidade Reguladora dos Serviços de Águas e Resíduos*. Lisbon: Instituto de Ciências Jurídico-Políticas/Centro de Investigação de Direito Público da Faculdade de Direito da Universidadede Lisboa; 2014. p. 8.

31. Roebben G, et al. *Towards a review of the EC recommendation for a definition of the term "nanomaterial": Part 2: Assessment of collected information concerning the experience with the definition (European Commission Joint Research Centre, 2014)*. 2014, Available at: <http://publications.jrc.ec.europa.eu/repository/bitstream/JRC91377/jrc_nm-def_report2_eur26744.pdf> [accessed 15.09.18].

32. ECHA. *Appendix R.6-1 for nanomaterials applicable to the guidance on QSARs and grouping of chemicals version 1.0*. Available at: <https://echa.europa.eu/documents/10162/23036412/appendix_r6_nanomaterials_en.pdf/71ad76f0-ab4c-fb04-acba-074cf045eaaa>; 2017 [accessed 15.09.18].

Index

Note: Page numbers followed by "*f*" and "*t*" refer to figures and tables, respectively.

B

C

H

I

N

V

W

X

Y

Z

Kolibri 360 GmbH
Ferdinand-Jühlke-Straße 7
99095 Erfurt, Deutschland
produktsicherheit@kolibri360.de